Tom Wolfe is the author of a dozen books, among them such contemporary classics as *The Electric Kool-Aid Acid Test, The Bonfire of the Vanities,* and *I Am Charlotte Simmons.* He lives in New York City.

Also by Tom Wolfe

The Kandy-Kolored Tangerine-Flake Streamline Baby

The Pump House Gang

The Electric Kool-Aid Acid Test

Radical Chic & Mau-Mauing the Flak Catchers

The Painted Word

Mauve Gloves & Madmen, Clutter & Vine

In Our Time

From Bauhaus to Our House

The Bonfire of the Vanities

A Man in Full

Hooking Up

I Am Charlotte Simmons

THE RIGHT STUFF

TOM WOLFE

THE RIGHT STUFF

PICADOR
Farrar, Straus and Giroux
New York

www.picadorusa.com

Picador® is a U.S. registered trademark and is used by Farrar, Straus and Giroux under license from Pan Books Limited.

For information on Picador Reading Group Guides, please contact Picador.
E-mail: readinggroupguides@picadorusa.com

ISBN-13: 978-0-312-42756-6
ISBN-10: 0-312-42756-5

First published in the United States by Farrar, Straus and Giroux

33

For Kailey Wong

CONTENTS

Foreword xi
I. The Angels 1
II. The Right Stuff 15
III. Yeager 33
IV. The Lab Rat 61
V. In Single Combat 85
VI. On the Balcony 102
VII. The Cape 127
VIII. The Thrones 141
IX. The Vote 169
X. Righteous Prayer 190
XI. The Unscrewable Pooch 214
XII. The Tears 238
XIII. The Operational Stuff 283
XIV. The Club 312
XV. The High Desert 331
Epilogue 349

FOREWORD

This book grew out of some ordinary curiosity. What is it, I wondered, that makes a man willing to sit up on top of an enormous Roman candle, such as a Redstone, Atlas, Titan, or Saturn rocket, and wait for someone to light the fuse? I decided on the simplest approach possible. I would ask a few astronauts and find out. So I asked a few in December of 1972 when they gathered at Cape Canaveral to watch the last mission to the moon, Apollo 17. I discovered quickly enough that none of them, no matter how talkative otherwise, was about to answer the question or even linger for more than a few seconds on the subject at the heart of it, which is to say, courage.

But I did sense that the answer was not to be found in any set of traits specific to the task of flying into space. The great majority of the astronauts who had flown the rockets had come from the ranks of test pilots. All but a few had been military test pilots, and even those few, such as Neil Armstrong, had been trained in the military. And it was this that led me to a rich and fabulous terrain that, in a literary sense, had remained as dark as the far side of the moon for more than half a century: military flying and the modern American officer corps.

Immediately following the First World War a certain fashion set in among writers in Europe and soon spread to their obedient colonial counterparts in the United States. War was looked upon as inherently monstrous, and those who waged it—namely, military officers—were looked upon as brutes and philistines. The tone was set by some brilliant novels; among them, *All Quiet on the Western Front, The Journey to the End of the Night,* and *The Good Soldier Schweik.* The only proper protagonist for a tale of war was an enlisted man, and he was to be presented not as a hero but as Everyman, as much a victim of war as any civilian. Any officer above the rank of second lieutenant was to be presented as a martinet or a fool, if not an outright villain, no matter whom he fought for. The old-fashioned tale of prowess and heroism was relegated to second- and third-rate forms of literature, ghost-written autobiographies and stories in pulp magazines on the order of *Argosy* and *Bluebook.*

Even as late as the 1930s the favorite war stories in the pulps concerned World War I pilots. One of the few scientific treatises ever written on the subject of bravery is *The Anatomy of Courage* by Charles Moran, who served as a doctor in the trenches for the British in World War I (and who was better known later as Lord Moran, personal physician to Winston Churchill). Writing in the 1920s, Moran predicted that in the wars of the future adventurous young men who sought glory in war would tend to seek it as pilots. In the twentieth century, he said, they would regard the military pilot as the quintessence of manly daring that the cavalryman had been in the nineteenth.

Serious treatment of the drama and psychology of this new pursuit, flying high-performance aircraft in battle, was left to the occasional pilot who could write, the most notable of them being Antoine de Saint-Exupéry. The literary world remained oblivious. Nevertheless, young men did exactly what Moran predicted. They became military officers so that they could fly, and they flew against astonishingly deadly odds. As late as 1970, I was to discover in an article by a military doctor in a medical journal a career Navy pilot faced a 23 percent likelihood of dying in an accident. This did not even include deaths in combat, which at that time, with the war in Vietnam in progress, were catastrophically high for Navy pilots. *The Right Stuff* became the story of why men were willing—willing?—*delighted!*—to take on such odds in this, an era literary people had long since characterized as the age of the anti-hero. Such was the psychological mystery that animated me in the writing of this book. And if there were those readers who were not interested in the exploration of space per se but who were interested in *The Right Stuff* nonetheless, perhaps it might have been because the mystery caught their imagination as well as mine.

Since this book was first published in 1979 I have enjoyed corresponding with many pilots and with many widows of pilots. Not all have written to pat me on the back, but almost all seemed grateful that someone had tried—and it had to be an outsider—to put into words certain matters that the very code of the pilot rules off-limits in conversation. These . . . matters . . . add up to one of the most extraordinary and most secret dramas of the twentieth century.

T.W.
August 1983

I. The Angels

Within five minutes, or ten minutes, no more than that, three of the others had called her on the telephone to ask her if she had heard that something had happened out there.

"Jane, this is Alice. Listen, I just got a call from Betty, and she said she heard something's happened out there. Have you heard anything?" That was the way they phrased it, call after call. She picked up the telephone and began relaying this same message to some of the others.

"Connie, this is Jane Conrad. Alice just called me, and she says something's happened . . ."

Something was part of the official Wife Lingo for tiptoeing blindfolded around the subject. Being barely twenty-one years old and new around here, Jane Conrad knew very little about this particular subject, since nobody ever talked about it. But the day was young! And what a setting she had for her imminent enlightenment! And what a picture she herself presented! Jane was tall and slender and had rich brown hair and high cheekbones and wide brown eyes. She looked a little like the actress Jean Simmons. Her father was a rancher in southwestern Texas. She had gone East to college, to Bryn Mawr, and had met her husband, Pete, at a debutante's party at the Gulph Mills Club in Philadelphia, when he was a senior at Princeton. Pete was a short, wiry, blond boy who joked around a lot. At any moment his face was likely to break into a wild grin revealing the gap between his front teeth. The Hickory Kid sort, he was; a Hickory Kid on the deb circuit, however. He had an air of energy, self-confidence, ambition, *joie de vivre.* Jane and Pete were married two days after he graduated from Princeton. Last year Jane gave birth to their first child, Peter. And today, here in Florida, in Jacksonville, in the peaceful year 1955, the sun shines through the pines outside, and the very air takes on the sparkle of the ocean. The ocean and a great mica-white beach are less than a mile away. Anyone driving by will see Jane's little house gleaming like a dream house in the pines. It is a brick house, but Jane and Pete painted the bricks white, so that it gleams in the sun against a great green screen of pine trees with a thousand little places where the sun peeks through. They painted the shutters black, which makes the white walls look even more brilliant. The house has only eleven hundred square feet of floor space, but Jane and Pete designed it themselves and that more than makes up for the size. A friend of theirs was the builder and gave them every possible break, so that it cost only eleven thousand dollars. Outside, the sun shines, and inside, the fever rises by the minute as five, ten, fifteen, and, finally, nearly all twenty of the wives join the circuit, trying to find out what has happened, which, in fact, means: to whose husband.

After thirty minutes on such a circuit—this is not an unusual morning around here—a wife begins to feel that the telephone is no longer located on a table or on the kitchen wall. It is exploding in her solar plexus. Yet it would be far worse right now to hear the front doorbell. The protocol is strict on that point, although written down nowhere. No woman is supposed to deliver the final news, and certainly not on the telephone. The matter mustn't be bungled!—that's the idea. No, a man should bring the news when the time comes, a man with some official or moral authority, a clergyman or a comrade of the newly deceased. Furthermore, he should bring the bad news in person. He should turn up at the front door and ring the bell and be standing there like a pillar of coolness and competence, bearing the bad news on ice, like a fish. Therefore, all the telephone calls from the wives were the frantic and portentous beating of the wings of the death angels, as it were. When the final news came, there would be a ring at the front door—a wife in this situation finds herself staring at the front door as if she no longer owns it or controls it—and outside the door would be a man . . . come to inform her that unfortunately something has happened out there, and her husband's body now lies incinerated in the swamps or the pines or the palmetto grass, "burned beyond recognition," which anyone who had been around an air base for very long (fortunately Jane had not) realized was quite an artful euphemism to describe a human body that now looked like an enormous fowl that has burned up in a stove, burned a blackish brown all over, greasy and blistered, fried, in a word, with not only the entire face and all the hair and the ears burned off, not to mention all the clothing, but also the *hands* and *feet*, with what remains of the arms and legs bent at the knees and elbows and burned into absolutely rigid angles, burned a greasy blackish brown like the bursting body itself, so that this husband, father, officer, gentleman, this *ornamentum* of some mother's eye, His Majesty the Baby of just twenty-odd years back, has been reduced to a charred hulk with wings and shanks sticking out of it.

My own husband—how could this be what they were talking about? Jane had heard the young men, Pete among them, talk about other young men who had "bought it" or "augered in" or "crunched," but it had never been anyone they knew, no one in the squadron. And in any event, the way they talked about it, with such breezy, slangy terminology, was the same way they talked about sports. It was as if they were saying, "He was thrown out stealing second base." And that was all! Not one word, not in print, not in conversation—not in this amputated language!—about an incinerated corpse from which a young man's spirit has vanished in an instant, from which all smiles, gestures, moods, worries, laughter, wiles, shrugs, tenderness, and loving looks—*you, my love!*—have disappeared like a sigh, while the terror consumes a cottage in the woods, and a young woman, sizzling with the fever, awaits her confirmation as the new widow of the day.

The next series of calls greatly increased the possibility that it was Pete to whom something had happened. There were only twenty men in the squadron, and soon nine or ten had been accounted for . . . by the fluttering reports of the death angels. Knowing that the word was out that an accident had occurred, husbands who could get to a telephone were calling home to say *it didn't happen to me.* This news, of course, was immediately fed to the fever. Jane's telephone would ring once more, and one of the wives would be saying:

"Nancy just got a call from Jack. He's at the squadron and he says something's happened, but he doesn't know what. He said he saw Frank D— take off about ten minutes ago with Greg in back, so they're all right. What have you heard?"

But Jane has heard nothing except that other husbands, and not hers, are safe and accounted for. And thus, on a sunny day in Florida, outside of the Jacksonville Naval Air Station, in a little white cottage, a veritable dream house, another beautiful young woman was about to be apprised of the *quid pro quo* of her husband's line of work, of the trade-off, as one might say, the subparagraphs of a contract written in no visible form. Just as surely as if she had the entire roster in front of her, Jane now realized that only two men in the squadron were unaccounted for. One was a pilot named Bud Jennings; the other was Pete. She picked up the telephone and did something that was much frowned on in a time of emergency. She called the squadron office. The duty officer answered.

"I want to speak to Lieutenant Conrad," said Jane. "This is Mrs. Conrad."

"I'm sorry," the duty officer said—and then his voice cracked. "I'm sorry . . . I . . ." He couldn't find the words! He was about to cry! "I'm—that's—I mean . . . he can't come to the phone!"

He can't come to the phone!

"It's very important!" said Jane.

"I'm sorry—it's impossible—" The duty officer could hardly get the words out because he was so busy gulping back sobs. *Sobs!* "He can't come to the phone."

"Why not? Where is he?"

"I'm sorry—" More sighs, wheezes, snuffling gasps. "I can't tell you that. I—I have to hang up now!"

And the duty officer's voice disappeared in a great surf of emotion and he hung up.

The duty officer! *The very sound of her voice was more than he could take!*

The world froze, congealed, in that moment. Jane could no longer calculate the interval before the front doorbell would ring and some competent long-faced figure would appear, some Friend of Widows and Orphans, who would inform her, officially, that Pete was dead.

Even out in the middle of the swamp, in this rot-bog of pine trunks, scum slicks, dead dodder vines, and mosquito eggs, even out in this great overripe sump, the smell of "burned beyond recognition" obliterated everything else. When airplane fuel exploded, it created a heat so intense that everything but the hardest metals not only *burned*—everything of rubber, plastic, celluloid, wood, leather, cloth, flesh, gristle, calcium, horn, hair, blood, and protoplasm—it not only burned, it gave up the ghost in the form of every stricken putrid gas known to chemistry. One could smell the horror. It came in through the nostrils and burned the rhinal cavities raw and penetrated the liver and permeated the bowels like a black gas until there was nothing in the universe, inside or out, except the stench of the char. As the helicopter came down between the pine trees and settled onto the bogs, the smell hit Pete Conrad even before the hatch was completely open, and they were not even close enough to see the wreckage yet. The rest of the way Conrad and the crewmen had to travel on foot. After a few steps the water was up to their knees, and then it was up to their armpits, and they kept wading through the water and the scum and the vines and the pine trunks, but it was nothing compared to the smell. Conrad, a twenty-five-year-old lieutenant junior grade, happened to be on duty as squadron safety officer that day and was supposed to make the on-site investigation of the crash. The fact was, however, that this squadron was the first duty assignment of his career, and he had never been at a crash site before and had never smelled any such revolting stench or seen anything like what awaited him.

When Conrad finally reached the plane, which was an SNJ, he found the fuselage burned and blistered and dug into the swamp with one wing sheared off and the cockpit canopy smashed. In the front seat was all that was left of his friend Bud Jennings. Bud Jennings, an amiable fellow, a promising young fighter pilot, was now a horrible roasted hulk—with no head. His head was completely gone, apparently torn off the spinal column like a pineapple off a stalk, except that it was nowhere to be found.

Conrad stood there soaking wet in the swamp bog, wondering what the hell to do. It was a struggle to move twenty feet in this freaking muck. Every time he looked up, he was looking into a delirium of limbs, vines, dappled shadows, and a chopped-up white light that came through the tree-tops—the ubiquitous screen of trees with a thousand little places where the sun peeked through. Nevertheless, he started wading back out into the muck and the scum, and the others followed. He kept looking up. Gradually he could make it out. Up in the treetops there was a pattern of broken limbs where the SNJ had come crashing through. It was like a tunnel through the treetops. Conrad and the others began splashing through the swamp, following the strange path ninety or a hundred feet above them. It took a sharp turn. That must have been where the wing broke off. The trail veered to one side and started downward. They kept looking up and wading through the muck. Then they stopped. There was a great green sap wound up there in the middle of a tree trunk. It was odd. Near the huge gash was . . . tree disease . . . some sort of brownish lumpy sac up in the branches, such as you see in trees infested by bagworms, and there were yellowish curds on the branches around it, as if the disease had caused the sap to ooze out and fester and congeal—except that it couldn't be sap because it was streaked with blood. In the next instant—Conrad didn't have to say a word. Each man could see it all. The lumpy sack was the cloth liner of a flight helmet, with the earphones attached to it. The curds were Bud Jennings's brains. The tree trunk had smashed through the cockpit canopy of the SNJ and knocked Bud Jennings's head to pieces like a melon.

In keeping with the protocol, the squadron commander was not going to release Bud Jennings's name until his widow, Loretta, had been located and a competent male death messenger had been dispatched to tell her. But Loretta Jennings was not at home and could not be found. Hence, a delay—and more than enough time for the other wives, the death angels, to burn with panic over the telephone lines. All the pilots were accounted for except the two who were in the woods, Bud Jennings and Pete Conrad. One chance in two, acey-deucey, one finger-two finger, and this was not an unusual day around here.

Loretta Jennings had been out at a shopping center. When she returned home, a certain figure was waiting outside, a man, a solemn Friend of Widows and Orphans, and it was Loretta Jennings who lost the game of odd and even, acey-deucey, and it was Loretta whose child (she was pregnant with a second) would have no father. It was this young woman who went through all the final horrors that Jane Conrad had imagined—*assumed!*—would be hers to endure forever. Yet this grim stroke of fortune brought Jane little relief.

On the day of Bud Jennings's funeral, Pete went into the back of the closet and brought out his bridge coat, per regulations. This was the most stylish item in the Navy officer's wardrobe. Pete had never had occasion to wear his before. It was a double-breasted coat made of navy-blue melton cloth and came down almost to the ankles. It must have weighed ten pounds. It had a double row of gold buttons down the front and loops for shoulder boards, big beautiful belly-cut collar and lapels, deep turnbacks on the sleeves, a tailored waist, and a center vent in back that ran from the waistline to the bottom of the coat. Never would Pete, or for that matter many other American males in the mid-twentieth century, have an article of clothing quite so impressive and aristocratic as that bridge coat. At the funeral the nineteen little Indians who were left—Navy boys!—lined up manfully in their bridge coats. They looked so young. Their pink, lineless faces with their absolutely clear, lean jawlines popped up bravely, correctly, out of the enormous belly-cut collars of the bridge coats. They sang an old Navy hymn, which slipped into a strange and lugubrious minor key here and there, and included a stanza added especially for aviators. It ended with: "O hear us when we lift our prayer for those in peril in the air."

Three months later another member of the squadron crashed and was burned beyond recognition and Pete hauled out the bridge coat again and Jane saw eighteen little Indians bravely going through the motions at the funeral. Not long after that, Pete was transferred from Jacksonville to the Patuxent River Naval Air Station in Maryland. Pete and Jane had barely settled in there when they got word that another member of the Jacksonville squadron, a close friend of theirs, someone they had had over to dinner many times, had died trying to take off from the deck of a carrier in a routine practice session a few miles out in the Atlantic. The catapult that propelled aircraft off the deck lost pressure, and his ship just dribbled off the end of the deck, with its engine roaring vainly, and fell sixty feet into the ocean and sank like a brick, and he vanished, *just like that.*

Pete had been transferred to Patuxent River, which was known in Navy vernacular as Pax River, to enter the Navy's new test-pilot school. This was considered a major step up in the career of a young Navy aviator. Now that the Korean War was over and there was no combat flying, all the hot young pilots aimed for flight test. In the military they always said "flight test" and not "test flying." Jet aircraft had been in use for barely ten years at the time, and the Navy was testing new jet fighters continually. Pax River was the Navy's prime test center.

Jane liked the house they bought at Pax River. She didn't like it as much as the little house in Jacksonville, but then she and Pete hadn't designed this one. They lived in a community called North Town Creek, six miles from the base. North Town Creek, like the base, was on a scrub-pine peninsula that stuck out into Chesapeake Bay. They were tucked in amid the pine trees. (Once more!) All around were rhododendron bushes. Pete's classwork and his flying duties were very demanding. Everyone in his flight test class, Group 20, talked about how difficult it was—and obviously loved it, because in Navy flying this was the big league. The young men in Group 20 and their wives were Pete's and Jane's entire social world. They associated with no one else. They constantly invited each other to dinner during the week; there was a Group party at someone's house practically every weekend; and they would go off on outings to fish or waterski in Chesapeake Bay. In a way they could not have associated with anyone else, at least not easily, because the boys could talk only about one thing: their flying. One of the phrases that kept running through the conversation was "pushing the outside of the envelope." The "envelope" was a flight-test term referring to the limits of a particular aircraft's performance, how tight a turn it could make at such-and-such a speed, and so on. "Pushing the outside," probing the outer limits, of the envelope seemed to be the great challenge and satisfaction of flight test. At first "pushing the outside of the envelope" was not a particularly terrifying phrase to hear. It sounded once more as if the boys were just talking about sports.

Then one sunny day a member of the Group, one of the happy lads they always had dinner with and drank with and went waterskiing with, was coming in for a landing at the base in an A3J attack plane. He let his airspeed fall too low before he extended his flaps, and the ship stalled out, and he crashed and was burned beyond recognition. And they brought out the bridge coats and sang about those in peril in the air and put the bridge coats away, and the Indians who were left talked about the accident after dinner one night. They shook their heads and said it was a damned shame, but he should have known better than to wait so long before lowering the flaps.

Barely a week had gone by before another member of the Group was coming in for a landing in the same type of aircraft, the A3J, making a ninety-degree turn to his final approach, and something went wrong with the controls, and he ended up with one rear stabilizer wing up and the other one down, and his ship rolled in like a corkscrew from 800 feet up and crashed, and he was burned beyond recognition. And the bridge coats came out and they sang about those in peril in the air and then they put the bridge coats away and after dinner one night they mentioned that the departed had been a good man but was inexperienced, and when the malfunction in the controls put him in that bad corner, he didn't know how to get out of it.

Every wife wanted to cry out: "Well, my God! The *machine* broke! What makes *any* of you think you would have come out of it any better!" Yet intuitively Jane and the rest of them knew it wasn't right even to suggest that. Pete never indicated for a moment that he thought any such thing could possibly happen to him. It seemed not only wrong but dangerous to challenge a young pilot's confidence by posing the question. And that, too, was part of the unofficial protocol for the Officer's Wife. From now on every time Pete was late coming in from the flight line, she would worry. She began to wonder if—no! *assume!*—he had found his way into one of those corners they all talked about so spiritedly, one of those little dead ends that so enlivened conversation around here.

Not long after that, another good friend of theirs went up in an F-4, the Navy's newest and hottest fighter plane, known as the Phantom. He reached twenty thousand feet and then nosed over and dove straight into Chesapeake Bay. It turned out that a hose connection was missing in his oxygen system and he had suffered hypoxia and passed out at the high altitude. And the bridge coats came out and they lifted a prayer about those in peril in the air and the bridge coats were put away and the little

Indians were incredulous. How could anybody fail to check his hose connections? And how could anybody be in such poor condition as to pass out *that quickly* from hypoxia?

A couple of days later Jane was standing at the window of her house in North Town Creek. She saw some smoke rise above the pines from over in the direction of the flight line. Just that, a column of smoke; no explosion or sirens or any other sound. She went to another room, so as not to have to think about it but there was no explanation for the smoke. She went back to the window. In the yard of a house across the street she saw a group of people . . . standing there and looking at her house, as if trying to decide what to do. Jane looked away—but she couldn't keep from looking out again. She caught a glimpse of *a certain figure* coming up the walkway toward her front door. She knew exactly who it was. She had had nightmares like this. And yet this was no dream. She was wide awake and alert. Never more alert in her entire life! Frozen, completely defeated by the sight, she simply waited for the bell to ring. She waited, but there was not a sound. Finally she could stand it no more. In real life, unlike her dream life, Jane was both too self-possessed and too polite to scream through the door: "Go away!" So she opened it. There was no one there, no one at all. There was no group of people on the lawn across the way and no one to be seen for a hundred yards in any direction along the lawns and leafy rhododendron roads of North Town Creek.

Then began a cycle in which she had both the nightmares and the hallucinations, continually. Anything could touch off an hallucination: a ball of smoke, a telephone ring that stopped before she could answer it, the sound of a siren, even the sound of trucks starting up (crash trucks!). Then she would glance out the window, and a certain figure would be coming up the walk, and she would wait for the bell. The only difference between the dreams and the hallucinations was that the scene of the dreams was always the little white house in Jacksonville. In both cases, the feeling that *this time it has happened* was quite real.

The star pilot in the class behind Pete's, a young man who was the main rival of their good friend Al Bean, went up in a fighter to do some power-dive tests. One of the most demanding disciplines in flight test was to accustom yourself to making precise readings from the control panel in the same moment that you were pushing the outside of the envelope. This young man put his ship into the test dive and was still reading out the figures, with diligence and precision and great discipline, when he augered straight into the oyster flats and was burned beyond recognition. And the bridge coats came out and they sang about those in peril in the air and the bridge coats were put away, and the little Indians remarked that the departed was a swell guy and a brilliant student of flying; a little too *much* of a student, in fact; he hadn't bothered to look out the window at the real world soon enough. Beano—Al Bean—wasn't quite so brilliant; on the other hand, he was still here.

Like many other wives in Group 20 Jane wanted to talk about the whole situation, the incredible series of fatal accidents, with her husband and the other members of the Group, to find out how they were taking it. But somehow the unwritten protocol forbade discussions of this subject, which was the fear of death. Nor could Jane or any of the rest of them talk, really *have a talk*, with anyone around the base. You could talk to another wife about being worried. But what good did it do? Who *wasn't* worried? You were likely to get a look that said: "*Why dwell on it?*" Jane might have gotten away with divulging the matter of the nightmares. But *hallucinations*? There was no room in Navy life for any such anomalous tendency as that.

By now the bad string had reached ten in all, and almost all of the dead had been close friends of Pete and Jane, young men who had been in their house many times, young men who had sat across from Jane and chattered like the rest of them about the grand adventure of military flying. And the survivors still sat around *as before*—with the same inexplicable exhilaration! Jane kept watching Pete for some sign that his spirit was cracking, but she saw none. He talked a mile a minute, kidded and joked, laughed with his Hickory Kid cackle. He always had. He still enjoyed the company of members of the group like Wally Schirra and Jim Lovell. Many young pilots were taciturn and cut loose with the strange fervor of this business only in the air. But Pete and Wally and Jim were not reticent; not in any situation. They loved to kid around. Pete called Jim Lovell "Shaky," because it was the last thing a pilot would want to be called. Wally Schirra was outgoing to the point of hearty; he loved practical jokes and dreadful puns, and so on. The three of them—*even in the midst of this bad string!*—would love to get on a subject such as accident-prone Mitch Johnson. Accident-prone Mitch Johnson, it seemed, was a Navy pilot whose life was in the hands of two angels, one of them bad and the other one good. The bad angel would put him into accidents that would have annihilated any ordinary pilot, and the good angel would bring him out of them without a scratch. Just the other day—this was the sort of story Jane would hear them tell—Mitch Johnson was coming in to land on a carrier. But he came in short, missed the flight deck, and crashed into the fantail, below the deck. There was a tremendous explosion, and the rear half of the plane fell into the water in flames. Everyone on the flight deck said, "Poor Johnson. The good angel was off duty." They were still debating how to remove the debris and his mortal remains when a phone rang on the bridge. A somewhat dopey voice said, "This is Johnson. Say, listen, I'm down here in the supply hold and the hatch is locked and I can't find the lights and I can't see a goddamned thing and I tripped over a cable and I think I hurt my leg." The officer on the bridge slammed the phone down, then vowed to find out what morbid sonofabitch could pull a phone prank at a time like this. Then the phone rang again, and the man with the dopey voice managed to establish the fact that he was, indeed, Mitch Johnson. The good angel had not left his side. When he smashed into the fantail, he hit some empty ammunition drums, and they cushioned the impact, leaving him groggy but not seriously hurt. The fuselage had blown to pieces; so he just stepped out onto the fantail and opened a hatch that led into the supply hold. It was pitch black in there, and there were cables all across the floor, holding down spare aircraft engines. Accident-prone Mitch Johnson kept tripping over these cables until he found a telephone. Sure enough, the one injury he had was a bruised shin from tripping over a cable. The man was accident-prone! Pete and Wally and Jim absolutely cracked up over stories like this. It was amazing. Great sports yarns! Nothing more than that.

A few days later Jane was out shopping at the Pax River commissary on Saunders Road, near the main gate to the base. She heard the sirens go off at the field, and then she heard the engines of the crash trucks start up. This time Jane was determined to keep calm. Every instinct made her want to rush home, but she forced herself to stay in the commissary and continue shopping. For thirty minutes she went through the motions of completing her shopping list. Then she drove home to North Town Creek. As she reached the house, she saw a figure going up the sidewalk. It was a man. Even from the back there was no question as to who he was. He had on a black suit, and there was a white band around his neck. It was her minister, from the Episcopal Church. She stared, and this vision did not come and go. The figure kept on walking up the front walk. She was not asleep now, and she was not inside her house glancing out the front window. She was outside in her car in front of her house. She was not dreaming, and she was not hallucinating, and the figure kept walking up toward her front door.

The commotion at the field was over one of the most extraordinary things that even veteran pilots had ever seen at Pax River. And they had all seen it, because practically the entire flight line had gathered out on the field for it, as if it had been an air show.

Conrad's friend Ted Whelan had taken a fighter up, and on takeoff there had been a structural failure that caused a hydraulic leak. A red warning light showed up on Whelan's panel, and he had a talk with the ground. It was obvious that the leak would cripple the controls before he could get the ship back down to the field for a landing. He would have to bail out; the only question was where and when, and so they had a talk about that. They decided that he should jump at 8,100 feet at such-and-such a speed, directly over the field. The plane would crash into the Chesapeake Bay, and he would float down to the field. Just as coolly as anyone could have asked for it, Ted Whelan lined the ship up to come across the field at 8,100 feet precisely and he punched out, ejected.

Down on the field they all had their faces turned up to the sky. They saw Whelan pop out of the cockpit. With his Martin-Baker seat-parachute rig strapped on, he looked like a little black geometric lump a mile and a half up in the blue. They watched him as he started dropping. Everyone waited for the parachute to open. They waited a few more seconds, and then they waited some more. The little shape was getting bigger and bigger and picking up tremendous speed. Then there came an unspeakable instant at which everyone on the field who knew anything about parachute jumps knew what was going to happen. Yet even for them it was an unearthly feeling, for no one had ever seen any such thing happen so close up, from start to finish, from what amounted to a grandstand seat. Now the shape was going so fast and coming so close it began to play tricks on the eyes. It seemed to stretch out. It became much bigger and hurtled toward them at a terrific speed, until they couldn't make out its actual outlines at all. Finally there was just a streaking black blur before their eyes, followed by what seemed like an explosion. Except that it was not an explosion; it was the tremendous *crack* of Ted Whelan, his helmet, his pressure suit, and his seat-parachute rig smashing into the center of the runway, precisely on target, right in front of the crowd; an absolute bull's-eye. Ted Whelan had no doubt been alive until the instant of impact. He had had about thirty seconds to watch the Pax River base and the peninsula and Baltimore County and continental America and the entire comprehensible world rise up to smash him. When they lifted his body up off the concrete, it was like a sack of fertilizer.

Pete took out the bridge coat again and he and Jane and all the little Indians went to the funeral for Ted Whelan. That it hadn't been Pete was not solace enough for Jane. That the preacher had not, in fact, come to her front door as the Solemn Friend of Widows and Orphans, but merely for a church call . . . had not brought peace and relief. That Pete still didn't show the slightest indication of thinking that any unkind fate awaited him no longer lent her even a moment's courage. The next dream and the next hallucination, and the next and the next, merely seemed more real. For she now *knew*. She now knew the subject and the essence of this enterprise, even though not a word of it had passed anybody's lips. She even knew why Pete—the Princeton boy she met at a deb party at the Gulph Mills Club!—would never quit, never withdraw from this grim business, unless in a coffin. And God knew, and she knew, there was a coffin waiting for each little Indian.

Seven years later, when a reporter and a photographer from *Life* magazine actually stood near her in her living room and watched her face, while outside, on the lawn, a crowd of television crewmen and newspaper reporters waited for a word, an indication, anything—perhaps a glimpse through a part in a curtain!—waited for some sign of what she felt—when one and all asked with their ravenous eyes and, occasionally, in so many words: "How do you feel?" and "Are you scared?"—America wants to know!—it made Jane want to laugh, but in fact she couldn't even manage a smile.

"Why ask *now*?" she wanted to say. But they wouldn't have had the faintest notion of what she was talking about.

II. The Right Stuff

What an extraordinary grim stretch that had been . . . and yet thereafter Pete and Jane would keep running into pilots from other Navy bases, from the Air Force, from the Marines, who had been through their own extraordinary grim stretches. There was an Air Force pilot named Mike Collins, a nephew of former Army Chief of Staff J. Lawton Collins. Mike Collins had undergone eleven weeks of combat training at Nellis Air Force Base, near Las Vegas, and in that eleven weeks twenty-two of his fellow trainees had died in accidents, which was an extraordinary rate of two per week. Then there was a test pilot, Bill Bridgeman. In 1952, when Bridgeman was flying at Edwards Air Force Base, sixty-two Air Force pilots died in the course of thirty-six weeks of training, an extraordinary rate of 1.7 per week. Those figures were for fighter-pilot trainees only; they did not include the test pilots, Bridgeman's own confreres, who were dying quite regularly enough.

Extraordinary, to be sure; except that every veteran of flying small high-performance jets seemed to have experienced these bad strings.

In time, the Navy would compile statistics showing that for a career Navy pilot, i.e., one who intended to keep flying for twenty years as Conrad did, there was a 23 percent probability that he would die in an aircraft accident. This did not even include combat deaths, since the military did not classify death in combat as accidental. Furthermore, there was a better than even chance, a 56 percent probability, to be exact, that at some point a career Navy pilot would have to eject from his aircraft and attempt to come down by parachute. In the era of jet fighters, ejection meant being exploded out of the cockpit by a nitroglycerine charge, like a human cannonball. The ejection itself was so hazardous—men lost knees, arms, and their lives on the rim of the cockpit or had the skin torn off their faces when they hit the "wall" of air outside—that many pilots chose to wrestle their aircraft to the ground rather than try it . . . and died that way instead.

The statistics were not secret, but neither were they widely known, having been eased into print rather obliquely in a medical journal. No pilot, and certainly no pilot's wife, had any need of the statistics in order to know the truth, however. The funerals took care of that in the most dramatic way possible. Sometimes, when the young wife of a fighter pilot would have a little reunion with the girls she went to school with, an odd fact would dawn on her: *they* have not been going to funerals. And then Jane Conrad would look at Pete . . . Princeton, Class of 1953 . . . Pete had already worn his great dark sepulchral bridge coat more than most boys of the Class of '53 had worn their tuxedos. How many of those happy young men had buried more than a dozen friends, comrades, and co-workers? (Lost through violent death in the execution of everyday duties.) At the time, the 1950's, students from Princeton took great pride in going into what they considered highly competitive, aggressive pursuits, jobs on Wall Street, on Madison Avenue, and at magazines such as *Time* and *Newsweek*. There was much fashionably brutish talk of what "dog-eat-dog" and "cutthroat" competition they found there; but in the rare instances when one of these young men died on the job, it was likely to be from choking on a chunk of Chateaubriand, while otherwise blissfully boiled, in an expense-account restaurant in Manhattan. How many would have gone to work, or stayed at work, on cutthroat Madison Avenue if there had been a 23 percent chance, nearly one chance in four, of dying from it? Gentlemen, we're having this little problem with chronic violent death . . .

And yet was there any basic way in which Pete (or Wally Schirra or Jim Lovell or any of the rest of them) was different from other college boys his age? There didn't seem to be, other than his love of flying. Pete's father was a Philadelphia stockbroker who in Pete's earliest years had a house in the Main Line suburbs, a limousine, and a chauffeur. The

Depression eliminated the terrific brokerage business, the house, the car, and the servants; and by and by his parents were divorced and his father moved to Florida. Perhaps because his father had been an observation balloonist in the First World War—an adventurous business, since the balloons were prized targets of enemy aircraft—Pete was fascinated by flying. He went to Princeton on the Holloway Plan, a scholarship program left over from the Second World War in which a student trained with midshipmen from the Naval Academy during the summers and graduated with a commission in the Regular Navy. So Pete graduated, received his commission, married Jane, and headed off to Pensacola, Florida, for flight training.

Then came the difference, looking back on it.

A young man might go into military flight training believing that he was entering some sort of technical school in which he was simply going to acquire a certain set of skills. Instead, he found himself all at once enclosed in a fraternity. And in this fraternity, even though it was military, men were not rated by their outward rank as ensigns, lieutenants, commanders, or whatever. No, herein the world was divided into those who had it and those who did not. This quality, this *it,* was never named, however, nor was it talked about in any way.

As to just what this ineffable quality was . . . well, it obviously involved bravery. But it was not bravery in the simple sense of being willing to risk your life. The idea seemed to be that any fool could do that, if that was all that was required, just as any fool could throw away his life in the process. No, the idea here (in the all-enclosing fraternity) seemed to be that a man should have the ability to go up in a hurtling piece of machinery and put his hide on the line and then have the moxie, the reflexes, the experience, the coolness, to pull it back in the last yawning moment—and then to go up again *the next day,* and the next day, and every next day, even if the series should prove infinite—and, ultimately, in its best expression, do so in a cause that means something to thousands, to a people, a nation, to humanity, to God. Nor was there *a test* to show whether or not a pilot had this righteous quality. There was, instead, a seemingly infinite series of tests. A career in flying was like climbing one of those ancient Babylonian pyramids made up of a dizzy progression of steps and ledges, a ziggurat, a pyramid extraordinarily high and steep; and the idea was to prove at every foot of the way up that pyramid that you were one of the elected and anointed ones who had *the right stuff* and could move higher and higher and even—ultimately, God willing, one day—that you might be able to join that special few at the very top, that elite who had the capacity to bring tears to men's eyes, the very Brotherhood of the Right Stuff itself.

None of this was to be mentioned, and yet it was acted out in a way that a young man could not fail to understand. When a new flight (i.e., a class) of trainees arrived at Pensacola, they were brought into an auditorium for a little lecture. An officer would tell them: "Take a look at the man on either side of you." Quite a few actually swiveled their heads this way and that, in the interest of appearing diligent. Then the officer would say: "One of the three of you is not going to make it!"—meaning, not get his wings. That was the opening theme, the *motif* of primary training. We already know that one-third of you do not have the right stuff—it only remains to find out who.

Furthermore, that was the way it turned out. At every level in one's progress up that staggeringly high pyramid, the world was once more divided into those men who had the right stuff to continue the climb and those who had to be *left behind* in the most obvious way. Some were eliminated in the course of the opening classroom work, as either not smart enough or not hardworking enough, and were left behind. Then came the basic flight instruction, in single-engine, propeller-driven trainers, and a few more—even though the military tried to make this stage easy—were washed out and left behind. Then came more demanding levels, one after the other, formation flying, instrument flying, jet training, all-weather flying, gunnery, and at each level more were washed out and left behind. By this point easily a third of the original candidates had been, indeed, eliminated . . . from the ranks of those who might prove to have the right stuff.

In the Navy, in addition to the stages that Air Force trainees went through, the neophyte always had waiting for him, out in the ocean, a certain grim gray slab; namely, the deck of an aircraft carrier; and with it perhaps the most difficult routine in military flying, carrier landings. He was shown films about it, he heard lectures about it, and he knew that carrier landings were hazardous. He first practiced touching down on the shape of a flight deck painted on an airfield. He was instructed to touch down and gun right off. This was safe enough—the shape didn't move, at least—but it could do terrible things to, let us say, the gyroscope of the soul. *That shape!—it's so damned small!* And more candidates were washed out and left behind. Then came the day, without warning, when those who remained were sent out over the ocean for the first of many days of reckoning with the slab. The first day was always a clear day with little wind and a calm sea. The carrier was so steady that it seemed, from up there in the air, to be resting on pilings, and the candidate usually made his first carrier landing successfully, with relief and even *élan*. Many young candidates looked like terrific aviators up to that very point—and it was not until they were actually standing on the carrier deck that they first began to wonder if they had the proper stuff, after all. In the training film the flight deck was a grand piece of gray geometry, perilous, to be sure, but an amazing abstract shape as one looks down upon it on the screen. And yet once the newcomer's two feet were on it . . . *Geometry*—my God, man, this is a . . . skillet! It *heaved*, it moved up and down underneath his feet, it pitched up, it pitched down, it rolled to port (this great beast *rolled!*) and it rolled to starboard, as the ship moved into the wind and, therefore, into the waves, and the wind kept sweeping across, sixty feet up in the air out in the open sea, and there were no railings whatsoever. This was a *skillet!*—a frying pan!—a short-order grill!—not gray but black, smeared with skid marks from one end to the other and glistening with pools of hydraulic fluid and the occasional jet-fuel slick, all of it still hot, sticky, greasy, runny, virulent from God knows what traumas—still ablaze!—consumed in detonations, explosions, flames, combustion, roars, shrieks, whines, blasts, horrible shudders, fracturing impacts, as little men in screaming red and yellow and purple and green shirts with black Mickey Mouse helmets over their ears skittered about on the surface as if for their very lives (you've said it now!), hooking fighter planes onto the catapult shuttles so that they can explode their afterburners and be slung off the deck in a red-mad fury with a *kaboom!* that pounds through the entire deck—a procedure that seems absolutely controlled, orderly, sublime, however, compared to what he is about to watch as aircraft return to the ship for what is known in the engineering stoicisms of the military as "recovery and arrest." To say that an F-4 was coming back onto this heaving barbecue from out of the sky at a speed of 135 knots . . . that might have been the truth in the training lecture, but it did not begin to get across the idea of what the newcomer saw from the deck itself, because it created the notion that perhaps the plane was gliding in. On the deck one knew differently! As the aircraft came closer and the carrier heaved on into the waves and the plane's speed did not diminish and the deck did not grow steady—indeed, it pitched up and down five or ten feet per greasy heave—one experienced a neural alarm that no lecture could have prepared him for: This is not an *airplane* coming toward me, it is a brick with some poor sonofabitch riding it *(someone much like myself!)*, and it is not *gliding*, it is *falling*, a thirty-thousand-pound brick, headed not for a stripe on the deck but for *me*—and with a horrible *smash!* it hits the skillet, and with a blur of momentum as big as a freight train's it hurtles toward the far end of the deck—another blinding storm!—another roar as the pilot pushes the throttle up to full military power and another smear of rubber screams out over the skillet—and this is nominal!—quite okay!—for a wire stretched across the deck has grabbed the hook on the end of the plane as it hit the deck tail down, and the smash was the rest of the fifteen-ton brute slamming onto the deck, as it tripped up, so that it is now straining against the wire at full throttle, in case it hadn't held and the plane had "boltered" off the end of the deck and had to struggle up into the air again. And already the Mickey Mouse helmets are running toward the fiery monster . . .

And the candidate, looking on, begins to *feel* that great heaving sunblazing deathboard of a deck wallowing in his own vestibular system—and suddenly he finds himself backed up against his own limits. He ends up going to the flight surgeon with so-called conversion symptoms. Overnight he develops blurred vision or numbness in his hands and feet or sinusitis so severe that he cannot tolerate changes in altitude. On one level the symptom is real. He really cannot see too well or use his fingers or stand the pain. But somewhere in his subconscious he knows it is a plea and a beg-off; he shows not the slightest concern (the flight surgeon notes) that the condition might be permanent and affect him in whatever life awaits him outside the arena of the right stuff.

Those who remained, those who qualified for carrier duty—and even more so those who later on qualified for *night* carrier duty—began to feel a bit like Gideon's warriors. *So many have been left behind!* The young warriors were now treated to a deathly sweet and quite unmentionable sight. They could gaze at length upon the crushed and wilted pariahs who had washed out. They could inspect those who did not have that righteous stuff.

The military did not have very merciful instincts. Rather than packing up these poor souls and sending them home, the Navy, like the Air Force and the Marines, would try to make use of them in some other role, such as flight controller. So the washout has to keep taking classes with the rest of his group, even though he can no longer touch an airplane. He sits there in the classes staring at sheets of paper with cataracts of sheer human mortification over his eyes while the rest steal looks at him . . . this man reduced to an ant, this untouchable, this poor sonofabitch. And in what test had he been found wanting? Why, it seemed to be nothing less than *manhood* itself. Naturally, this was never mentioned, either. Yet there it was. *Manliness, manhood, manly courage* . . . there was something ancient, primordial, irresistible about the challenge of this stuff, no matter what a sophisticated and rational age one might think he lived in.

Perhaps because it could not be talked about, the subject began to take on superstitious and even mystical outlines. A man either had it or he didn't! There was no such thing as having *most* of it. Moreover, it could blow at any seam. One day a man would be ascending the pyramid at a terrific clip, and the next—bingo!—he would reach his own limits in the most unexpected way. Conrad and Schirra met an Air Force pilot who had had a great pal at Tyndall Air Force Base in Florida. This man had been the budding ace of the training class; he had flown the hottest fighter-style trainer, the T–38, like a dream; and then he began the routine step of being checked out in the T–33. The T–33 was not nearly as hot an aircraft as the T–38; it was essentially the old P–80 jet fighter. It had an exceedingly small cockpit. The pilot could barely move his shoulders. It was the sort of airplane of which everybody said, "You don't get into it, you *wear* it." Once inside a T–33 cockpit this man, this budding ace, developed claustrophobia of the most paralyzing sort. He tried everything to overcome it. He even went to a psychiatrist, which was a serious mistake for a military officer if his superiors learned of it. But nothing worked. He was shifted over to flying jet transports, such as the C–135. Very demanding and necessary aircraft they were, too, and he was still spoken of as an excellent pilot. But as everyone knew—and, again, it was never explained in so many words—only those who were assigned to fighter squadrons, the "fighter jocks," as they called each other with a self-satisfied irony, remained in the true fraternity. Those assigned to transports were not humiliated like washouts—*somebody* had to fly those planes—nevertheless, they, too, had been *left behind* for lack of the right stuff.

Or a man could go for a routine physical one fine day, feeling like a million dollars, and be grounded for *fallen arches*. It happened!—just like that! (And try raising them.) Or for breaking his wrist and losing only *part* of its mobility. Or for a minor deterioration of eyesight, or for any of hundreds of reasons that would make no difference to a man in an ordinary occupation. As a result all fighter jocks began looking upon doctors as their natural enemies. Going to see a flight surgeon was a no-gain proposition; a pilot could only hold his own or lose in the doctor's office. To be grounded for a medical reason was no humiliation, looked at objectively. But it was a humiliation, nonetheless!—for it meant you no longer had that indefinable, unutterable, integral stuff. (It could blow at *any* seam.)

All the hot young fighter jocks began trying to test the limits themselves in a superstitious way. They were like believing Presbyterians of a century before who used to probe their own experience to see if they were truly among *the elect*. When a fighter pilot was in training, whether in the Navy or the Air Force, his superiors were continually spelling out strict rules for him, about the use of the aircraft and conduct in the sky. They repeatedly forbade so-called hot-dog stunts, such as outside loops, buzzing, flat-hatting, hedgehopping and flying under bridges. But somehow one got the message that the man who truly *had* it could ignore those rules—not that he should make a point of it, but that he *could*—and that after all there was only one way to find out—and that in some strange unofficial way, peeking through his fingers, his instructor halfway expected him to challenge all the limits. They would give a lecture about how a pilot should never fly without a good solid breakfast—eggs, bacon, toast, and so forth—because if he tried to fly with his blood-sugar level too low, it could impair his alertness. Naturally, the next day every hot dog in the unit would get up and have a breakfast consisting of one cup of black coffee and take off and go up into a vertical climb until the weight of the ship exactly canceled out the upward thrust of the engine and his air speed was zero, and he would hang there for one thick adrenal instant—and then fall like a rock, until one of three things happened: he keeled over nose first and regained his aerodynamics and all was well, he went into a spin and fought his way out of it, or he went into a spin and had to eject or crunch it, which was always supremely possible.

Likewise, "hassling"—mock dogfighting—was strictly forbidden, and so naturally young fighter jocks could hardly wait to go up in, say, a pair of F–100s and start the duel by making a pass at each other at 800 miles an hour, the winner being the pilot who could slip in behind the other one and get locked in on his tail ("wax his tail"), and it was not uncommon for some eager jock to try too tight an outside turn and have his engine flame out, whereupon, unable to restart it, he has to eject . . . and he shakes his fist at the victor as he floats down by parachute and his million-dollar aircraft goes *kaboom!* on the palmetto grass or the desert floor, and he starts thinking about how he can get together with the other guy back at the base in time for the two of them to get their stories straight before the investigation: "I don't know what happened, sir. I was pulling up after a target run, and it just flamed out on me." Hassling was forbidden, and hassling that led to the destruction of an aircraft was a serious court-martial offense, and the man's superiors knew that the engine hadn't *just flamed out*, but every unofficial impulse on the base seemed to be saying: "Hell, we wouldn't give you a nickel for a pilot who hasn't done some crazy rat-racing like that. It's all part of the right stuff."

The other side of this impulse showed up in the reluctance of the young jocks to admit it when they had maneuvered themselves into a bad corner they couldn't get out of. There were two reasons why a fighter pilot hated to declare an emergency. First, it triggered a complex and very public chain of events at the field: all other incoming flights were held up, including many of one's comrades who were probably low on fuel; the fire trucks came trundling out to the runway like yellow toys (as seen from way up there), the better to illustrate one's hapless state; and the bureaucracy began to crank up the paper monster for the investigation that always followed. And second, to declare an emergency, one first had to reach that conclusion in his own mind, which to the young pilot was the same as saying: "A minute ago I still *had* it—now I need your help!" To have a bunch of young fighter pilots up in the air thinking this way used to drive flight controllers crazy. They would see a ship beginning to drift off the radar, and they couldn't rouse the pilot on the microphone for anything other than a few meaningless mumbles, and they would know he was probably out there with engine failure at a low altitude, trying to reignite by lowering his auxiliary generator rig, which had a little propeller that was supposed to spin in the slipstream like a child's pinwheel.

"Whiskey Kilo Two Eight, do you want to declare an emergency?"

This would rouse him!—to say: "Negative, negative, Whiskey Kilo Two Eight is not declaring an emergency."

Kaboom. Believers in the right stuff would rather crash and burn.

One fine day, after he had joined a fighter squadron, it would dawn on the young pilot exactly how the losers in the great fraternal competition were now being left behind. Which is to say, not by instructors or other superiors or by failures at prescribed levels of competence, but by death. At this point the essence of the enterprise would begin to dawn on him. Slowly, step by step, the ante had been raised until he was now involved in what was surely the grimmest and grandest gamble of manhood. Being a fighter pilot—for that matter, simply taking off in a single-engine jet fighter of the Century series, such as an F-102, or any of the military's other marvelous bricks with fins on them—presented a man, on a perfectly sunny day, with more ways to get himself killed than his wife and children could imagine in their wildest fears. If he was barreling down the runway at two hundred miles an hour, completing the takeoff run, and the board started lighting up red, should he (a) abort the takeoff (and try to wrestle with the monster, which was gorged with jet fuel, out in the sand beyond the end of the runway) or (b) eject (and hope that the goddamned human cannonball trick works at zero altitude and he doesn't shatter an elbow or a kneecap on the way out) or (c) continue the takeoff and deal with the problem aloft (knowing full well that the ship may be on fire and therefore seconds away from exploding)? He would have one second to sort out the options and act, and this kind of little workaday decision came up all the time. Occasionally a man would look coldly at the binary problem he was now confronting every day—Right Stuff/Death—and decide it wasn't worth it and voluntarily shift over to transports or reconnaissance or whatever. And his comrades would wonder, for a day or so, what evil virus had invaded his soul . . . as they left him behind. More often, however, the reverse would happen. Some college graduate would enter Navy aviation through the Reserves, simply as an alternative to the Army draft, fully intending to return to civilian life, to some waiting profession or family business; would become involved in the obsessive business of ascending the ziggurat pyramid of flying; and, at the end of his enlistment, would astound everyone back home and very likely himself as well by signing up for another one. What on earth got into him? He couldn't explain it. After all, the very words for it had been amputated. A Navy study showed that two-thirds of the fighter pilots who were rated in the top rungs of their groups—i.e., the hottest young pilots—reenlisted when the time came, and practically all were college graduates. By this point, a young fighter jock was like the preacher in *Moby Dick* who climbs up into the pulpit on a rope ladder and then pulls the ladder up behind him; except the pilot could not use the words necessary to express the vital lessons. Civilian life, and even home and hearth, now seemed not only far away but far *below*, back down many levels of the pyramid of the right stuff.

A fighter pilot soon found he wanted to associate only with other fighter pilots. Who else could understand the nature of the little proposition (right stuff/death) they were all dealing with? And what other subject could compare with it? It was riveting! To talk about it in so many words was forbidden, of course. The very words *death, danger, bravery, fear* were not to be uttered except in the occasional specific instance or for ironic effect. Nevertheless, the subject could be adumbrated in *code* or *by example.* Hence the endless evenings of pilots huddled together talking about flying. On these long and drunken evenings (the bane of their family life) certain theorems would be propounded and demonstrated—and all by *code* and *example.* One theorem was: There are no *accidents* and no fatal flaws in the machines; there are only pilots with the wrong stuff. (I.e., blind Fate can't kill me.) When Bud Jennings crashed and burned in the swamps at Jacksonville, the other pilots in Pete Conrad's squadron said: *How could he have been so stupid?* It turned out that Jennings had gone up in the SNJ with his cockpit canopy opened in a way that was expressly forbidden in the manual, and carbon monoxide had been sucked in from the exhaust, and he passed out and crashed. All agreed that Bud Jennings was a good guy and a good pilot, but his epitaph on the ziggurat was: *How could he have been so stupid?* This seemed shocking at first, but by the time Conrad had reached the end of that bad string at Pax River, he was capable of his own corollary to the theorem: viz., no single factor ever killed a pilot; there was always a chain of mistakes. But what about Ted Whelan, who fell like a rock from 8,100 feet when his parachute failed? Well, the parachute was merely part of the chain: first, someone should have caught the structural defect that resulted in the hydraulic leak that triggered the emergency; second, Whelan did not check out his seat-parachute rig, and the drogue failed to separate the main parachute from the seat; but even after those two mistakes, Whelan had fifteen or twenty seconds, as he fell, to disengage himself from the seat and open the parachute manually. Why just stare at the scenery coming up to smack you in the face! And everyone nodded. (He failed—but I wouldn't have!) Once the theorem and the corollary were understood, the Navy's statistics about one in every four Navy aviators dying meant nothing. The figures were averages, and averages applied to those with average stuff.

A riveting subject, especially if it were one's own hide that was on the line. Every evening at bases all over America, there were military pilots huddled in officers clubs eagerly cutting the right stuff up in coded slices so they could talk about it. What more compelling topic of conversation was there in the world? In the Air Force there were even pilots who would ask the tower for priority landing clearance so that they could make the beer call on time, at 4 p.m. sharp, at the Officers Club. They would come right out and state the reason. The drunken rambles began at four and sometimes went on for ten or twelve hours. Such conversations! They diced that righteous stuff up into little bits, bowed ironically to it, stumbled blindfolded around it, groped, lurched, belched, staggered, bawled, sang, roared, and feinted at it with self-deprecating humor. Nevertheless!—they never mentioned it by name. No, they used the approved codes, such as: "Like a jerk I got myself into a hell of a corner today." They told of how they "lucked out of it." To get across the extreme peril of his exploit, one would use certain oblique cues. He would say, "I looked over at Robinson"—who would be known to the listeners as a non-com who sometimes rode backseat to read radar—"and he wasn't talking any more, he was just staring at the radar, like this, giving it that *zombie* look. Then I *knew* I was in trouble!" Beautiful! Just right! For it would also be known to the listeners that the non-coms advised one another: "*Never* fly with a lieutenant. *Avoid* captains and majors. Hell, man, do yourself a favor: don't fly with anybody below colonel." Which in turn said: "Those young bucks shoot dice with death!" And yet once in the air the non-com had his own standards. He was determined to remain as outwardly cool as the pilot, so that when the pilot did something that truly petrified him, he would say nothing; instead, he would turn silent, catatonic, like a zombie. Perfect! *Zombie*. There you had it, compressed into a single word, all of the foregoing. I'm a hell of a pilot! I shoot dice with death! And now all you fellows know it! And I haven't spoken of that unspoken stuff even once!

The talking and drinking began at the beer call, and then the boys would break for dinner and come back afterward and get more wasted and more garrulous or else more quietly fried, drinking good cheap PX booze until 2 a.m. The night was young! Why not get the cars and go out for a little proficiency run? It seemed that every fighter jock thought himself an ace driver, and he would do anything to obtain a hot car, especially a sports car, and the drunker he was, the more convinced he would be about his driving skills, as if the right stuff, being indivisible, carried over into any enterprise whatsoever, under any conditions. A little proficiency run, boys! (There's only one way to find out!) And they would roar off in close formation from, say, Nellis Air Force Base, down Route 15, into Las Vegas, barreling down the highway, rat-racing, sometimes four abreast, jockeying for position, piling into the most listless curve in the desert flats as if they were trying to root each other out of the groove at the Rebel 500—and then bursting into downtown Las Vegas with a rude fraternal roar like the Hell's Angels—and the natives chalked it up to youth and drink and the bad element that the Air Force attracted. They knew nothing about the right stuff, of course.

More fighter pilots died in automobiles than in airplanes. Fortunately, there was always some kindly soul up the chain to certify the papers "line of duty," so that the widow could get a better break on the insurance. That was okay and only proper because somehow the system itself had long ago said *Skol!* and *Quite right!* to the military cycle of Flying & Drinking and Drinking & Driving, as if there were no other way. Every young fighter jock knew the feeling of getting two or three hours' sleep and then waking up at 5:30 a.m. and having a few cups of coffee, a few cigarettes, and then carting his poor quivering liver out to the field for another day of flying. There were those who arrived not merely hungover but still drunk, slapping oxygen tank cones over their faces and trying to burn the alcohol out of their systems, and then going up, remarking later: "I don't *advise* it, you understand, but it *can* be done." (Provided you have the right stuff, you miserable pudknocker.)

Air Force and Navy airfields were usually on barren or marginal stretches of land and would have looked especially bleak and Low Rent to an ordinary individual in the chilly light of dawn. But to a young pilot there was an inexplicable bliss to coming out to the flight line while the sun was just beginning to cook up behind the rim of the horizon, so that the whole field was still in shadow and the ridges in the distance were in silhouette and the flight line was a monochrome of Exhaust Fume Blue, and every little red light on top of the water towers or power stanchions looked dull, shriveled, congealed, and the runway lights, which were still on, looked faded, and even the landing lights on a fighter that had just landed and was taxiing in were no longer dazzling, as they would be at night, and looked instead like shriveled gobs of candlepower out there—and yet it was beautiful, exhilarating!—for he was revved up with adrenalin, anxious to take off before the day broke, to burst up into the sunlight over the ridges before all those thousands of comatose souls down there, still dead to the world, snug in home and hearth, even came to their senses. To take off in an F–100 at dawn and cut in the afterburner and hurtle twenty-five thousand feet up into the sky so suddenly that you felt not like a bird but like a trajectory, yet with full control, full control of *five tons* of thrust, all of which flowed from your will and through your fingertips, with the huge engine right beneath you, so close that it was as if you were riding it bareback, until you leveled out and went supersonic, an event registered on earth by a tremendous cracking boom that shook windows, but up here only by the fact that you now felt utterly free of the earth—to describe it, even to wife, child, near ones and dear ones, seemed impossible. So the pilot kept it to himself, along with an even more indescribable . . . an even more sinfully inconfessable . . . feeling of superiority, appropriate to him and to his kind, lone bearers of the right stuff.

From *up here* at dawn the pilot looked down upon poor hopeless Las Vegas (or Yuma, Corpus Christi, Meridian, San Bernardino, or Dayton) and began to wonder: How can all of them down there, those poor souls who will soon be waking up and trudging out of their minute rectangles and inching along their little noodle highways toward whatever slots and grooves make up their everyday lives—how could they live like that, with such earnestness, if they had the faintest idea of what it was like up here in this righteous zone?

But of course! Not only the washed-out, grounded, and dead pilots had been left behind—but also all of those millions of sleepwalking souls who never even attempted the great gamble. The entire world below . . . *left behind.* Only at this point can one begin to understand just how big, how titanic, the ego of the military pilot could be. The world was used to enormous egos in artists, actors, entertainers of all sorts, in politicians, sports figures, and even journalists, because they had such familiar and convenient ways to show them off. But that slim young man over there in uniform, with the enormous watch on his wrist and the withdrawn look on his face, that young officer who is so shy that he can't even open his mouth unless the subject is flying—that young pilot—well, my friends, his ego is even *bigger!*—so big, it's *breathtaking!* Even in the 1950's it was difficult for civilians to comprehend such a thing, but *all* military officers and many enlisted men tended to feel superior to civilians. It was really quite ironic, given the fact that for a good thirty years the rising business classes in the cities had been steering their sons away from the military, as if from a bad smell, and the officer corps had never been held in lower esteem. Well, career officers returned the contempt in trumps. They looked upon themselves as men who lived by higher standards of behavior than civilians, as men who were the bearers and protectors of the most important values of American life, who maintained a sense of discipline while civilians abandoned themselves to hedonism, who maintained a sense of honor while civilians lived by opportunism and greed. Opportunism and greed: there you had your much-vaunted corporate business world. Khrushchev was right about one thing: when it came time to hang the capitalist West, an American businessman would sell him the rope. When the showdown came—and the showdowns always came—not all the wealth in the world or all the sophisticated nuclear weapons and radar and missile systems it could buy would take the place of those who had the uncritical willingness to face danger, those who, in short, had the right stuff.

In fact, the feeling was so righteous, so exalted, it could become religious. Civilians seldom understood this, either. There was no one to teach them. It was no longer the fashion for serious writers to describe the glories of war. Instead, they dwelt upon its horrors, often with cynicism or disgust. It was left to the occasional pilot with a literary flair to provide a glimpse of the pilot's self-conception in its heavenly or spiritual aspect. When a pilot named Robert Scott flew his P–43 over Mount Everest, quite a feat at the time, he brought his hand up and snapped a salute to his fallen adversary. He thought he had *defeated* the mountain, surmounting all the forces of nature that had made it formidable. And why not? "God is my co-pilot," he said—that became the title of his book—and he meant it. So did the most gifted of all the pilot authors, the Frenchman Antoine de Saint-Exupéry. As he gazed down upon the world . . . from up there . . . during transcontinental flights, the good Saint-Ex saw civilization as a series of tiny fragile patches clinging to the otherwise barren rock of Earth. He felt like a lonely sentinel, a protector of those vulnerable little oases, ready to lay down his life in their behalf, if necessary; a saint, in short, true to his name, flying up here at the right hand of God. The good Saint-Ex! And he was not the only one. He was merely the one who put it into words most beautifully and anointed himself before the altar of the right stuff.

There were many pilots in their thirties who, to the consternation of their wives, children, mothers, fathers, and employers, volunteered to go active in the reserves and fly in combat in the Korean War. In godforsaken frozen Chosen! But it was simple enough. Half of them were fliers who had trained during the Second World War and had never seen combat. It was well understood—and never said, of course—that no one could reach the top of the pyramid without going into combat.

The morale of foot soldiers in the Korean War was so bad it actually reached the point where officers were prodding men forward with gun barrels and bayonets. But in the air—it was Fighter Jock Heaven! Using F–86s mainly, the Air Force was producing aces, pilots who had shot down five planes or more, as fast as the Koreans and Chinese could get their Soviet MiG–15s up to fight them. By the time the fighting was stopped, there were thirty-eight Air Force aces, and they had accounted for a total of 299.5 kills. Only fifty-six F–86s were lost. High spirits these lads had. They chronicled their adventures with a good creamy romanticism such as nobody in flying had treated himself to since the days of Lufbery, Frank Luke, and von Richthofen in the First World War. Colonel Harrison R. Thyng, who shot down five MiGs in Korea (and eight German and Japanese planes in the Second World War), glowed like Excalibur when he described his Fourth Fighter-Interceptor Wing: "Like olden knights the F–86 pilots ride up over North Korea to the Yalu River, the sun glinting off silver aircraft, contrails streaming behind, as they challenge the numerically superior enemy to come on up and fight." Lances and plumes! *I'm a knight!* Come on up and fight! Why hold back! Knights of the Right Stuff!

When a pilot named Gus Grissom (whom Conrad, Schirra, Lovell, and the others would meet later on) first went to Korea, the Air Force used to take the F–86 jocks out to the field before dawn, in the dark, in buses, and the pilots who had not been shot at by a MiG in air-to-air combat had to stand up. At first Grissom couldn't believe it and then he couldn't bear it—those bastards sitting down were *the only ones with the right stuff!* The next morning, as they rumbled out there in the dark, he was sitting down. He had gone up north toward the Yalu on the first day and had it out with some howling supersonic Chinee just so he could have a seat on the bus. Even at the level of combat, the main thing was not to be *left behind.*

Combat had its own infinite series of tests, and one of the greatest sins was "chattering" or "jabbering" on the radio. The combat frequency was to be kept clear of all but strategically essential messages, and all unenlightening comments were regarded as evidence of funk, of the wrong stuff. A Navy pilot (in legend, at any rate) began shouting, "I've got a MiG at zero! A MiG at zero!"—meaning that it had maneuvered in behind him and was locked in on his tail. An irritated voice cut in and said, "Shut up and die like an aviator." One had to be a Navy pilot to appreciate the final nuance. A good Navy pilot was a real *aviator*; in the Air Force they merely had pilots and not precisely the proper stuff.

No, the tests were never-ending. And in the periods between wars a man's past successes in combat did not necessarily keep him at the top of the heavenly pyramid. By the late 1950's there was yet another plateau to strive for. On that plateau were men who had flown in combat in the Second World War or Korea and had then gone on to become test pilots in the new age of jet and rocket engines. Not every combat pilot could make the climb. Two of the great aces of the Second World War, Richard I. Bong and Don Gentile, tried it but didn't have the patience for the job. They only wanted to go up and poke holes in the sky; and presently they were just part of combat history. Of course, by now, thanks to the accident of age, you began to find young men who had reached the exalted level of test pilot without ever having had a chance to fight in combat. One was Pete Conrad, who was just graduating, with the survivors of Group 20, to the status of full-fledged test pilot at Pax River. Like every Navy test pilot, Conrad was proud of Pax River and its reputation. Out loud every true Navy aviator insisted that Pax River was the place . . . and inwardly knew it really wasn't. For every military pilot knew where the apex of the great ziggurat was located. You could point it out on a map. The place was Edwards Air Force Base in the high desert 150 miles northeast of Los Angeles. Everyone knew who resided there, too, although their actual status was never put into words. Not only that, everyone knew the name of the individual who ranked foremost in the Olympus, the ace of all the aces, as it were, among the true brothers of the right stuff.

III. Yeager

Anyone who travels very much on airlines in the United States soon gets to know the voice of *the airline pilot* . . . coming over the intercom . . . with a particular drawl, a particular folksiness, a particular down-home calmness that is so exaggerated it begins to parody itself (nevertheless!—it's reassuring) . . . the voice that tells you, as the airliner is caught in thunderheads and goes bolting up and down a thousand feet at a single gulp, to check your seat belts because "it might get a little choppy" . . . the voice that tells you (on a flight from Phoenix preparing for its final approach into Kennedy Airport, New York, just after dawn): "Now, folks, uh . . . this is the captain . . . ummmm . . . We've got a little ol' red light up here on the control panel that's tryin' to tell us that the *lan*din' gears're not . . . uh . . . *lock*in' into position when we lower 'em . . . Now . . . *I* don't believe that little ol' red light knows what it's *talk*in' about—I believe it's that little ol' red light that iddn' workin' right" . . . faint chuckle, long pause, as if to say, *I'm not even sure all this is really worth going into—still, it may amuse you* . . . "*But* . . . I guess to play it by the rules, we oughta *hum*or that little ol' light . . . so we're gonna take her down to about, oh, two or three hundred feet over the runway at Kennedy, and the folks down there on the ground are gonna see if they caint give us a *visual* inspection of those ol' landin' gears"—with which he is obviously on intimate ol'-buddy terms, as with every other working part of this mighty ship—"and if I'm right . . . they're gonna tell us everything is copa*cet*ic all the way aroun' an' we'll jes take her on in" . . . and, after a couple of low passes over the field, the voice returns: "Well, folks, those folks down there on the ground—it must be too early for 'em or somethin'—I 'spect they still got the *sleep*ers in their eyes . . . 'cause they say they caint tell if those ol' landin' gears are all the way down or not . . . But, you know, up here in the cockpit we're convinced they're all the way down, so we're jes gonna take her on in . . . And oh" . . . *(I almost forgot)* . . . "while we take a little swing out over the ocean an' empty some of that surplus fuel we're not gonna be needin' anymore—that's what you might be seein' comin' out of the wings—our lovely little ladies . . . if they'll be so kind . . . they're gonna go up and down the aisles and show you how we do what we call 'assumin' the position' " . . . another faint chuckle *(We do this so often, and it's so much fun, we even have a funny little name for it)* . . . and the stewardesses, a bit grimmer, by the looks of them, than *that voice,* start telling the passengers to take their glasses off and take the ballpoint pens and other sharp objects out of their pockets, and they show them *the position,* with the head lowered . . . while down on the field at Kennedy the little yellow emergency trucks start roaring across the field—and even though in your pounding heart and your sweating palms and your broiling brainpan you *know* this is a critical moment in your life, you still can't quite bring yourself to be*lieve* it, because if it were . . . how could *the captain,* the man who knows the actual situation most intimately . . . how could he keep on drawlin' and chucklin' and driftin' and lollygaggin' in that particular voice of his—

Well!—who doesn't know that voice! And who can forget it!—even after he is proved right and the emergency is over.

That particular voice may sound vaguely Southern or Southwestern, but it is specifically Appalachian in origin. It originated in the mountains of West Virginia, in the coal country, in Lincoln County, so far up in the hollows that, as the saying went, "they had to pipe in daylight." In the late 1940's and early 1950's this up-hollow voice drifted down from on high, from over the high desert of California, down, down, down, from the upper reaches of the Brotherhood into all phases of American aviation. It was amazing. It was *Pygmalion* in reverse. Military pilots and then, soon, airline pilots, pilots from Maine and Massachusetts and the Dakotas and Oregon and everywhere else, began to talk in that poker-hollow West Virginia drawl, or as close to it as they could bend their native accents. It was the drawl of the most righteous of all the possessors of the right stuff: Chuck Yeager.

Yeager had started out as the equivalent, in the Second World War, of the legendary Frank Luke of the 27th Aero Squadron in the First. Which is to say, he was the boondocker, the boy from the back country, with only a high-school education, no credentials, no cachet or polish of any sort, who took off the feed-store overalls and put on a uniform and climbed into an airplane and lit up the skies over Europe.

Yeager grew up in Hamlin, West Virginia, a town on the Mud River not far from Nitro, Hurricane, Whirlwind, Salt Rock, Mud, Sod, Crum, Leet, Dollie, Ruth, and Alum Creek. His father was a gas driller (drilling for natural gas in the coalfields), his older brother was a gas driller, and he would have been a gas driller had he not enlisted in the Army Air Force in 1941 at the age of eighteen. In 1943, at twenty, he became a flight officer, i.e., a non-com who was allowed to fly, and went to England to fly fighter planes over France and Germany. Even in the tumult of the war Yeager was somewhat puzzling to a lot of other pilots. He was a short, wiry, but muscular little guy with dark curly hair and a tough-looking face that seemed (to strangers) to be saying: "You best not be lookin' me in the eye, you peckerwood, or I'll put four more holes in your nose." But that wasn't what was puzzling. What was puzzling was the way Yeager talked. He seemed to talk with some older forms of English elocution, syntax, and conjugation that had been preserved uphollow in the Appalachians. There were people up there who never said they disapproved of anything, they said: "I don't hold with it." In the present tense they were willing to *help* out, like anyone else; but in the past tense they only *holped.* "H'it weren't nothin' I hold with, but I holped him out with it, anyways."

In his first eight missions, at the age of twenty, Yeager shot down two German fighters. On his ninth he was shot down over German-occupied French territory, suffering flak wounds; he bailed out, was picked up by the French underground, which smuggled him across the Pyrenees into Spain disguised as a peasant. In Spain he was jailed briefly, then released, whereupon he made it back to England and returned to combat during the Allied invasion of France. On October 12, 1944, Yeager took on and shot down five German fighter planes in succession. On November 6, flying a propeller-driven P–51 Mustang, he shot down one of the new jet fighters the Germans had developed, the Messerschmitt–262, and damaged two more, and on November 20 he shot down four FW–190s. It was a true Frank Luke–style display of warrior fury and personal prowess. By the end of the war he had thirteen and a half kills. He was twenty-two years old.

In 1946 and 1947 Yeager was trained as a test pilot at Wright Field in Dayton. He amazed his instructors with his ability at stunt-team flying, not to mention the unofficial business of hassling. That plus his up-hollow drawl had everybody saying, "He's a natural-born stick 'n' rudder man." Nevertheless, there was something extraordinary about it when a man so young, with so little experience in flight test, was selected to go to Muroc Field in California for the XS–1 project.

Muroc was up in the high elevations of the Mojave Desert. It looked like some fossil landscape that had long since been left behind by the rest of terrestrial evolution. It was full of huge dry lake beds, the biggest being Rogers Lake. Other than sagebrush the only vegetation was Joshua trees, twisted freaks of the plant world that looked like a cross between cactus and Japanese bonsai. They had a dark petrified green color and horribly crippled branches. At dusk the Joshua trees stood out in silhouette on the fossil wasteland like some arthritic nightmare. In the summer the temperature went up to 110 degrees as a matter of course, and the dry lake beds were covered in sand, and there would be windstorms and sandstorms right out of a Foreign Legion movie. At night it would drop to near freezing, and in December it would start raining, and the dry lakes would fill up with a few inches of water, and some sort of putrid prehistoric shrimps would work their way up from out of the ooze, and sea gulls would come flying in a hundred miles or more from the ocean, over the mountains, to gobble up these squirming little throwbacks. A person had to see it to believe it: flocks of sea gulls wheeling around in the air out in the middle of the high desert in the dead of winter and grazing on antediluvian crustaceans in the primordial ooze.

When the wind blew the few inches of water back and forth across the lake beds, they became absolutely smooth and level. And when the water evaporated in the spring, and the sun baked the ground hard, the lake beds became the greatest natural landing fields ever discovered, and also the biggest, with miles of room for error. That was highly desirable, given the nature of the enterprise at Muroc.

Besides the wind, sand, tumbleweed, and Joshua trees, there was nothing at Muroc except for two quonset-style hangars, side by side, a couple of gasoline pumps, a single concrete runway, a few tarpaper shacks, and some tents. The officers stayed in the shacks marked "barracks," and lesser souls stayed in the tents and froze all night and fried all day. Every road into the property had a guardhouse on it manned by soldiers. The enterprise the Army had undertaken in this godforsaken place was the development of supersonic jet and rocket planes.

At the end of the war the Army had discovered that the Germans not only had the world's first jet fighter but also a rocket plane that had gone 596 miles an hour in tests. Just after the war a British jet, the Gloster Meteor, jumped the official world speed record from 469 to 606 in a single day. The next great plateau would be Mach 1, the speed of sound, and the Army Air Force considered it crucial to achieve it first.

The speed of sound, Mach 1, was known (thanks to the work of the physicist Ernst Mach) to vary at different altitudes, temperatures, and wind speeds. On a calm 60-degree day at sea level it was about 760 miles an hour, while at 40,000 feet, where the temperature would be at least sixty below, it was about 660 miles an hour. Evil and baffling things happened in the transonic zone, which began at about .7 Mach. Wind tunnels choked out at such velocities. Pilots who approached the speed of sound in dives reported that the controls would lock or "freeze" or even alter their normal functions. Pilots had crashed and died because they couldn't budge the stick. Just last year Geoffrey de Havilland, son of the famous British aircraft designer and builder, had tried to take one of his father's DH 108s to Mach 1. The ship started buffeting and then disintegrated, and he was killed. This led engineers to speculate that the shock waves became so severe and unpredictable at Mach 1, no aircraft could survive them. They started talking about "the sonic wall" and "the sound barrier."

So this was the task that a handful of pilots, engineers, and mechanics had at Muroc. The place was utterly primitive, nothing but bare bones, bleached tarpaulins, and corrugated tin rippling in the heat with caloric waves; and for an ambitious young pilot it was perfect. Muroc seemed like an outpost on the dome of the world, open only to a righteous few, closed off to the rest of humanity, including even the Army Air Force brass of command control, which was at Wright Field. The commanding officer at Muroc was only a colonel, and his superiors at Wright did not relish junkets to the Muroc rat shacks in the first place. But to pilots this prehistoric throwback of an airfield became . . . shrimp heaven! the rat-shack plains of Olympus!

Low Rent Septic Tank Perfection . . . yes; and not excluding those traditional essentials for the blissful hot young pilot: Flying & Drinking and Drinking & Driving.

Just beyond the base, to the southwest, there was a rickety windblown 1930's-style establishment called Pancho's Fly Inn, owned, run, and bartended by a woman named Pancho Barnes. Pancho Barnes wore tight white sweaters and tight pants, after the mode of Barbara Stanwyck in *Double Indemnity*. She was only forty-one when Yeager arrived at Muroc, but her face was so weatherbeaten, had so many hard miles on it, that she looked older, especially to the young pilots at the base. She also shocked the pants off them with her vulcanized tongue. Everybody she didn't like was an old bastard or a sonofabitch. People she liked were old bastards and sonsabitches, too. "I tol' 'at ol' bastard to get 'is ass on over here and I'd g'im a drink." But Pancho Barnes was anything but Low Rent. She was the granddaughter of the man who designed the old Mount Lowe cable-car system, Thaddeus S. C. Lowe. Her maiden name was Florence Leontine Lowe. She was brought up in San Marino, which adjoined Pasadena and was one of Los Angeles' wealthiest suburbs, and her first husband—she was married four times—was the pastor of the Pasadena Episcopal Church, the Rev. C. Rankin Barnes. Mrs. Barnes seemed to have few of the conventional community interests of a Pasadena matron. In the late 1920's, by boat and plane, she ran guns for Mexican revolutionaries and picked up the nickname Pancho. In 1930 she broke Amelia Earhart's air-speed record for women. Then she barnstormed around the country as the featured performer of "Pancho Barnes's Mystery Circus of the Air." She always greeted her public in jodhpurs and riding boots, a flight jacket, a white scarf, and a white sweater that showed off her terrific Barbara Stanwyck chest. Pancho's desert Fly Inn had an airstrip, a swimming pool, a dude ranch corral, plenty of acreage for horseback riding, a big old guest house for the lodgers, and a connecting building that was the bar and restaurant. In the barroom the floors, the tables, the chairs, the walls, the beams, the bar were of the sort known as extremely weatherbeaten, and the screen doors kept banging. Nobody putting together such a place for a movie about flying in the old days would ever dare make it as dilapidated and generally go-to-hell as it actually was. Behind the bar were many pictures of airplanes and pilots, lavishly autographed and inscribed, badly framed and crookedly hung. There was an old piano that had been dried out and cracked to the point of hopeless desiccation. On a good night a huddle of drunken aviators could be heard trying to bang, slosh, and navigate their way through old Cole Porter tunes. On average nights the tunes were not that good to start with. When the screen door banged and a man walked through the door into the saloon, every eye in the place checked him out. If he wasn't known as somebody who had something to do with flying at Muroc, he would be eyed like some lame goddamned mouseshit sheepherder from *Shane.*

The plane the Air Force wanted to break the sound barrier with was called the X–I at the outset and later on simply the X–I. The Bell Aircraft Corporation had built it under an Army contract. The core of the ship was a rocket of the type first developed by a young Navy inventor, Robert Truax, during the war. The fuselage was shaped like a 50-caliber bullet—an object that was known to go supersonic smoothly. Military pilots seldom drew major test assignments; they went to highly paid civilians working for the aircraft corporations. The prime pilot for the X–I was a man whom Bell regarded as the best of the breed. This man looked like a movie star. He looked like a pilot from out of *Hell's Angels.* And on top of everything else there was his name: Slick Goodlin.

The idea in testing the X–I was to nurse it carefully into the transonic zone, up to seven-tenths, eight-tenths, nine-tenths the speed of sound (.7 Mach, .8 Mach, .9 Mach) before attempting the speed of sound itself, Mach 1, even though Bell and the Army already knew the X–I had the rocket power to go to Mach 1 and beyond, if there *was* any *beyond.* The consensus of aviators and engineers, after Geoffrey de Havilland's death, was that the speed of sound was an absolute, like the firmness of the earth. The sound barrier was a farm you could buy in the sky. So Slick Goodlin began to probe the transonic zone in the X–I, going up to .8 Mach. Every time he came down he'd have a riveting tale to tell. The buffeting, it was so fierce—and the listeners, their imaginations aflame, could practically see poor Geoffrey de Havilland disintegrating in midair. And the goddamned aerodynamics—and the listeners got a picture of a man in ballroom pumps skidding across a sheet of ice, pursued by bears. A controversy arose over just how much bonus Slick Goodlin should receive for assaulting the dread Mach 1 itself. Bonuses for contract test pilots were not unusual; but the figure of $150,000 was now bruited about. The Army balked, and Yeager got the job. He took it for $283 a month, or $3,396 a year; which is to say, his regular Army captain's pay.

The only trouble they had with Yeager was in holding him back. On his first powered flight in the X–I he immediately executed an unauthorized zero-g roll with a full load of rocket fuel, then stood the ship on its tail and went up to .85 Mach in a vertical climb, also unauthorized. On subsequent flights, at speeds between .85 Mach and .9 Mach, Yeager ran into most known airfoil problems—loss of elevator, aileron, and rudder control, heavy trim pressures, Dutch rolls, pitching and buffeting, the lot—yet was convinced, after edging over .9 Mach, that this would all get better, not worse, as you reached Mach 1. The attempt to push beyond Mach 1—"breaking the sound barrier"—was set for October 14, 1947. Not being an engineer, Yeager didn't believe the "barrier" existed.

October 14 was a Tuesday. On Sunday evening, October 12, Chuck Yeager dropped in at Pancho's, along with his wife. She was a brunette named Glennis, whom he had met in California while he was in training, and she was such a number, so striking, he had the inscription "Glamorous Glennis" written on the nose of his P–51 in Europe and, just a few weeks back, on the X–I itself. Yeager didn't go to Pancho's and knock back a few because two days later the big test was coming up. Nor did he knock back a few because it was the weekend. No, he knocked back a few because night had come and he was a pilot at Muroc. In keeping with the military tradition of Flying & Drinking, that was what you did, for no other reason than that the sun had gone down. You went to Pancho's and knocked back a few and listened to the screen doors banging and to other aviators torturing the piano and the nation's repertoire of Familiar Favorites and to lonesome mouse-turd strangers wandering in through the banging doors and to Pancho classifying the whole bunch of them as old bastards and miserable peckerwoods. That was what you did if you were a pilot at Muroc and the sun went down.

So about eleven Yeager got the idea that it would be a hell of a kick if he and Glennis saddled up a couple of Pancho's dude-ranch horses and went for a romp, a little rat race, in the moonlight. This was in keeping with the military tradition of Flying & Drinking and Drinking & Driving, except that this was prehistoric Muroc and you rode horses. So Yeager and his wife set off on a little proficiency run at full gallop through the desert in the moonlight amid the arthritic silhouettes of the Joshua trees. Then they start racing back to the corral, with Yeager in the lead and heading for the gateway. Given the prevailing conditions, it being nighttime, at Pancho's, and his head being filled with a black sandstorm of many badly bawled songs and vulcanized oaths, he sees too late that the gate has been closed. Like many a hard-driving midnight pilot before him, he does not realize that he is not equally gifted in the control of all forms of locomotion. He and the horse hit the gate, and he goes flying off and lands on his right side. His side hurts like hell.

The next day, Monday, his side still hurts like hell. It hurts every time he moves. It hurts every time he breathes deep. It hurts every time he moves his right arm. He knows that if he goes to a doctor at Muroc or says anything to anybody even remotely connected with his superiors, he will be scrubbed from the flight on Tuesday. They might even go so far as to put some other miserable peckerwood in his place. So he gets on his motorcycle, an old junker that Pancho had given him, and rides over to see a doctor in the town of Rosamond, near where he lives. Every time the goddamned motorcycle hits a pebble in the road, his side hurts like a sonofabitch. The doctor in Rosamond informs him he has two broken ribs and he tapes them up and tells him that if he'll just keep his right arm immobilized for a couple of weeks and avoid any physical exertion or sudden movements, he should be all right.

Yeager gets up before daybreak on Tuesday morning—which is supposed to be the day he tries to break the sound barrier—and his ribs still hurt like a sonofabitch. He gets his wife to drive him over to the field, and he has to keep his right arm pinned down to his side to keep his ribs from hurting so much. At dawn, on the day of a flight, you could hear the X–1 screaming long before you got there. The fuel for the X–1 was alcohol and liquid oxygen, oxygen converted from a gas to a liquid by lowering its temperature to 297 degrees below zero. And when the lox, as it was called, rolled out of the hoses and into the belly of the X–1, it started boiling off and the X–I started steaming and screaming like a teakettle. There's quite a crowd on hand, by Muroc standards . . . perhaps nine or ten souls. They're still fueling the X–I with the lox, and the beast is wailing.

The X–I looked like a fat orange swallow with white markings. But it was really just a length of pipe with four rocket chambers in it. It had a tiny cockpit and a needle nose, two little straight blades (only three and a half inches thick at the thickest part) for wings, and a tail assembly set up high to avoid the "sonic wash" from the wings. Even though his side was throbbing and his right arm felt practically useless, Yeager figured he could grit his teeth and get through the flight—except for one specific move he had to make. In the rocket launches, the X–I, which held only two and a half minutes' worth of fuel, was carried up to twenty-six thousand feet underneath a B–29. At seven thousand feet, Yeager was to climb down a ladder from the bomb bay of the B–29 to the open doorway of the X–I, hook up to the oxygen system and the radio microphone and earphones, and put his crash helmet on and prepare for the launch, which would come at twenty-five thousand feet. This helmet was a homemade number. There had never been any such thing as a crash helmet before, except in stunt flying. Throughout the war pilots had used the old skin-tight leather helmet-and-goggles. But the X–I had a way of throwing the pilot around so violently that there was danger of getting knocked out against the walls of the cockpit. So Yeager had bought a big leather football helmet—there were no plastic ones at the time—and he butchered it with a hunting knife until he carved the right kind of holes in it, so that it would fit down over his regular flying helmet and the earphones and the oxygen rig. Anyway, then his flight engineer, Jack Ridley, would climb down the ladder, out in the breeze, and shove into place the cockpit door, which had to be lowered out of the belly of the B–29 on a chain. Then Yeager had to push a handle to lock the door airtight. Since the X–I's cockpit was minute, you had to push the handle with your right hand. It took quite a shove. There was no way you could move into position to get enough leverage with your left hand.

Out in the hangar Yeager makes a few test shoves on the sly, and the pain is so incredible he realizes that there is no way a man with two broken ribs is going to get the door closed. It is time to confide in somebody, and the logical man is Jack Ridley. Ridley is not only the flight engineer but a pilot himself and a good old boy from Oklahoma to boot.

He will understand about Flying & Drinking and Drinking & Driving through the goddamned Joshua trees. So Yeager takes Ridley off to the side in the tin hangar and says: Jack, I got me a little ol' problem here. Over at Pancho's the other night I sorta . . . dinged my goddamned ribs. Ridley says, Whattya mean . . . *dinged?* Yeager says, Well, I guess you might say I damned near like to . . . *broke* a coupla the sonsabitches. Whereupon Yeager sketches out the problem he foresees.

Not for nothing is Ridley the engineer on this project. He has an inspiration. He tells a janitor named Sam to cut him about nine inches off a broom handle. When nobody's looking, he slips the broomstick into the cockpit of the X–I and gives Yeager a little advice and counsel.

So with that added bit of supersonic flight gear Yeager went aloft.

At seven thousand feet he climbed down the ladder into the X–I's cockpit, clipped on his hoses and lines, and managed to pull the pumpkin football helmet over his head. Then Ridley came down the ladder and lowered the door into place. As Ridley had instructed, Yeager now took the nine inches of broomstick and slipped it between the handle and the door. This gave him just enough mechanical advantage to reach over with his left hand and whang the thing shut. So he whanged the door shut with Ridley's broomstick and was ready to fly.

At 26,000 feet the B–29 went into a shallow dive, then pulled up and released Yeager and the X–I as if it were a bomb. Like a bomb it dropped and shot forward (at the speed of the mother ship) at the same time. Yeager had been launched straight into the sun. It seemed to be no more than six feet in front of him, filling up the sky and blinding him. But he managed to get his bearings and set off the four rocket chambers one after the other. He then experienced something that became known as the ultimate sensation in flying: "booming and zooming." The surge of the rockets was so tremendous, forced him back into his seat so violently, he could hardly move his hands forward the few inches necessary to reach the controls. The X–I seemed to shoot straight up in an absolutely perpendicular trajectory, as if determined to snap the hold of gravity via the most direct route possible. In fact, he was only climbing at the 45-degree angle called for in the flight plan. At about .87 Mach the buffeting started.

On the ground the engineers could no longer see Yeager. They could only hear . . . that poker-hollow West Virginia drawl.

"Had a mild buffet there . . . jes the usual instability . . ."

Jes the usual instability?

Then the X-1 reached the speed of .96 Mach, and that incredible caint-hardlyin' aw-shuckin' drawl said:

"Say, Ridley . . . make a note here, will ya?" *(if you ain't got nothin' better to do)* ". . . elevator effectiveness *re*gained."

Just as Yeager had predicted, as the X-1 approached Mach 1, the stability improved. Yeager had his eyes pinned on the machometer. The needle reached .96, fluctuated, and went off the scale.

And on the ground they heard . . . that voice:

"Say, Ridley . . . make another note, will ya?" *(if you ain't too bored yet)* ". . . there's somethin' wrong with this ol' machometer . . ." (faint chuckle) ". . . it's gone kinda screwy on me . . ."

And in that moment, on the ground, they heard a boom rock over the desert floor—just as the physicist Theodore von Kármán had predicted many years before.

Then they heard Ridley back in the B-29: "If it is, Chuck, we'll fix it. Personally I think you're seeing things."

Then they heard Yeager's poker-hollow drawl again:

"Well, I guess I am, Jack . . . And I'm still goin' upstairs like a bat."

The X-1 had gone through "the sonic wall" without so much as a bump. As the speed topped out at Mach 1.05, Yeager had the sensation of shooting straight through the top of the sky. The sky turned a deep purple and all at once the stars and the moon came out—and the sun shone at the same time. He had reached a layer of the upper atmosphere where the air was too thin to contain reflecting dust particles. He was simply looking out into space. As the X-1 nosed over at the top of the climb, Yeager now had seven minutes of . . . Pilot Heaven . . . ahead of him. He was going faster than any man in history, and it was almost silent up here, since he had exhausted his rocket fuel, and he was so high in such a vast space that there was no sensation of motion. He was master of the sky. His was a king's solitude, unique and inviolate, above the dome of the world. It would take him seven minutes to glide back down and land at Muroc. He spent the time doing victory rolls and wing-over-wing aerobatics while Rogers Lake and the High Sierras spun around below.

On the ground they had understood the code as soon as they heard Yeager's little exchange with Ridley. The project was secret, but the radio exchanges could be picked up by anyone within range. The business of the "screwy machometer" was Yeager's deadpan way of announcing that the X–1's instruments indicated Mach 1. As soon as he landed, they checked out the X–1's automatic recording instruments. Without any doubt the ship had gone supersonic. They immediately called the brass at Wright Field to break the tremendous news. Within two hours Wright Field called back and gave some firm orders. A top security lid was being put on the morning's events. That the press was not to be informed went without saying. But neither was anyone else, anyone at all, to be told. Word of the flight was not to go beyond the flight line. And even among the people directly involved—who were there and knew about it, anyway—there was to be no celebrating. Just what was on the minds of the brass at Wright is hard to say. Much of it, no doubt, was a simple holdover from wartime, when every breakthrough of possible strategic importance was kept under wraps. That was what you did—you shut up about them. Another possibility was that the chief at Wright had never quite known what to make of Muroc. There was some sort of weird ribald aerial tarpaper mad-monk squadron up on the roof of the desert out there . . .

In any case, by mid-afternoon Yeager's tremendous feat had become a piece of thunder with no reverberation. A strange and implausible stillness settled over the event. Well . . . there was not supposed to be any celebration, but come nightfall . . . Yeager and Ridley and some of the others ambled over to Pancho's. After all, it was the end of the day, and they were pilots. So they knocked back a few. And they had to let Pancho in on the secret, because Pancho had said she'd serve a free steak dinner to any pilot who could fly supersonic and walk in here to tell about it, and they had to see the look on *her* face. So Pancho served Yeager a big steak dinner and said they were a buncha miserable peckerwoods all the same, and the desert cooled off and the wind came up and the screen doors banged and they drank some more and bawled some songs over the cackling dry piano and the stars and the moon came out and Pancho screamed oaths no one had ever heard before and Yeager and Ridley roared and the old weatherbeaten bar boomed and the autographed pictures of a hundred dead pilots shook and clattered on the frame wires and the faces of the living fell apart in the reflections, and by and by they all left and stumbled and staggered and yelped and bayed for glory before the arthritic silhouettes of the Joshua trees. Shit!—there was no one to tell except for Pancho and the goddamned Joshua trees!

Over the next five months Yeager flew supersonic in the X-1 more than a dozen times, but still the Air Force insisted on keeping the story secret. *Aviation Week* published a report of the flights late in December (without mentioning Yeager's name) provoking a minor debate in the press over whether or not *Aviation Week* had violated national security—and *still* the Air Force refused to publicize the achievement until June of 1948. Only then was Yeager's name released. He received only a fraction of the publicity that would have been his had he been presented to the world immediately, on October 14, 1947, as the man who "broke the sound barrier." This dragged-out process had curious effects.

In 1952 a British movie called *Breaking the Sound Barrier*, starring Ralph Richardson, was released in the United States, and its promoters got the bright idea of inviting the man who had actually done it, Major Charles E. Yeager of the U.S. Air Force, to the American premiere. So the Air Force goes along with it and Yeager turns up for the festivities. When he watches the movie, he's stunned. He can't believe what he's seeing. Far from being based on the exploits of Charles E. Yeager, *Breaking the Sound Barrier* was inspired by the death of Geoffrey de Havilland in his father's DH 108. At the end of the movie a British pilot solves the mystery of "the barrier" by *reversing the controls* at the critical moment during a power dive. The buffeting is tearing his ship to pieces, and every rational process in his head is telling him to *pull back* on the stick to keep from crashing—and he *pushes it down* instead . . . and zips right through Mach 1 as smooth as a bird, regaining full control!

Breaking the Sound Barrier happened to be one of the most engrossing movies about flying ever made. It seemed superbly realistic, and people came away from it sure of two things: it was an Englishman who had broken the sound barrier, and he had done it by reversing the controls in the transonic zone.

Well, after the showing they bring out Yeager to meet the press, and he doesn't know where in the hell to start. To him the whole goddamned picture is outrageous. He doesn't want to get mad, because this thing has been set up by Air Force P.R. But he is not happy. In as calm a way as he can word it on the spur of the moment, he informs one and all that the picture is an utter shuck from start to finish. The promoters respond, a bit huffily, that this picture is not, after all, a documentary. Yeager figures, well, anyway, that settles that. But as the weeks go by, he discovers an incredible thing happening. He keeps running into people who think he's the first *American* to break the sound barrier . . . and that he learned how to *reverse the controls* and zip through from the Englishman who did it first. The last straw comes when he gets a call from the Secretary of the Air Force.

"Chuck," he says, "do you mind if I ask you something? Is it true that you broke the sound barrier by reversing the controls?"

Yeager is stunned by this. The Secretary—*the Secretary!*—of the U.S. Air Force!

"No, sir," he says, "that is . . . not correct. Anyone who reversed the controls going transonic would be dead."

Yeager and the rocket pilots who soon joined him at Muroc had a hard time dealing with publicity. On the one hand, they hated the process. It meant talking to reporters and other fruit flies who always hovered, eager for the juice . . . and invariably got the facts screwed up . . . *But that wasn't really the problem, was it!* The real problem was that reporters violated the invisible walls of the fraternity. They blurted out questions and spoke boorish words about . . . all the unspoken things!—about fear and bravery (they would say the words!) and how you *felt* at such-and-such a moment! It was obscene! They presumed a knowledge and an intimacy they did not have and had no right to. Some aviation writer would sidle up and say, "I hear Jenkins augered in. That's too bad." *Augered in!*—a phrase that belonged exclusively to the fraternity!—coming from the lips of this *ant* who was *left behind* the moment Jenkins made his first step up the pyramid long, long ago. It was repulsive! But on the other hand . . . one's healthy pilot ego loved the glory—wallowed in it!—lapped it up!—no doubt about it! The Pilot Ego—ego didn't come any bigger! The boys wouldn't have minded the following. They wouldn't have minded appearing once a year on a balcony over a huge square in which half the world is assembled. They wave. The world roars its approval, its applause, and breaks into a sustained thirty-minute storm of cheers and tears (moved by my righteous stuff!). And then it's over. All that remains is for the wife to paste the clippings in the scrapbook.

A little adulation on the order of the Pope's; that's all the True Brothers at the top of the pyramid really wanted.

Yeager received just about every major decoration and trophy that was available to test pilots, but the Yeager legend grew not in the press, not in public, but within the fraternity. As of 1948, after Yeager's flight was made public, every hot pilot in the country knew that Muroc was what you aimed for if you wanted to reach the top. In 1947 the National Security Act, Title 10, turned the Army Air Force into the U.S. Air Force, and three years later Muroc Army Air Base became Edwards Air Force Base, named for a test pilot, Glenn Edwards, who had died testing a ship with no tail called the Flying Wing. So now the magic word became *Edwards*. You couldn't keep a really hot, competitive pilot away from Edwards. Civilian pilots (almost all of whom had been trained in the military) could fly for the National Advisory Committee for Aeronautics (NACA) High Speed Center at Edwards, and some of the rocket pilots did that: Scott Crossfield, Joe Walker, Howard Lilly, Herb Hoover, and Bill Bridgeman, among them. Pete Everest, Kit Murray, Iven Kincheloe, and Mel Apt joined Yeager as Air Force rocket pilots. There was a constant rivalry between NACA and the Air Force to push the rocket planes to their outer limits. On November 20, 1953, Crossfield, in the D–558-2, raised the speed record to Mach 2. Three weeks later Yeager flew the X–IA to Mach 2.4. The rocket program was quickly running out of frontiers within the atmosphere; so NACA and the Air Force began planning a new project, with a new rocket plane, the X–15, to probe altitudes as high as fifty miles, which was well beyond anything that could still be called "air."

My God!—to be a part of Edwards in the late forties and early fifties!—even to be on the ground and hear one of those incredible explosions from 35,000 feet somewhere up there in the blue over the desert and know that some True Brother had commenced his rocket launch . . . in the X–1, the X–1A, the X–2, the D–558-1, the horrible XF–92A, the beautiful D–558-2 . . . and to know that he would soon be at an altitude, in the thin air at the edge of space, where the stars and the moon came out at noon, in an atmosphere so thin that the ordinary laws of aerodynamics no longer applied and a plane could skid into a flat spin like a cereal bowl on a waxed Formica counter and then start tumbling, not spinning and not diving, but tumbling, end over end like a brick . . . In those planes, which were like chimneys with little razor-blade wings on them, you had to be "afraid to panic," and that phrase was no joke. In the skids, the tumbles, the spins, there was, truly, as Saint-Exupéry had said, only one thing you could let yourself think about: *What do I do next?* Sometimes at Edwards they used to play the tapes of pilots going into the final dive, the one that killed them, and the man would be tumbling, going end over end in a fifteen-ton length of pipe, with all aerodynamics long gone, and not one prayer left, and he knew it, and he would be screaming into the microphone, but not for Mother or for God or the nameless spirit of Ahor, but for one last hopeless crumb of information about the loop: "I've tried A! I've tried B! I've tried C! I've tried D! Tell me what else I can try!" And then that truly spooky click on the machine. *What do I do next?* (In this moment when the Halusian Gulp is opening?) And everybody around the table would look at one another and nod ever so slightly, and the unspoken message was: Too bad! There was a man with the right stuff. There was no national mourning in such cases, of course. Nobody outside of Edwards knew the man's name. If he were well liked, he might get one of those dusty stretches of road named for him on the base. He was probably a junior officer doing all this for four or five thousand a year. He owned perhaps two suits, only one of which he dared wear around people he didn't know. But none of that mattered!—not at Edwards—not in the Brotherhood.

What made it truly beautiful (for a True Brother!) was that for a good five years Edwards remained primitive and Low Rent, with nothing out there but the bleached prehistoric shrimp terrain and the rat shacks and the blazing sun and the thin blue sky and the rockets sitting there moaning and squealing before dawn. Not even Pancho's changed—except to become more gloriously Low Rent. By 1949 *the girls* had begun turning up at Pancho's in amazing numbers. They were young, lovely, juicy, frisky—and there were so many of them, at all hours, every day of the week! And they were not prostitutes, despite the accusations made later. They were just . . . well, just young juicy girls in their twenties with terrific young conformations and sweet cupcakes and loamy loins. They were sometimes described with a broad sweep as "stewardesses," but only a fraction of them really were. No, they were lovely young things who arrived as mysteriously as the sea gulls who sought the squirming shrimp. They were moist labial piping little birds who had somehow learned that at this strange place in the high Mojave lived the hottest young pilots in the world and that this was *where things were happening.* They came skipping and screaming in through the banging screen doors at Pancho's—and it completed the picture of Pilot Heaven. There was no other way to say it. Flying & Drinking and Drinking & Driving and Driving & Balling. The pilots began calling the old Fly Inn dude ranch "Pancho's Happy Bottom Riding Club," and there you had it.

All of this was fraternal bliss. No pilot was shut off from it because he was "in the public eye." Not even the rocket aces were isolated like stars. Most of them also performed the routine flight-test chores. Some of Yeager's legendary exploits came when he was merely a supporting player, flying "chase" in a fighter plane while another pilot flew the test aircraft. One day Yeager was flying chase for another test pilot at 20,000 feet when he noticed the man veering off in erratic maneuvers. As soon as he reached him on the radio, he realized the man was suffering from hypoxia, probably because an oxygen hose connection had come loose. Some pilots in that state became like belligerent drunks—prior to losing consciousness. Yeager would tell the man to check his oxygen system, he'd tell him to go to a lower altitude, and the man kept suggesting quaint anatomical impossibilities for Yeager to perform on himself. So Yeager hit upon a ruse that only he could have pulled off. "Hey," he said, "I got me a problem here, boy. I caint keep this thing running even on the emergency system. She just flamed out. Follow me down." He started descending, but his man stayed above him, still meandering. So Yeager did a very un-Yeager-like thing. He *yelled* into the microphone!

He yelled: "Look, my dedicated young scientist—*follow me down!*" The change in tone—*Yeager yelling!*—penetrated the man's impacted hypoxic skull. *My God! The fabled Yeager! He's yelling—Yeager's yelling!—to me for help! Jesus H. Christ!* And he started following him down. Yeager knew that if he could get the man down to 12,000 feet, the oxygen content of the air would bring him around, which it did. *Hey! What happened?* After he landed, he realized he had been no more than a minute or two from passing out and punching a hole in the desert. As he got out of the cockpit, an F–86 flew overhead and did a slow roll sixty feet off the deck and then disappeared across Rogers Lake. That was Yeager's signature.

Yeager was flying chase one day for Bill Bridgeman, the prime pilot for one of the greatest rocket planes, the Douglas Skyrocket, when the ship went into a flat spin followed by a violent tumble. Bridgeman fought his way out of it and regained stability, only to have his windows ice up. This was another common danger in rocket flights. He was out of fuel, so that he was now faced with the task of landing the ship both deadstick and blind. At this point Yeager drew alongside in his F–86 and became his eyes. He told Bridgeman every move to make every foot of the way down . . . as if he knew that ol' Skyrocket like the back of his hand . . . and this was jes a little ol' fishin' trip on the Mud River . . . and there was jes the two of 'em havin' a little poker-hollow fun in the sun . . . and that lazy lollygaggin' chucklin' driftin' voice was still purrin' away . . . the very moment Bridgeman touched down safely. You could almost hear Yeager saying to Bridgeman, as he liked to do:

"How d'ye hold with rockets now, son?"

That was what you thought of when you saw the F–86 do a slow roll sixty feet off the deck and disappear across Rogers Lake.

Yeager had just turned thirty. Bridgeman was thirty-seven. It didn't dawn on him until later that Yeager always called him *son*. At the time it had seemed perfectly natural. Somehow Yeager was like the big daddy of the skies over the dome of the world. In keeping with the eternal code, of course, for anyone to have suggested any such thing would have been to invite hideous ridicule. There were even other pilots with enough Pilot Ego to believe that *they* were actually better than this drawlin' hot dog. But no one would contest the fact that as of that time, the 1950's,

Chuck Yeager was at the top of the pyramid, number one among all the True Brothers.

And *that voice* . . . started drifting down from on high. At first the tower at Edwards began to notice that all of a sudden there were an awful lot of test pilots up there with West Virginia drawls. And pretty soon there were an awful lot of fighter pilots up there with West Virginia drawls. The air space over Edwards was getting so caint-hardly supercool day by day, it was terrible. And then that lollygaggin' poker-hollow air space began to spread, because the test pilots and fighter pilots from Edwards were considered the pick of the litter and had a cachet all their own, wherever they went, and other towers and other controllers began to notice that it was getting awfully drawly and down-home up there, although they didn't know exactly why. And then, because the military is the training ground for practically all airline pilots, it spread further, until airline passengers all over America began to hear that awshuckin' driftin' gone-fishin' Mud River voice coming from the cockpit . . . "Now, folks, uh . . . this is the captain . . . ummmm . . . We've got a little ol' red light up here on the control panel that's tryin' to tell us that the *lan*din' gears're not . . . uh . . . *lock*in' into position . . ."

But so what! What could possibly go wrong! We've obviously got a man up there in the cockpit who doesn't have a nerve in his body! He's a block of ice! He's made of 100 percent righteous victory-rolling True Brotherly stuff.

Yeager quit testing rocket planes in 1954 and returned to strictly military flying. First he went to Okinawa to test a Soviet MiG–15 that a North Korean defector, a pilot named Kim Sok No, had arrived in, giving the Air Force its first opportunity to study this fabled craft. American pilots used to come back from the Yalu River saying that the MiG–15 was so hot you could put your F–86 in a power dive and the MiG would fly outside loops around you all the way down. Yeager took the MiG–15 up to 50,000 feet and then down to 12,000 feet in a power dive without even so much as an instruction manual to go by. He found that it would outclimb and outaccelerate the F–86, but that the F–86 had a higher top speed in both level flight and in dives. The MiG–15 was good but not exactly a superfighter that should strike terror in the heart of the West. Yeager had to chuckle. Some things never changed. You let any fighter jock talk about the enemy aircraft and he'll tell you it's the hottest thing that ever left the ground. After all, it made him look just that much better when he waxed the bandit's tail. Then Yeager went to Germany to fly F–86s and to train the American combat squadrons there in a special air-alert system. By October 4, 1957, he was back in the United States, at George Air Force Base, about fifty miles southeast of Edwards, commanding a squadron of F–100s, when the Soviet Union launched the rocket that put a 184-pound artificial satellite called Sputnik I into orbit around the earth.

Yeager was not terribly impressed. The thing was so goddamned small. The idea of an artificial earth satellite was not novel to anyone who had been involved in the rocket program at Edwards. By now, ten years after Yeager had first flown a rocket faster than Mach 1, rocket development had reached the point where the idea of unmanned satellites such as Sputnik I was taken for granted. Two years ago, 1955, the government had published a detailed description of the rockets that would be used to launch a small satellite in late 1957 or early 1958 as part of the United States' contribution to the International Geophysical Year. Engineers for NACA and the Air Force and several aircraft companies were already designing manned spacecraft as the logical extension of the X series. The preliminary design section of North American Aviation had working drawings and most of the specifications for a fifteen-ton ship called the X–15B, a winged craft that would be launched by three enormous rockets, each with 415,000 pounds of thrust, whereupon the ship's two pilots would take over with the X–15B's own 75,000-pound engine, make three or more orbits of the earth, reenter the atmosphere, and land on a dry lake bed at Edwards like any other pilots in the X series. This was no mere dream. North American was already manufacturing a ship almost as ambitious: namely the X–15. Scott Crossfield was in training to fly it. The X–15 was designed to achieve an altitude of 280,000 feet, just above fifty miles, which was generally regarded as the boundary where all trace of atmosphere ended and "space" began. Within a month after the launching of Sputnik I, North American's chief engineer, Harrison Storms, was in Washington with a completely detailed proposal for the X–15B project. His turned out to be one among 421 proposals for manned spacecraft that had been submitted to NACA and the Defense Department. The Air Force was interested in a rocket-glider craft, similar to the X–15B, that would be called the X–20 or Dyna-Soar, for "dynamic soaring"; an Air Force rocket, the Titan, which was under development, would provide the 500,000 pounds of thrust that would be required. Naturally the pilots of the X–15B or the X–20 or whatever—the first Americans and possibly the first men in the world to go into space—would come from Edwards. At Edwards you had men like Crossfield, Iven Kincheloe, and Joe Walker, who had already flown rockets many times.

So what was the big deal about Sputnik I? The problem was already on the way to being solved.

That was the way it looked to Yeager and to everybody involved in the X series at Edwards. It was hard to realize how Sputnik I looked to the rest of the country and particularly to politicians and the press . . . and other technological illiterates with influence . . . It was hard to realize that Sputnik I, if not the MiG–15, would strike terror in the heart of the West.

After two weeks, however, the situation was obvious: a colossal panic was underway, with congressmen and newspapermen leading a huge pack that was baying at the sky where the hundred-pound Soviet satellite kept beeping around the world. In their eyes Sputnik I had become the second momentous event of the Cold War. The first had been the Soviet development of the atomic bomb in 1953. From a purely strategic standpoint, the fact that the Soviets had the rocket power to launch Sputnik I meant that they now also had the capacity to deliver the bomb on an intercontinental ballistic missile. The panic reached far beyond the relatively sane concern for tactical weaponry, however. Sputnik I took on a magical dimension—among highly placed persons especially, judging by opinion surveys. It seemed to dredge up primordial superstitions about the influence of heavenly bodies. It gave birth to a modern, i.e., technological, astrology. Nothing less than *control of the heavens* was at stake. It was Armageddon, the final and decisive battle of the forces of good and evil. Lyndon Johnson, who was the Senate majority leader, said that whoever controlled "the high ground" of space would control the world. This phrase, "the high ground," somehow caught hold. "The Roman Empire," said Johnson, "controlled the world because it could build roads. Later—when it moved to sea—the British Empire was dominant because it had ships. In the air age we were powerful because we had airplanes. Now the Communists have established a foothold in outer space." *The New York Times*, in an editorial, said the United States was now in a "race for survival." The panic became more and more apocalyptic. Nothing short of doom awaited the loser, now that the battle had begun. When the Soviets shot a Sputnik called *Mechta* into a heliocentric orbit, the House Select Committee on Astronautics, headed by House Speaker John McCormack, said that the United States faced the prospect of "national extinction" if it did not catch up with the Soviet space program. "It cannot be overemphasized that the survival of the free world—indeed, all the world—is caught up in the stakes." The public, according to the Gallup poll, was not all that alarmed. But McCormack, like a great many powerful people, genuinely believed in the notion of "controlling the high ground." He was genuinely convinced that the Soviets would send up space platforms from which they could drop nuclear bombs at will, like rocks from a highway overpass.

The Soviet program gave off an aura of sorcery. The Soviets released practically no figures, pictures, or diagrams. And no names; it was revealed only that the Soviet program was guided by a mysterious individual known as "the Chief Designer." But his powers were indisputable! Every time the United States announced a great space experiment, the Chief Designer accomplished it first, in the most startling fashion. In 1955 the United States announces plans to launch an artificial earth satellite by early 1958. The Chief Designer startles the world by doing it in October 1957. The United States announces plans to send a satellite into orbit around the sun in March of 1959. The Chief Designer does it in January 1959. The fact that the United States went ahead and successfully conducted such experiments on schedule, as announced, impressed no one—and Americans least of all. In a marvelously morose novel of the future called *We*, completed in 1921, the Russian writer Evgeny Zamyatin describes a gigantic "fire-breathing, electric" rocket ship that is poised to "soar into cosmic space" in order to "subjugate the unknown beings on other planets, who may still be living in the primitive condition of freedom"—all this in the name of "the Benefactor," ruler of "the One State." This omnipotent spaceship is called the Integral, and its designer is known only as "D–503, Builder of the Integral." In 1958 and early 1959, as magical success followed magical success, that was the way Americans, the leaders even more so than the followers, began to look on the Soviet space program. It was a thing of misty but stupendous dimensions . . . the mighty Integral . . . with an anonymous but omnipotent Chief Designer . . . Builder of the Integral. Within the federal government and throughout the education bureaucracies rose a cry for a complete overhauling of American education in order to catch up with the new generation, the new dawn, of socialist scientists, out of which had come geniuses like the Chief Designer (Builder of the Integral!) and his assistants.

The panic was greatly exacerbated by the figure of Nikita Khrushchev, who now emerged as the new Stalin in terms of his autocratic rule of the Soviet Union. Khrushchev was a type whom Americans could readily understand and fear. He was the burly, hearty, crude but shrewd farmboy capable of grinning with barnyard humor one moment and of tormenting small animals the next. After Sputnik I Khrushchev became the wicked master of mocking the United States for its incompetence. Two months after Sputnik I the Navy tried to launch the first American satellite with a Vanguard rocket. The first nationally televised *countdown* began . . . "Ten, nine, eight . . ." Then . . . "Ignition!" A mighty surge of noise and flames. The rocket lifts—some six inches. The first stage, bloated with fuel, explodes, and the rest of the rocket sinks into the sand beside the launch platform. It appears to sink very slowly, like a fat old man collapsing into a Barcalounger. The sight is absolutely ludicrous, if one is in the mood for a practical joke. Oh, Khrushchev had fun with that, all right! This picture—the big buildup, the dramatic countdown, followed by the exploding cigar—was unforgettable. It became *the* image of the American space program. The press broke into a hideous cackle of national self-loathing, with the headline KAPUTNIK! being the most inspired rendition of the mood.

The rocket pilots at Edwards simply could not understand what sort of madness possessed everybody. They watched in consternation as a war effort mentality took over. Catch up! On all fronts! That was the imperative. They could scarcely believe the outcome of a meeting held in Los Angeles in March of 1958. This was an emergency meeting (*what* emergency?) of government, military, and aircraft industry leaders to discuss the possibility of getting a man into space before the Russians. Suddenly there was no longer time for orderly progress. To put an X–15B or an X–20 into orbit, with an Edwards rocket pilot aboard, would require rockets that were still three or four years away from delivery. So a so-called quick and dirty approach was seized upon. Using available rockets such as the Redstone (70,000 pounds of thrust) and the just-developed Atlas (367,000 pounds), they would try to launch not a flying ship but a pod, a container, a *capsule,* with a man in it. The man would not be a pilot; he would be a human cannonball. He would not be able to alter the course of the capsule in the slightest. The capsule would go up like a cannonball and come down like a cannonball, splashing into the ocean, with a parachute to slow it down and spare the life of the human specimen inside. The job was assigned to NACA, the National Advisory Committee for Aeronautics, which was converted into NASA, the National Aeronautics and Space Administration. The project was called Project Mercury.

The capsule approach was the brainchild of a highly regarded Air Force research physician, Brigadier General Don Flickinger. The Air Force named it the MISS project, for "man in space soonest." The man in the MISS capsule would be an aero-medical test subject and little more. In fact, in the first flights, as Flickinger envisioned it, the capsule would contain a chimpanzee. Mercury was a slightly modified version of MISS, and so naturally enough Flickinger became one of the five men in charge of selecting Project Mercury's astronauts, as they would be called. The fact that NASA would soon be choosing men to go into space had not been made public, but Scott Crossfield was aware of it. Shortly after the Sputnik I launching, Crossfield, Flickinger, and seven others had been named to an emergency committee on "human factors and training" for space flight. Crossfield had also worked closely with Flickinger when he was testing pressure suits at Wright-Patterson Air Force Base in preparation for the X–15 project. Now Crossfield approached Flickinger and told him he was interested in becoming an astronaut. Flickinger liked Crossfield and admired him. And he told him: "Scotty, don't even bother applying, because you'll only be turned down. You're too independent." Crossfield was the most prominent of the rocket pilots, now that Yeager was no longer at Edwards, and he had as well developed an ego as any of Edwards' fabled jocks, and he was one of the most brilliant of all the pilots when it came to engineering. Flickinger seemed to be telling him that Project Mercury just wasn't suited for the righteous brethren of yore, the veterans of those high desert rat-shack broomstick days when there were no chiefs and no Indians and the pilot huddled in the hangar with the engineer and then went out and took the beast up and lit the candle and reached for the stars and rode his chimney and landed it on the lake bed and made it to Pancho's in time for beer call. When Flickinger explained to him that the first flight of the Mercury system would be made by a chimpanzee . . . well, Crossfield wasn't even particularly interested any more. Nor were most of the other pilots who were in line to fly the X–15. A *monkey's gonna make the first flight.* That was what you started hearing. Astronaut meant "star voyager," but in fact the poor devil would be a guinea pig for the study of the effects of weightlessness on the body and the central nervous system. As the brethren knew, NASA's original civil-service job specifications for Mercury astronaut did not even require that the star voyager be a pilot of any description whatsoever. Just about any young male college graduate with experience in a physically dangerous pursuit would do, so long as he was under five feet eleven and could fit into the Mercury capsule. The announcement calling for volunteers did mention test pilots as being among the types of men who might qualify, but it also mentioned submarine crew members, parachute jumpers, arctic explorers, mountain climbers, deep sea divers, even scuba divers, combat veterans, and, for that matter, mere veterans of combat training, and men who had served as test subjects for acceleration and atmospheric pressure tests, such as the Air Force and Navy had been running. The astronaut would not be expected to *do* anything; he only had to be able to take it.

NASA was ready to issue the call when the President himself, Eisenhower, stepped in. He foresaw bedlam. Every lunatic in the U.S.A. would volunteer for this thing. Every dingaling in the U.S. Congress would be touting a favorite son. It would be chaos. The selection process might take months, and the inevitable business of security clearances would take a few more. Late in December Eisenhower directed that NASA select the astronauts from among the 540 military test pilots already on duty, even though they were rather overqualified for the job. The main thing was that their records were immediately available, they already had security clearances, and they could be ordered to Washington at a moment's notice. The specifications were that they be under five feet eleven and no older than thirty-nine and that they be graduates of test-pilot schools, with at least 1,500 hours of flying time and experience in jets, and that they have bachelor's degrees "or the equivalent." One hundred and ten of the pilots fit the profile. There were men on the NASA selection committee who wondered if the pool was big enough. They figured that they would be lucky if one test pilot in ten volunteered. Even that wouldn't be quite enough, because they were looking for twelve astronaut candidates. They only needed six for the flights themselves, but they assumed that at least half the candidates would drop out because of the frustration of training to become passive guinea pigs in an automated capsule.

After all, they already knew how the leading test pilots at Edwards felt. North American had rolled out the first X–15 in the fall of 1958, and Crossfield and his colleagues, Joe Walker and Iven Kincheloe, had become absorbed in the assignment. Joe Walker was NASA's prime pilot for the project, and Kincheloe was prime pilot for the Air Force. Kincheloe had set the world altitude record of 126,000 feet in the X–2, and the Air Force envisioned him as the new Yeager . . . and then some. Kincheloe was a combat hero and test pilot from out of a dream, blond, handsome, powerful, bright, supremely ambitious and yet popular with all who worked with him, including other pilots. There was absolutely no ceiling on his future in the Air Force. Then one perfectly sunny day he was making a routine takeoff in an F–104 and the panel lit up red and he had *that one second* in which to decide whether or not to punch out at an altitude of about fifty feet . . . a choice complicated by the fact that the F–104's seat ejected straight down, out of the belly . . . and so he tried to roll the ship over and eject upside down, but he went out sideways and was killed. His backup, Major Robert White, took his place in the X–15 project. Joe Walker's backup was a former Navy fighter pilot named Neil Armstrong. Crossfield, White, Walker, Armstrong—they no longer had time to even think about Project Mercury. Project Mercury did not mean the end of the X–15 program. Not at all. The testing of the X–15 would proceed, in order to develop a true spacecraft, a ship that a pilot could fly into space and fly back down through the atmosphere for a landing. Much was made of the fact that the X–15 would "land with dignity" rather than splash down in the water like the proposed Mercury capsule. Press interest in the X–15 had become tremendous, because it was the country's sole existing "spaceship." Reporters had started writing about Kincheloe as "Mr. Space," since he was the one who held the altitude record. After his death they hung the title on Crossfield. It was a bother . . . but a fellow could learn to live with it . . . In any case, Project Mercury, the human cannonball approach, looked like a Larry Lightbulb scheme, and it gave off the funk of panic. Any pilot who went into it would no longer be a pilot. He would be a laboratory animal wired up from skull to rectum with medical sensors. The rocket pilots had fought this medical crap every foot of the way. Scott Crossfield had reluctantly allowed them to wire him for heartbeat and respiration in rocket flights but had refused to let them insert a rectal thermometer. The pilots who signed up to crawl into the Mercury capsule—the *capsule,* everybody noted, not the *ship*—would be called "astronauts." But, in fact, they would be lab rabbits with wires up the tail and everywhere else. Nobody in his right mind would hang his hide out over the edge for ten or fifteen years and ascend the pyramid and finally reach the dome of the world, Edwards . . . only to end up like that: a lab rabbit curled up motionless in a *capsule* with his little heart pitter-patting and a wire up the kazoo.

Some of the most righteous of the brethren weren't even eligible for the preliminary screening for Project Mercury. Yeager was young enough—still only thirty-five—but had never attended college. Crossfield and Joe Walker were civilians. Not that any of them gave a damn . . . at the time. The commanding officer at Edwards passed the word around that he wanted his top boys, the test pilots in Fighter Ops, to avoid Project Mercury because it would be a ridiculous waste of talent; they would just become "Spam in a can." This phrase "Spam in a can" became very popular at Edwards as the nickname for Project Mercury.

IV. The Lab Rat

Pete conrad, being an alumnus of Princeton and the Philadelphia Main Line, had the standard E.S.A. charm and command of the proprieties. E.S.A. was 1950's Princeton club code for "Eastern Socially Attractive." E.S.A. qualities served a man well in the Navy, where refinement in the officer ranks was still valued. Yet Conrad remained, at bottom, the Hickory Kid. He had the same combination of party manners and Our Gang scrappiness that his wife, Jane, had found attractive when she met him six years before. Now, in 1959, at the age of twenty-eight, Conrad was still just as wirily built, five feet six and barely 140 pounds, still practically towheaded, and he had the same high-pitched nasal voice, the same collegiate cackle when he laughed, and the same Big Weekend grin that revealed the gap between his two front teeth. Nevertheless, people gave him room. There was an old-fashioned Huck Finn hickory-stick don't-cross-that-line-or-I'll-crawl-you streak in him. Unlike a lot of pilots, he tended to say exactly what was on his mind when aroused. He couldn't stand being trifled with. Consequently he seldom was.

That was Conrad. Add the normal self-esteem of the healthy young fighter jock making his way up the mighty ziggurat . . . and the lab rat's revolt was probably in the cards from the beginning.

The survivors of Group 20's *bad string* had just completed their flight-test training when the orders arrived. Conrad received them, and so did Wally Schirra and Jim Lovell. "Shaky" Lovell—he was stuck with the nickname Conrad had given him—had finished first in the training class. The orders were marked "top secret." That already had half the base talking, of course. There was nothing like issuing top-secret orders for a whole batch of officers in the same outfit to make the grapevine start lunging about like a live wire. They were supposed to report to a certain room at the Pentagon disguised as civilians.

So on the appointed Monday morning, February 2, Conrad, along with Schirra and Lovell, arrives at the Pentagon and presents his orders and files into a room with thirty-four other young men, most of them with crew cuts and all of them with lean lineless faces and suntans and the unmistakable cocky rolling gait of fighter jocks, not to mention the pathetic-looking civilian suits and the enormous wristwatches. The wristwatches had about two thousand calibrations on them and dials for recording everything short of the sound of enemy guns. These terrific wristwatches were practically fraternal insignia among the pilots. Thirty-odd young souls wearing Robert Hall clothes that cost about a fourth as much as their watches: in the year 1959 this just had to be a bunch of military pilots trying to disguise themselves as civilians.

Once inside the room, the boys realized that they were part of a secret gathering of military test pilots from all over the country. That was rather righteous stuff. Two of the highest ranking engineers of the National Aeronautics and Space Administration, Abe Silverstein and George Low, started briefing them. They had been brought to Washington, they were told, because NASA needed volunteers for suborbital and orbital flights above the earth's atmosphere in Project Mercury. The project had the highest national priority, comparable to that of a crash program in wartime. NASA intended to put astronauts into space by mid-1960, fifteen months from now.

A pilot could tell, if he listened carefully to the briefing, that an astronaut on a Project Mercury flight would do none of the things that comprised *flying* a ship: he would not take it aloft, control its flight, or land it. In short, he would be a passenger. The propulsion, guidance, and landing would all be determined automatically by the ground. Yet the slender engineer, Low, went out of his way to show that the astronaut would exercise some forms of control. He would have "altitude control," for example. In fact, this meant only that the astronaut could make the capsule yaw, pitch, or roll by means of little hydrogen-peroxide thrusters, just as you could rock a seat on a Ferris wheel but couldn't change its orbit or direction in the slightest. But when a capsule was put into earth orbit, said Low, controlling the altitude would be essential for bringing the capsule back in through the atmosphere. Otherwise, the vehicle would burn up, and the astronaut with it. If the automatic control system malfunctioned, then the astronaut would have to take over on the manual or the fly-by-wire. In the fly-by-wire system the apparatus of the automatic system could be commandeered by the astronaut for manual control. The astronaut might also have to override the automatic system, in the case of a malfunction, to fire the retrorockets to reduce the capsule's speed and bring it out of orbit. *Retrofire! Fly-by-wire!* It was as if you really would be flying the thing. The stocky engineer, Silverstein, told them that obviously the Mercury flights might be hazardous. The first men to go into space would be running considerable risk. Therefore, the astronauts would be chosen on a strictly volunteer basis; and if a man did not volunteer, that fact would not be entered on his record or held against him in any way.

The message had a particular ring to it; but coming, as it did, from a civilian, it took a while for it to register.

Conrad and the rest of the Pax River contingent were staying at the Marriott motel near the Pentagon, and after dinner they got together in one of the rooms and had a long discussion. Schirra was there, and Lovell, and Alan Shepard, a veteran test pilot who had recently been reassigned from Pax River to a staff position in Norfolk, and a few others. What they talked about was not space travel, the future of the galaxy, or even the problems of riding a rocket into earth orbit. No, they talked about a rather more urgent matter: what this Project Mercury might do to your Navy career.

Wally Schirra had a lot of reservations about the thing, and Conrad and everybody else listened. Schirra was farther up the pyramid than anybody else in the room. Alan Shepard had more flight test experience, but he had never been in combat. Schirra, at thirty-five, had an outstanding combat record and was the sort of man who was obviously going places in the Navy. He had graduated from the Naval Academy, and his wife, Jo, was the stepdaughter of Admiral James Holloway, former commander of the Pacific Theater in the Second World War. Wally had been on ninety combat missions in Korea and shot down two MiGs. He had been chosen for the initial testing of the Sidewinder air-to-air missile out at China Lake, California, he had tested the F–4H for the Navy at Edwards itself—all this before joining Group 20 to complete his flight test training. Wally was quite popular. He was a stocky fellow with a big wide open face who was given to pranks, cosmic winks, fast cars, and all other ways of "maintaining an even strain," to use a Schirraism. He was a practical joker of the amiable sort. He would call up and say, "Hey, you gotta come over here! You'll never guess what I caught in the woods . . . A mongoose! I'm not kidding—a *mon*goose! You gotta see this thing!" And it would sound so incredible, you'd go over and take a look. Up on a table Wally would have a box that looked as if it had been converted into a cage, and he'd say: "Here, I'll open the top a little, so you can see him. But don't put your hand in, because he'll take it off for you. This baby is *vicious*." You'd lean down to take a look and—bango!—the lid flies open and this huge gray streak springs toward your face—and, well, my God, veteran aviators would recoil in terror, dive for the deck—and only then realize that the gray streak was some sort of foxtail rig and the whole thing was a jack-in-the-box, Schirra-style. It was a broad joke, strictly speaking, but the delight Wally took in such things came in a wave, a wave so big that it swept you along in spite of yourself. A smile about a foot wide would spread over his face and his cheekbones would well up into a pair of cherub bellies, St. Nicholas-style, and an incredible rocking-druid laugh would come shaking and rumbling up from his rib cage, and he'd say: "Gotcha!" Schirra's "gotchas" were famous. Wally was one of those people who didn't mind showing their emotions, happiness, rage, frustration, whatever. But in the air he was as cool as they made them. His father had been an ace in the First World War, shooting down five German planes, and both his father and his mother had done barnstorm stunt flying after the war. For all his cutting up, Wally was absolutely serious about his career. And that was his mood now that this "astronaut" business confronted them.

There were some obvious problems. One, Project Mercury was a civilian program; two, NASA had not yet developed the rockets or the capsule to carry it out; three, it involved no flying, at least not in the sense a pilot used that word. The Mercury capsule was not a ship but a can. Not only did it involve no *flying*, there wasn't even a window to look out of. There wasn't even a hatch you could egress from like a man; it would take a crew of swabbos with lug wrenches to get you out of the thing. It was a *can*. Suppose you volunteered and got tied up in the project for two or three years, and then the whole thing fizzled? That was entirely possible, because this rocket-and-capsule system was novel and had a lot of Rube Goldberg stuff in it. Any test pilot who had ever been to the Society of Experimental Test Pilots convention, to one of those sessions where they show movies of Great Ideas that never made it out of the test stage, would know what he meant . . . the Sea Dart, a ten-ton jet fighter that was supposed to take off and land on water skis (up on the screen it keeps hopping up out of the waves, like a rock skipping across a pond, and the audience roars with laughter) . . . the single-engine plane, with a 25-foot propeller, that was propped up on its tail to take off vertically, like a hummingbird (it hangs in the air, suspended at forty feet, tail down, its engine churning furiously, not realizing that it has turned from an airplane into a cockeyed helicopter, and the audience roars with laughter) . . . In the history of flight these well-meant farces took place all the time. And where would you be then? You would be three years behind in flight test. You would be three years off the line in the general jockeying for promotion. You would be giving up whatever brownie points you had built up over the past four or five years. For somebody like Wally this was no joke. He was at the point in his career where you really start to climb—or you go off on some ill-advised tangent. He was in line to command his own fighter squadron. That was the route to the top, to admiral rank, for a Navy flier.

They talked for a long time. Someone Schirra's age had more to lose than Conrad, who was only twenty-eight. But as every officer knew, it was never too early to screw up your career in the Navy by getting involved in what was known, with some sarcasm, as "innovative duty."

From the beginning George Low and others in the NASA hierarchy had been afraid that the pilots would react in precisely this way. As a result, they were amazed. They had briefed thirty-five test pilots on Monday,

February 2, Conrad, Schirra, Lovell, and Alan Shepard among them, and another thirty-four the following Monday; and of the total of sixty-nine, fifty-six volunteered to become astronauts. They now had so many volunteers they didn't even call in the remaining forty-one men who fit the profile. Why bother? They already had fifty-six grossly overqualified volunteers. Not only that, the men seemed so gung-ho about the project, they figured they could get by with seven astronauts instead of twelve.

Pete Conrad had ended up volunteering, and so had Jim Lovell. In fact, every man who had been in that room at the motel had volunteered, including Wally Schirra, who had been the most dubious of all. And why? That was a good question. Despite all the pondering, all the discussions, all the career-wrestling, all the toting up of pros and cons, none of them could give you a very clear-cut answer. The matter had not been decided by sheer logic. Somehow, in that briefing in the inner room at the Pentagon, Silverstein and Low had hit every button just right. It was as if they possessed a blueprint of the way the fighter jock was wired.

"The highest national priority" . . . "hazardous undertaking" . . . "strictly volunteer" . . . so hazardous that "if you don't volunteer, it won't be held against you" . . . And they had all gotten the signal, subliminally, in the solar plexus. They were being presented with the Cold War version of *the dangerous mission.* One of the maxims that was drilled into all career officers went: *Never refuse a combat assignment.* Moreover, there was the business of "the first men to go into space." *The first men to go into space.* Well . . . suppose it happened just that way? The rocket aces at Edwards, from their eminence, might be able to look down upon the whole scheme. But within the souls of the rest of the fighter jocks who came to the Pentagon was triggered a motivation that overrode all strictly logical career considerations: I must not get . . . *left behind.*

That feeling was magnified by the public reaction. No sooner had the first group of men been briefed than the news that NASA was looking for Mercury astronauts made its way into the press. From the beginning the reporters and broadcasters dealt with the subject in tones of awe. It was the awe that one has of an impending death-defying stunt. The question of whether an astronaut was a pilot or a mere guinea pig never entered into it for a moment, so far as the press was concerned. "Are they really looking for somebody to go into space on top of a rocket?" That was the question and the only one that seemed to matter. To almost anyone who had followed NASA's efforts on television, the odds against the successful launch of an American into space seemed absolutely dreadful. For fourteen months now the Eisenhower Administration had adopted the strategy of openly publicizing its attempts to catch up with the Russians—and so people were being treated to the sight of the rockets at Cape Canaveral and Wallops Island, Virginia, either blowing up on the launch pad in the most ignominious, if briefly hilarious, fashion or else heading off on crazy trajectories, toward downtown Orlando instead of outer space, in which case they had to be blown up by remote control. Well, not all of them, of course, for the United States had succeeded in putting up some small satellites, mere "oranges," as Nikita Khrushchev liked to put it, in his cruel colorful farmboy way, as compared to the 1,000-pound Sputniks the mighty Integral was sending around the earth loaded with dogs and other experimental animals. But the only obvious American talent was for blowing up. They had many names, these rockets, Atlas, Navaho, Little Joe, Jupiter, but they all blew up.

Conrad, like Schirra or any other test pilot, did not look at the TV footage in the same light, however. What people were seeing on television were, in fact, ordinary test events. Blown engines were par for the course in testing aircraft prototypes and were inevitable in testing an entirely new propulsion system such as jet or rocket engines. It had happened at Muroc in testing the engine of the second American jet fighter, the XP–80. Obviously you didn't send a man up with an engine until it had attained a certain level of reliability. The only thing unusual about the testing of big rocket engines like the Navaho and the Atlas was that so much of it was televised and that these normal test events came across as colossal "failures." They were not even radical engines. The rocket engines that had been used in the X–1 project and all of the X projects that followed employed the same basic power plants as the Atlas, the Jupiter, and the other rockets NASA was working with. They used the same fuel, liquid oxygen. The X-project rockets, inevitably, had *blown* in the testing stage, but had been made reliable in the end. No rocket pilot had ever had an engine blow up under him in flight, although one, Skip Ziegler, had died when an X–2 had exploded while it was still attached to the B–50 that was supposed to launch it. To pilots who had been through *bad strings* at Pax River or Edwards, it was hard to see how the risk would be any greater than in testing the Century series of jet fighters. Just think of a beast like the F–102 . . . or the F–104 . . . or the F–105 . . .

When Pete talked to Jane about Project Mercury, she was all for it! If he wanted to volunteer, then he should by all means do so. The thought of Pete riding a NASA rocket did not fill her with horror. On the contrary. Although she never quite put it this way to Pete, she felt that anything would be better, safer, saner than for him to continue flying high-performance jet fighters for the Navy. At the very least, astronaut training would take him away from that. As for rocket flights themselves, how could they possibly be any more dangerous than flying every day at Pax River? What rocket pilot's wife had ever been to more funerals than the wives of Group 20?

Albuquerque, home of the Lovelace Clinic, was a dirty red sod-hut tortilla highway desert city that was remarkably short on charm, despite the Mexican touch here and there. But career officers were used to dreary real estate. That was what they inhabited in America, especially if they were fliers. No, it was Lovelace itself that began to get everybody's back up. Lovelace was a fairly new private diagnostic clinic, somewhat like the Mayo Clinic, doing "aerospace-medical" work for the government, among other things. Lovelace had been founded by Randy Lovelace—W. Randolph Lovelace II—who had served along with Crossfield and Flickinger on the committee on "human factors" in space flight. The chief of the medical staff at Lovelace was a recently retired general of the Air Force medical corps, Dr. A. H. Schwichtenberg. He was General Schwichtenberg to everybody at Lovelace. The operation took itself very seriously. The candidates for astronaut would be given their physical testing here. Then they would go to Wright-Patterson Air Force Base in Dayton for psychological and stress testing. It was all very hush-hush. Conrad went to Lovelace in a group of only six men, once more in their ill-fitting mufti and terrific watches, apparently so that they would blend in with the clinic's civilian patients. They had been warned that the tests at Lovelace and Wright-Patterson would be more exacting and strenuous than any they had ever taken. It was not the tests *per se*, however, that made every self-respecting fighter jock, early in the game, begin to hate Lovelace.

Military pilots were veterans of physical examinations, but in addition to all the usual components of "the complete physical," the Lovelace doctors had devised a series of novel tests involving straps, tubes, hoses, and needles. They would put a strap around your head, clamp some sort of instrument over your eyes—and then stick a hose in your ear and pump cold water into your ear canal. It would make your eyeballs flutter. It was an unpleasant, disorienting sensation, although not painful. If you wanted to know what it was all about, the Lovelace doctors and technicians, in their uncompromising white smocks, indicated that you really didn't need to know, and that was that.

What really made Conrad feel that something *eccentric* was going on here, however, was the business of the electrode in the thumb muscle. They brought him into a room and strapped his hand down to a table, palm up. Then they brought out an ugly-looking needle attached to an electrical wire. Conrad didn't like needles in the first place, and this one looked like a monster. *Hannh?*—they drove the needle into the big muscle at the base of his thumb. It hurt like a bastard. Conrad looked up as if to say, "What the hell's going on?" But they weren't even looking at him. They were looking—at the meter. The wire from the needle led to what looked like a doorbell. They pushed the buzzer. Conrad looked down, and his hand—his own goddamned hand!—was balling up into a fist and springing open and balling up into a fist and springing open and balling up into a fist and springing open and balling up into a fist and springing open at an absolutely furious rate, faster than he could have ever made it do so on its own, and there seemed to be nothing that he, with his own mind and his own central nervous system, could do to stop his own hand or even slow it down. The Lovelace doctors in their white smocks, with their reflectors on their heads, were having a hell of a time for themselves . . . with *his* hand . . . They were reading the meter and scribbling away on their clipboards at a jolly rate.

Afterward Conrad said, "What was that for?"

A doctor looked up, distractedly, as if Conrad were interrupting an important train of thought.

"I'm afraid there's no simple way to explain it to you," he said. "There's nothing for you to worry about."

It was then that it began to dawn on Conrad, first as a feeling rather than as a fully formed thought: "Lab rats."

It went on like that. The White Smocks gave each of them a test tube and said they wanted a sperm count. *What do you mean?* Place your sperm in the tube. *How?* Through ejaculation. *Just like that?* Masturbation is the customary procedure. *What!* The best results seem to be obtained through fantasization, accompanied by masturbation, followed by ejaculation. *Where, f'r chrissake?* Use the bathroom. A couple of the boys said things such as, "Well, okay, I'll do it if you'll send a nurse in with me—to help me along if I get stuck." The White Smocks looked at them as if they were schoolboys making obscene noises. This got the pilots' back up, and a couple of them refused, flat out. But by and by they gave in, and so now you had the ennobling prospect of half a dozen test pilots padding off one by one to the head in their skivvies to jack off for the Lovelace Clinic, Project Mercury, and America's battle for the heavens. Sperm counts were supposed to determine the density and motility of the sperm. What this had to do with a man's fitness to fly on top of a rocket or anywhere else was incomprehensible. Conrad began to get the feeling that it wasn't just him and his brother lab rats who didn't know what was going on. He now had the suspicion that the Reflector Heads didn't know, either. They had somehow gotten *carte blanche* to try out any goddamned thing they could think up—and that was what they were doing, whether there was any logic to it or not.

Each candidate was to deliver two stool specimens to the Lovelace laboratory in Dixie cups, and days were going by and Conrad had been unable to egest even one, and the staff kept getting after him about it. Finally he managed to produce a single bolus, a mean hard little ball no more than an inch in diameter and shot through with some kind of seeds, whole seeds, undigested. Then he remembered. The first night in Albuquerque he had gone to a Mexican restaurant and eaten a lot of jalapeño peppers. They were jalapeño seeds. Even in the turd world this was a pretty miserable-looking *objet.* So Conrad tied a red ribbon around the goddamned thing, with a bow and all, and put it in the Dixie cup and delivered it to the lab. Curious about the ribbons that flopped out over the lip of the cup, the technicians all peered in. Conrad broke into his full cackle of mirth, much the way Wally might have. No one was swept up in the joke, however. The Lovelace staffers looked at the beribboned bolus, and then they looked at Conrad . . . as if he were a bug on the windshield of the pace car of medical progress.

One of the tests at Lovelace was an examination of the prostate gland. There was nothing exotic about this, of course; it was a standard part of the complete physical for men. The doctor puts a rubber sleeve on a finger and slips the finger up the subject's rectum and presses the prostate, looking for signs of swelling, infection, and so on. But several men in Conrad's group had come back from the prostate examination gasping with pain and calling the doctor a sadistic little pervert and worse. He had prodded the prostate with such force a couple of them had passed blood.

Conrad goes into the room, and sure enough, the man reams him so hard the pain brings him to his knees.

"What the hell!—"

Conrad comes up swinging, but an orderly, a huge monster, immediately grabs him, and Conrad can't move. The doctor looks at him blankly, as if he's a vet and Conrad's a barking dog.

The probings of the bowels seemed to be endless, full proctosigmoidoscope examinations, the works. These things were never pleasant; in fact, they were a bit humiliating, involving, as they did, various things being shoved up your tail. The Lovelace Clinic specialty seemed to be the exacting of maximum indignity from each procedure. The pilots had never run into anything like this before. Not only that, before each ream-out you had to report to the clinic at seven o'clock in the morning and give yourself an enema. *Up yours!* seemed to be the motto of the Lovelace Clinic—and they even made you do it to yourself. So Conrad reports at seven one morning and gives himself the enema. He's supposed to undergo a lower gastrointestinal tract examination that morning. In the so-called lower G.I. examination, barium is pumped into the subject's bowels; then a little hose with a balloon on the end of it is inserted in the rectum, and the balloon is inflated, blocking the canal to keep the barium from forcing its way out before the radiologist can complete his examination. After the examination, like everyone who has ever been through the procedure, Conrad now feels as if there are eighty-five pounds of barium in his intestines and they are about to explode. The Smocks inform him that there is no john on this floor. He's supposed to pick up the tube that is coming out of his rectum and follow an orderly, who will lead him to a john two floors below. On the tube there is a clamp, and he can release the clamp, deflating the balloon, at the proper time. *It's unbelievable!* To try to walk, with this explosive load sloshing about in your pelvic saddle, is agony. Nevertheless, Conrad picks up the tube and follows the orderly. Conrad has on only the standard bed patient's tunic, the angel robes, open up the back. The tube leading out of his tail to the balloon gizmo is so short that he has to hunch over to about two feet off the floor to carry it in front of him. His tail is now, as the saying goes, flapping in the breeze, with a tube coming out of it. The orderly has on red cowboy boots. Conrad is intensely aware of that fact, because he is now hunched over so far that his eyes hit the orderly at about calf level. He's hunched over, with his tail in the breeze, scuttling like a crab after a pair of red cowboy boots. Out into a corridor they go, an ordinary public corridor, the full-moon hunchback and the red cowboy boots, amid men, women, children, nurses, nuns, the lot. The red cowboy boots are beginning to trot along like mad. The orderly is no fool. He's been through this before. He's been through the whole disaster. He's seen the explosions. Time is of the essence. There's a hunchback stick of dynamite behind him. To Conrad it becomes more incredible every step of the way. They actually have to go down an elevator—full of sane people—and do their crazy tango through another public hallway—agog with normal human beings—before finally reaching the goddamned john.

Later that day Conrad received, once more, instructions to report to the clinic at seven the next morning to give himself an enema. The next thing the people in the administrative office of the clinic knew, a small but enraged young man was storming into the office of General Schwichtenberg himself, waving a great flaccid flamingo-pink enema bag and hose like some sort of obese whip. As he waved it, it gurgled.

The enema bag came slamming down on the general's desk. It landed with a tremendous *plop* and then began gurgling and sighing.

"General Schwichtenberg," said Conrad, "you're looking at a man who has given himself his last enema. If you want enemas from me, from now on you can come get 'em yourself. You can take this bag and give it to a nurse and send her over—"

Just you—

"—and let her do the honors. I've given myself my last enema. Either things shape up around here, or I ship out."

The general stared at the great flamingo bag, which lay there heaving and wheezing on his desk, and then he stared at Conrad. The general seemed appalled . . . All the same it wouldn't do anybody any good, least of all the Lovelace Clinic, if one of the candidates pulled out, firing broadsides at the operation. The general started trying to mollify this vision of enema rage.

"Now, Lieutenant," he said, "I know this hasn't been pleasant. This is probably the toughest examination you'll ever have to go through in your life, but as you know, it's for a project of utmost importance. The project needs men like yourself. You have a compact build, and every pound saved in Project Mercury can be critical."

And so forth and so on. He kept spraying Conrad's fire.

"All the same, General, I've given myself my last enema."

Word of the Enema Bag Showdown spread rapidly among the other candidates, and they were delighted to hear about it. Practically all of them had wanted to do something of the sort. It wasn't just that the testing procedures were unpleasant; the entire atmosphere of the testing constituted an affront. There was something . . . decidedly *out of joint* about it. Pilots and doctors were natural enemies, of course, at least as pilots saw it. The flight surgeon was pretty much kept *in his place* in the service. His only real purpose was to tend to pilots and keep 'em flying. He was an attendant to the pilots' vital stuff. In fact, flight surgeons were encouraged to fly back-seat with fighter pilots from time to time, so as to understand what stresses and righteous stuff the job entailed. Regardless of how much he thought of himself, no flight surgeon dared position himself *above* the pilots in his squadron in the way he conducted himself before them: i.e., it was hard for him to be a consummate panjandrum, the way the typical civilian doctor was.

But at Lovelace, in the testing for Project Mercury, the natural order was turned upside down. These people not only did not treat them as righteous pilots, they did not treat them as pilots of any sort. They never even alluded to the fact that they were pilots. An irksome thought was beginning to intrude. In the competition for *astronaut* the kind of stuff you were made of as a *pilot* didn't count for a goddamned thing. They were looking for a certain type of animal who registered bingo on the meter. You wouldn't win this competition in the air. If you won it, it would be right here on the examination table in the land of the rubber tubes.

Yes, the boys were delighted when Conrad finally told off General Schwichtenberg. Attaboy, Pete! At the same time, they were quite content to let the credit for the Lab Rat Revolt fall to Conrad and to him alone.

At Wright-Patterson Air Force Base, where they went for the psychological and stress testing, the air of secrecy was even more pronounced than at Lovelace. At Wright-Patterson they went through the testing in groups of eight. They were billeted off to themselves at the BOQ, the Bachelor Officers Quarters. If they had to call for anything on the base, they were not to refer to themselves by name. Instead, each of them had a number. Conrad was "Number 7." If he needed a car to take him from one place to another, to keep an appointment, he was supposed to ring up the car pool and say only: "This is Number Seven. I need a car . . ."

The testing, on the other hand, seemed—at first—more like what a self-respecting fighter jock might expect. They gave the candidate an oxygen mask and a partial-pressure suit and put him in an air-pressure chamber and reduced the pressure until an altitude of 65,000 feet was simulated. It made one feel as if his entire body were being squeezed by thongs, and he had to force his breath out in order to bring new oxygen into his lungs. Part of the stress was in the fact that they didn't tell him how long he had to stay there. They put each man in a small, pitch-black, windowless, soundproofed room—a "sensory deprivation chamber"—and locked the door, again without telling him how long he would have to stay there. It turned out to be three hours. They strapped each man into a huge human milkshake apparatus that vibrated the body at tremendous amplitudes and bombarded it with high-energy sound, some of it at excruciating frequencies. They put each man at the console of a machine called "the idiot box." It was like a simulator or a trainer. There were fourteen different signals that the candidate was supposed to respond to in different ways by pressing buttons or throwing switches; but the lights began lighting up so fast no human being could possibly keep up with them. This appeared to be not only a test of reaction times but of perseverance or ability to cope with frustration.

No, there was nothing wrong with tests of this sort. Nevertheless, the atmosphere around them was a bit . . . *off.* Psychiatrists were running the show at Wright-Patterson. Every inch of the way there were psychiatrists and psychologists standing over you taking notes and giving you little jot'em'n'dot'em tests. Before they put you in the Human Milkshake, some functionary in a white smock would present you with a series of numbered dots on a piece of paper on a clipboard and you were supposed to take a pencil and connect the dots so that the numbers beside them added up to certain sums. Then when you got out of the machine, the White Smock character would give you the same test again, presumably to see if the physical experience had impaired your ability to calculate. And that was all right, too. But they also had people staring at the candidate the whole time and taking notes. They took notes in little spiral notebooks. Every gesture you made, every tic, twitch, smile, stare, frown, every time you rubbed your nose—there was some White Smock standing by jotting it down in a notebook.

One of the most assiduous of the monitors was a psychologist, a woman named Dr. Gladys J. Loring—as Conrad could tell from the nameplate on her smock. Gladys J. Loring was beginning to annoy him intensely. Every time he turned around she seemed to be standing there staring at him, without a word, staring at him with utter White Smock detachment, as if he were a frog, a rabbit, a rat, a gerbil, a guinea pig, or some other lab animal, scribbling furiously in her notebook. For days she had been watching him, and they had never even been introduced. One day Conrad suddenly looked her straight in the eye and said: "Gladys! What . . . are . . . you . . . writing . . . in . . . your . . . notebook!"

Dr. Gladys J. Loring looked at him as if he were a flatworm. All she did was make another notation of the specimen's behavior in her notebook.

To fighter jocks it was bad enough to have doctors of any sort as your final judges. To find psychologists and psychiatrists positioned above you in this manner was irritating in the extreme. Military pilots, almost to a man, perceived psychiatry as a pseudo-science. They regarded the military psychiatrist as the modern and unusually bat-brained version of the chaplain. But the shrink could be dealt with. You just turned on the charm—lit up the halo of the right stuff—and did some prudent lying.

In the interviews for this job of "astronaut," as in other situations, the psychiatrists would get on the subject of the hazards of the assignment, the unknowns, the potentially high risk, and then gauge the candidate's reaction. As all heads-up pilots knew, this called for "second-convolution" thinking. It was a mistake to say anything along the lines of: *Oh, I rather enjoy risks, I enjoy hanging my hide out over the edge, day in and day out, for that is what makes me superior to other men.* The psychiatrists always interpreted that as a reckless love of danger, an irrational impulse associated with the late-Freudian concept of "the death wish." The proper response—heard more than once during that week at Wright-Patterson—was to say: "Oh, I don't regard Project Mercury as a particularly high-risk proposition, certainly not compared to the routine test work I've been doing for [the Air Force, the Navy, the Marines]. Since this project has a high national priority, I'm sure that the safety precautions will be far more thorough and reliable than they were on something like the [F–100F, F–102, F–104, F–4B], when I had *that one* in the test stage." (Very slight smile and roll of the eyeballs.) Beautiful stuff! This showed that you were a rational test pilot, as concerned about safety as any sensible professional . . . while at the same time getting across the idea that you had been routinely risking your life and were so used to it, had such righteous stuff, that riding a rocket seemed like a vacation by comparison. That created the Halo Effect. Offhand allusions to derring-do would have the psychiatrists looking at you with big wide eyes, like little boys.

Conrad knew all this as well as the rest of them. He knew exactly how the prudent officer should deal with these people. It was hard not to know. Every night the boys got together in the BOQ and regaled each other with stories of how they had lied their heads off or otherwise diligently subverted the inquiries of the shrinks. Conrad's problem was that somewhere along the way the Hickory Kid always took over and had to add *a wink or two* for good measure.

In one test the interviewer gave each candidate a blank sheet of paper and asked him to study it and describe what he saw in it. There was no one right response in this sort of test, because it was designed to force the candidate to free-associate in order to see where his mind wandered. The test-wise pilot knew that the main thing was to stay on dry land and not go swimming. As they described with some relish later on in the BOQ, quite a few studied the sheet of paper and then looked the interviewer in the eye and said, "All I see is a blank sheet of paper." This was not a "correct" answer, since the shrinks probably made a note of "inhibited imaginative capacity" or some goddamned thing, but neither did it get you in trouble. One man said, "I see a field of snow." Well, you might get away with that, as long as you didn't go any further . . . as long as you did not thereupon start ruminating about freezing to death or getting lost in the snow and running into bears or something of that sort. But Conrad . . . well, the man is sitting across the table from Conrad and gives him the sheet of paper and asks him to study it and tell him what he sees. Conrad stares at the piece of paper and then looks up at the man and says in a wary tone, as if he fears a trick: "But it's upside down."

This so startles the man, he actually leans across the table and looks at this absolutely blank sheet of paper to see if it's true—and only after he is draped across the table does he realize that he has been had. He looks at Conrad and smiles a smile of about 33 degrees Fahrenheit.

This was *not* the way to produce the Halo Effect.

In another test they showed the candidates pictures of people in various situations and asked them to make up stories about them. One of the pictures they showed Conrad was a piece of American Scene Okie Realism, apparently from the Depression years. You could see a poor sunken hookwormy sharecropper in bib overalls trying to push a rusty plow through some eroded ground that was more gully than topsoil, aided by a mule with all his ribs showing, while off to one side the man's sallow hollow-socketed pellagra-ravaged wife with a swollen eight-month belly covered by a dress made from a fertilizer sack leans up against their shack to catch her breath or else to prop up the side wall. Conrad looks at the picture and says, "Well, you can tell that this man is a nature lover. He not only tills the soil, he appreciates the scenery, as you can tell by the way he is looking off toward the mountains, the better to observe the way the pale blue of the range in the distance harmonizes with the purple haze of the hills near his beloved homestead"—and on and on in this fashion until, at long last, it dawns on the interviewer that this wiry wiseacre who chatters away with a gap between his front teeth . . . is sending him up, him and his whole test.

This did not create the Halo Effect, either.

Oh, Conrad was rolling now. He was beginning to have a good time. But he had one piece of unfinished business. That night he called up the car pool.

"This is Number Seven," he said. "Number Seven needs a car to go to the PX."

The next day, after the heat-chamber test, in which he spent three hours shut up in a cubicle heated to 130 degrees, Conrad was rubbing the sweat off the end of his nose when he looked up—and sure enough, Dr. Gladys J. Loring was right there, making note of the event in her spiral notebook with a ballpoint pen. Conrad reached into the pocket of his pants . . . and came up with a spiral notebook and a ballpoint pen just like hers.

"Gladys!" he said. She looked up. She was startled. Conrad started scribbling in his notebook and then looked at her again. "Aha! You touched your ear, Gladys! We call that inhibition of the exhibitionism!" More scribbling in the notebook. "Oh-oh! Lowering of the eyes, Gladys! Repressed hypertrophy of the latency! I'm sorry, but it has to go in the report!"

Word of how the flatworm turned . . . how the lab rat had risen up . . . how Pavlov's dog rang Pavlov's bell and took notes on it . . . oh, word of all this circulated quickly, too, and everyone, from Number 1 to Number 8, was quite delighted. There was no indication, however, then or later, that Dr. Gladys Loring was amused in the slightest.

When Scott Carpenter called home to California from Wright-Patterson in the evenings—and he was always careful to take advantage of the lower evening rates—his wife, Rene, was usually in the living room. They had a house in Garden Grove, a town near Disneyland. The focal point of the living room was a three-piece sectional sofa that had a great teardrop-shaped monkeypod coffee table in front of it and a monkeypod end table at this end and another monkeypod end table at the other end. A great deal was summed up by those three great showy slabs of monkeypod wood with the yellow-brown grain streaks curling this way and that. Every officer and his wife in the U.S. Navy in the year 1959 understood the Monkeypod Life.

Scott was a lieutenant, which meant that his pay, including subsistence and housing allowance, was only about $7,200 a year, plus some extra flight pay. Young officers and their wives realized from the outset, of course, that abysmal pay was one of the realities of a service career. There were other forms of compensation: the opportunity to fly, which Scott loved; the status of a Naval officer; the community of the squadron (when one was in the mood for it); a certain sense of mission (if you were feeling good about yourself) that civilians lacked—extras, such as flight pay and a housing allowance, and *the goodies*. Given the dreadfully low pay, the goodies, which were usually trivial by ordinary standards, tended to take on an overblown importance. That was why so many young military couples in the late 1950's had living rooms dominated, ruled, oppressed, enslaved by the most bizarre furniture imaginable: Chinese k'ang tables with entire village scenes carved into the tops of them in deep relief, squads of high-backed Turkish chairs that could have swallowed a ballroom, Korean divans with wood frames so ornately inlaid with mother-of-pearl that the entire room seemed to be grinning hideously, Spanish armoires so gross, so gloomy, so overbearing that the mere sight of them brought conversation to a halt . . . and the flamboyant monkeypod. For one of the goodies was the opportunity to buy hand-carved wooden furniture cheaply when you were assigned to the far ends of the earth. That was your opportunity to furnish the living room at last!—and the military would ship it back to the States free of charge. Of course, your choice was limited by the local taste. In Korea you settled for mother-of-pearl or Chinese baroque. And in Hawaii, where Scott and Rene had been assigned, there was always the monkeypod.

In a department store in Hawaii a first-rate finished monkeypod coffee table cost about $150. A modest sum, one might say; but if your base pay was only $7,200 a year, that meant you had only forty-eight such sums to last you for an entire year. And Scott and Rene had four children! But unfinished slabs of this amazing wood with its flaming yellow grain were available for as little as nine dollars. If you were willing to spend twenty-four hours sanding, rubbing, oiling, and polishing them, and another ten or twenty constructing legs or frames, you could save $140. Fortunately, Rene had a good sense of design and was even able to use the monkeypod with finesse—a rare accomplishment in the Monkeypod Life.

Both Scott and Rene had been brought up in Boulder, Colorado. Scott was top drawer by Boulder social standards, such as they were. He was descended from the first white settlers in the state. His mother's father, Victor Noxon, owned and edited a newspaper, the Boulder *Miner-Journal*. Scott's parents had separated when he was only three, and his mother contracted tuberculosis and was confined to a sanitarium for long stretches, so that Scott lived in Victor Noxon's house and was, as it turned out, raised by him. Rene met Scott at the University of Colorado and dropped out in her sophomore year to marry him. They spent practically that entire first year on the ski slopes. They were an extraordinarily good-looking couple, both blond, trim, athletic, high-spirited, outgoing, the sort of couple you seldom actually saw outside the Lucky Strike ads. Many wives of fighter pilots would end up looking on helplessly as their husbands grew more and more distant, a fact they would acknowledge in what were meant as lighthearted remarks, such as: "I'm only his mistress—he's married to an airplane." Often she would be overstating their intimacy; the actual mistress would be someone she didn't know about. Scott, on the contrary, was completely devoted to Rene and their two sons and two daughters. Many nights, during the testing for Project Mercury, Scott wrote Rene long letters, some of them ten or fifteen pages, rather than run up more telephone bills. He kept trying to reassure her that he wasn't getting involved in anything reckless. One night he wrote: "Most of all, don't worry. You know what is uppermost in my mind and that I wouldn't needlessly jeopardize what we have together." He was determined, he said, to live long enough "to make love to you as a grandmother."

All along, his feelings on this score had been so profound that he had done an extraordinary thing eight years before. After finishing his basic flight training at Pensacola and his advanced training at Corpus Christi, he had voluntarily chosen to fly multi-engine PBY–4 patrol planes rather than fighter planes. He didn't even like PBY–4s. He could barely stand to fly them. Who could? They were big slow awkward trucks. Yet he had stepped down off the great ziggurat pyramid here at the first major plateau. If the subject ever came up, he would say: "I did it out of allegiance to my family," meaning that patrol planes didn't leave as many widows behind. The Korean War was just beginning, and Scott ended up flying P2V patrol planes up and down the Pacific mainland. Naturally this was completely outside the big league for Navy pilots in the war. Any truly righteous aviator wanted to be assigned, on loan, to an Air Force fighter squadron for aerial combat over North Korea. But reconnaissance had its own hazards and trials, and Scott was considered extremely proficient at it; so much so that after the war he was brought to Patuxent River and trained as a test pilot.

Nevertheless, Scott had forsaken the righteous competition. Of his own accord! Out of allegiance to his family! Perhaps it had to do with his memories of how his own family had been disrupted during his early childhood. Well, that was something for psychiatrists to speculate about, and no doubt they were doing so. In his first psychiatric interview at Wright-Patterson, Scott himself had opened the session. He asked the first question. He said to the man: "How many children do you have? I have four."

Scott was surprised when he found himself among the thirty-two finalists in the competition for astronaut. He had bowed out of the big-league competition long ago. He had only two hundred hours of flying time in jets; and most of that he had accumulated in the course of flight test training. The other candidates all seemed to have fifteen hundred to twenty-five hundred. Yet here he was. Not only that, as the days had gone by, first at Lovelace and now at Wright-Pat, his prospects had begun to look promising. It was amazing.

One thing Scott had going for him was his superb physical condition, although at the outset he would have never believed that sheer physical condition could be of any vital importance. He had been a gymnast at the University of Colorado and had terrific shoulders with the deltoid muscles bulging out in high relief, a thick strong neck, an absolutely lean and perfectly formed chest, like a South Sea pearl diver's—and, in fact, he had done a great deal of scuba diving—and his torso tapered down like Captain America's in the comic strip. Others complained the whole time, but the tests at Lovelace and Wright-Patterson didn't bother Scott in the slightest. Each one was a moment of triumph for him.

One night Scott called, and Rene could tell that he was exceptionally pleased over how things were going. It seemed that there was a test of lung capacity. The candidate sat at a table and blew into a tube. The tube led into an instrument with a column of mercury. The idea was to see how long you could hold the column of mercury up to a designated level with the pressure of your breath. The record, they were informed, was ninety-one seconds. Scott—as he excitedly told Rene that night—knew from years of undersea swimming that after your lungs feel completely out of air and every signal in your central nervous system predicts disaster if you hold your breath an instant longer, you actually have a substantial reserve supply of oxygen in your system. It is the buildup of carbon dioxide in the lungs, not the absolute depletion of oxygen, that signals the emergency. Scott forced himself to hold his breath, through all the early signals, while he counted slowly to one hundred, with the idea of surpassing the mark of ninety-one seconds. He counted very slowly, as it turned out, and held the column of mercury up for 171 seconds, almost doubling the record.

Another candidate in Scott's group also broke the old record by holding the mercury up 150 seconds. He was a Marine pilot named John Glenn. Scott had known Glenn slightly when both were at Pax River during Scott's flight test training. Glenn had set a cross-country speed record, Los Angeles to New York, of three hours and twenty-three minutes in an F8U fighter plane in July 1957. The two of them had hit it off immediately at Wright-Pat, partly, perhaps, because they appeared to be the two pacesetters in the testing. Scott had broken five records in all, and Glenn was usually his runner-up. One day they overheard one of the doctors saying to another: "Let's call Washington and tell them about these two guys."

Dr. Gladys J. Loring and the others were astounded by Scott's performance, and not merely by such matters as his lung capacity, either; for many of the tests were not physical tests in the ordinary sense but, rather, tests of perseverance and one's willingness to push oneself beyond the usual limits of human endurance.

Scott Carpenter did not mind the note-taking of Dr. Gladys J. Loring in the slightest. Scribble away! Scott was in his element.

Conrad was back home in North Town Creek, back at Pax River, when the letter from NASA arrived. He knew he had not played it very smoothly during the testing. He had compared notes with Wally Schirra, and it turned out that when Wally's group had gone through Lovelace, they had been just as ticked off at the way the place was run as Conrad and his group had been. Wally had even led his own little rebellion when they wanted their goddamned stool samples. One afternoon they had told Wally and the boys to avoid all highly seasoned food that evening, because they wanted to take stool samples the next day. So the whole group headed off to the Mexican section of Albuquerque, and they had picked out the rankest-looking restaurant they could find and fired up their innards with every red-mad dish they had ever heard of and hosed it down with plenty of good cheap rank Mexican beer. The jalapeño peppers had even gotten into the act! One of the boys had discovered a bowl of jalapeño pepper sauce on the table, a fiery reddish-brown concoction, and had poured it into a Dixie cup and presented it to the lab technicians as if he had a ferocious case of diarrhea—and had laughed his head off when the first sultry cloud of jalapeño aroma nearly wiped them out. But that was as far as he went—the lab technician level. He let the good General Schwichtenberg remain reserved and serene. That was the way Wally himself would have done it. He always knew where the outside of the envelope was, even when it came to pranks.

No, Conrad knew he had poked a few unfortunate holes in the old envelope . . . Still, he had spoken his mind before and it had never hurt him. His career in the Navy had gone up on a steady curve. He had never been left behind. So he opened the letter.

From the very first line he knew the rest. The letter noted that he had been among the finalists in the selection process and said he was to be commended for that. Alas, it went on, he had not been one of the seven chosen for the assignment, but NASA and one & all were grateful to him for volunteering and so forth and so on.

Well, there it was, your classic Dear John Letter. Even though in his detached moments he realized that he had perhaps screwed the pooch here and there, it was hard to believe. He had been left behind. This was something that had never happened to him in the nearly six years he had been moving up the great invisible ziggurat.

A couple of days later he found out that Wally had made it. Wally was one of the seven. So was Alan Shepard, who had also been in that room at the Marriott.

Well, what the hell. Wally himself had given them all a lot of pretty sound reasons why a man shouldn't feel too goddamned unlucky if he *didn't* get involved in this Rube Goldberg *capsule* business. It was probably all for the best. Project Mercury was a civilian enterprise and slightly wacky when you got right down to it. They hadn't even chosen *pilots,* f'r chrissake. Jim Lovell had been ranked number one in Group 20 at Pax River, and he hadn't been chosen, either. They had all been lab rats from beginning to end. It was a good thing someone had set the record straight . . .

Still! It was incredible! He had been . . . *left behind!*

Not too long afterward, Conrad was told that across the master sheet on top of his file at Wright-Patterson had been written: "Not suitable for long-duration flight."

V. In Single Combat

They bubble, they boil, they steam and scream, they rumble, and then they boil some more in the most excited way. This sound of boiling voices was exactly like the sound an actor hears backstage before the curtain goes up on a play that everyone—*tout le monde*—must attend. Once there, everyone starts chattering away, out of the sheer excitement of being there at all, of being *where things are happening*, until everybody's beaming face is boiling away with words and grins and laughs that burst out whether or not anything the least bit funny has been said.

As he was not much of an actor, however, this was the sort of sound that terrified Gus Grissom. He was only moments away from the part he was likely to be worst at, and these people were all waiting on the other side of the curtain. At 2 p.m. the curtain was pulled back, and he had to walk onto the stage.

A sheet of light hit Gus and the others, and the boiling voices dropped down to a rumble, or a buzz, and then you could make them out. There appeared to be hundreds of them, packed in shank to flank, sitting, standing, squatting. Some of them were up on a ladder that was propped against the wall under one of the huge lights. Some of them had cameras with the most protuberant lenses, and they had a way of squatting and crawling at the same time, like the hunkered-down beggars you saw all over the Far East. The lights were on for television crews. This building was the Dolley Madison House, at the northeast corner of Lafayette Square, just a few hundred yards from the White

House. It had been converted into NASA's Washington headquarters, and this room was the ballroom, which they used for press conferences, and it was not nearly big enough for all these people. The little beggar figures were crawling all over it.

The NASA people steered Gus and the other six to seats at a long table on the stage. The table had a felt cloth over it. They put Gus in a seat at the middle of the table, and sticking up over the felt right in front of him was the needle nose of a miniature escape tower on top of a model of the Mercury capsule mounted on an Atlas rocket. The model was evidently propped up against the other side of the table so that the press could see it. A man from NASA named Walter Bonney got up, a man with a jolly-sounding voice, and he said: "Ladies and gentlemen, may I have your attention, please. The rules of this briefing are very simple. In about sixty seconds we will give you the announcement that you have all been waiting for: the names of the seven volunteers who will become the Mercury astronaut team. Following the distribution of the kit—and this will be done as speedily as possible—those of you who have p.m. deadline problems had better dash for your phones. We will have about a ten- or twelve-minute break during which the gentlemen will be available for picture taking."

Some men from over on the sides appeared and began handing out folders, and people were rushing up and grabbing these kits and bolting from the room. Bonney pointed to the seven of them sitting there at the table and said: "Gentlemen, these are the astronaut volunteers. Take your pictures as you will, gentlemen."

And now began a very odd business. Without another word, all these grim little crawling beggar figures began advancing toward them, elbowing and hipping one another out of the way, growling and muttering, but never looking at each other, since they had their cameras screwed into their eye sockets and remained concentrated on Gus and the six other pilots at the table in the most obsessive way, like a swarm of root weevils which, no matter how much energy they might expend in all directions trying to muscle one another out of the way, keep their craving beaks homed in on the juicy stuff that the whole swarm has sensed—until they were all over them, within inches of their faces in some cases, poking their mechanical beaks into everything but their belly buttons. Yet this by itself was not what made the moment so strange. It was something else on top of it. There was such frantic excitement—and their names had not even been mentioned! Yet it didn't matter in the slightest! They didn't care whether he was Gus Grissom or Joe Blow! They were ravenous for his picture all the same! They were crawling all over him and the other six as if they were creatures of tremendous value and excitement, real prizes.

Ravenous they were!—these swarming photographers who could grunt but not speak, who crawled over them for a full fifteen minutes. Nevertheless, who was dying for the press conference to start? Anytime Gus had to tell total strangers how he felt about anything, it made him uncomfortable; and the thought of doing so publicly, in this room, in front of several hundred people, made him extremely uncomfortable. Gus came from the sort of background where, to put it mildly, glibness was not encouraged. Back in Mitchell, Indiana, his father had been a railroad worker. His mother used to take the family to the Church of Christ, a Protestant denomination that was so fundamentalist no musical instruments were allowed in the church, not even a piano. The human voice raising thanks to God was music enough. Not that Gus was much of a singer, either, however. His public incantations consisted mainly of Hoosier gus gruffisms. He was a short man with sloping shoulders, a compact build, black crew-cut hair, and black bushy eyebrows, a broad nose, and a face given to very dour looks. The only time Gus felt like talking was when he was with other pilots, particularly at beer call. Then he became another human being. His sleepy eyes lit up to about 200 watts. A crazy confident grin took over his mouth. He would start talking a streak and drinking a lake and, when the midnight madness struck, getting into his hot rod and sucking the surrounding countryside up his two exhaust pipes. Flying & Drinking and Drinking & Driving, of course. Gus was one of those young men, quite common in the United States, actually, who would fight you down to the last unbroken bone over an insult to the gray little town they came from or the grim little church they fidgeted in all those years—while at the same time, in some hidden corner of the soul, they prostrated themselves daily in thanksgiving to the things that had gotten them the hell out of there. In Gus's case those things had been hot rods and, now, airplanes.

Gus had flown a hundred missions in combat in Korea and had won the Distinguished Flying Cross, after breaking formation to chase away a MiG–15 that was about to jump one of his outfit's reconnaissance planes, although somehow, during that great duck shoot over Korea, to his regret, he had never succeeded in downing an enemy plane. After the war he had done all-weather testing of fighter planes at Wright-Patterson and was highly thought of there. So far he had not reached the big league, which was being prime pilot for testing a new fighter, preferably at Edwards. But Gus had every confidence in himself; which is to say, he was a typical fighter jock heading up the pyramid. He already sensed that winning out in the competition for Project Mercury was a tremendous achievement for him, even if it remained to be seen just how far this advanced a man up the ziggurat.

There was just one odd thing about the situation. Last night, at Langley Field, near Newport News, Virginia, Gus had met the other six pilots who had won out. Two of the Air Force pilots were from Edwards. That was to be expected; Edwards was the big league. But one of them, Gordon Cooper, was a man Gus had known at Wright-Pat at one point, and Cooper was not in Fighter Ops at Edwards. The very hottest pilots at Edwards, of course, were in the rocket-plane projects, the X-series. The best line-test pilots were in Fighter Ops as prime pilots in the testing of aircraft such as the Century series of jet fighters. That was what the other Edwards pilot, Deke Slayton, had been involved in. But Cooper—Cooper had graduated from Test Pilot School and was officially a test pilot, but he had been involved mainly in engineering. Not only that, there was this fellow from the Navy, Scott Carpenter. He seemed to be a likable sort—but he had never been in a fighter squadron. He had been flying multi-engine propeller planes and had only two hundred hours in jets. What did this say about the business of being selected as a Mercury astronaut?

Finally, the NASA people were shooing the photographers away from the table, and the head of NASA, a man with big smooth jowls named T. Keith Glennan, got up and said: "Ladies and gentlemen, today we are introducing to you and to the world these seven men who have been selected to begin training for orbital space flight. These men, the nation's Project Mercury astronauts, are here after a long and perhaps unprecedented series of evaluations which told our medical consultants and scientists of their superb adaptability to their upcoming flight."

And it was probably noticed by no one other than the seven pilots themselves that he mentioned only their *adaptability*. He had nothing to say, not a word, about their prowess or standing as pilots.

"It is my pleasure," said Glennan, "to introduce to you—and I consider it a very real honor, gentlemen—Malcolm S. Carpenter, Leroy G. Cooper, John H. Glenn, Jr., Virgil I. Grissom, Walter M. Schirra, Jr., Alan B. Shepard, Jr., and Donald K. Slayton . . . the nation's Mercury Astronauts!"

With that, applause erupted, applause of the most fervent sort, amazing applause. Reporters rose to their feet, applauding as if they had come for no other reason. Smiles of weepy and grateful sympathy washed across their faces. They gulped, they cheered, as if this were one of the most inspiring moments of their lives. Even some of the photographers straightened up from out of their beggar's crouches and let their cameras dangle from their straps, so that they could use their hands for clapping.

But for what?

Once the reporters and photographers got hold of themselves again, men from NASA, the Air Force, and the Navy got up and testified to how terrifically the seven of them had done on all the tests at Lovelace and Wright-Patterson—yet *not one word* was uttered about any ability or experience they might have had *as pilots*. The tone of the thing, the angle, didn't improve with the questions from the reporters. The first reporter who raised his hand wanted to know from each of them whether his wife and children had "had anything to say about this."

Wife and children?

Most of them, Gus included, dealt with this question in typical military-pilot fashion. Which is to say, they managed to get out something brief, obvious, abstract, and above all safe and impersonal. But when it becomes the turn of the guy sitting on Gus's left, John Glenn, the only Marine in the group—it's hard to believe. This guy starts *turning on the charm!* He has a regular little speech on the subject.

"I don't think any of us could really go on with something like this," he says, "if we didn't have pretty good backing at home, really. My wife's attitude toward this has been the same as it has been all along through my flying. If it is what I want to do, she is behind it, and the kids are, too, a hundred percent."

What the hell was he talking about? *I don't think any of us could really go on with something like this* . . . *What* possible difference could a *wife's attitude* make about an opportunity for a giant step up the great ziggurat? What was with this guy? It kept on in that fashion. Some reporter gets up and asks them all to tell about their religious affiliations (re*lig*ious affiliations?)—and Glenn tees off again.

"I am a Presbyterian," he says, "a Protestant Presbyterian, and I take my religion very seriously, as a matter of fact." He starts telling them about all the Sunday schools he has taught at and the church boards he has served on and all the church work that he and his wife and his children have done. "I was brought up believing that you are placed on earth here more or less with sort of a fifty-fifty proposition, and this is what I still believe. We are placed here with certain talents and capabilities. It is up to each of us to use those talents and capabilities as best you can. If you do that, I think there is a power greater than any of us that will place the opportunities in our way, and if we use our talents properly, we will be living the kind of life we should live."

Jesus Christ—share it, brother. You can see the boys cutting glances from either end of the table up at this flying churchman Gus is sitting next to. They're seated in alphabetical order, with Scott Carpenter at one end and Deke Slayton at the other and Glenn in the middle. What can anybody say as a follow-up to this man and his speeches about the Wife and the Children and the Family and Sunday School and God? What can you do, say that as a matter of fact you can get along just as well without any of them as long as they'll let you fly? That didn't seem very prudent. (Turn on the halo—and lie!) You could see these pilots struggling to put up enough chips to stay in the God & Family game with this pious Marine named Glenn.

When it was Gus's turn, he said: "I consider myself religious. I am a Protestant and belong to the Church of Christ. I am not real active in church, as Mr. Glenn is"—*Mister* Glenn, he calls him—"but I consider myself a good Christian still." Deke Slayton says: "As far as my religious faith is concerned, I am a Lutheran, and I go to church periodically."

One of the Navy pilots, Alan Shepard, says: "I am not a member of any church. I attend the Christian Science Church regularly." And so it went. It was a struggle.

God . . . Family . . . the only thing that Glenn hadn't wrapped them all up in was Country, and so he took care of that, too. He gave a nice little speech that started with Orville and Wilbur Wright standing on a hill at Kitty Hawk, North Carolina, tossing a coin to see which one would take the first airplane flight, and then he tied that in with the first space flight. "I think we are very fortunate," he said, "that we have, should we say, been blessed with the talents that have been picked for something like this." (No one had said a word about *talents,* however.) "I think we would be most remiss in our duty," he went on, "if we didn't make the fullest use of our talents in volunteering for something that is as important as this is to our country and to the world in general right now. This can mean an awful lot to this country, of course."

This guy had the halo turned on at all times! Glenn had all the verbal skills that Gus lacked, and yet he didn't seem glib or smooth about it. He looked like a balding and slightly tougher version of the cutest-looking freckle-faced country boy you ever saw. He had a snub nose, light-hazel eyes, reddish-blond hair, a terrific smile, and thousands of freckles. He had the sunniest face in ten counties. He was also one of the best-known pilots in the Marines. He had flown in combat in both the Second World War and the Korean War and had won many medals, including five DFCs, and two years ago, in 1957, he had made the first coast-to-coast non-stop supersonic flight. On the basis of that accomplishment, he had been invited onto a TV show, *Name That Tune,* with a child singer, Eddie Hodges, as his partner, and he had beamed the freckled smile on TV and just charmed the hell out of everybody. The two of them were on the show several weeks.

Well . . . hell . . . maybe he was sincere, after all. God knew that for any pilot to get involved in that much Sunday school and that many church boards and good works, he'd have to be a true believer and a half. Perhaps he even meant it about Wife & Family . . . which would make him an even rarer breed of fighter pilot. If anybody asked Gus—like right now—if he were religious, a family man, and a patriot, he would say yes, he was religious, and yes, he was a family man, and yes, he was a patriot.

But the firmest conviction of the three was about being a patriot. When Gus said he would gladly ride a Mercury rocket for the sake of his country, he meant it. Alan Shepard said the same thing and there was not the slightest doubt that he meant it. And there was no doubt that Glenn meant it, no matter how he went on about it. That was one of the inexpressible things about being a military flying officer. You *meant* it! You were among the few who had "the uncritical willingness to face danger"! There was an exhilaration to this that few civilians could possibly comprehend! No, Gus was a patriot, and he had his hundred combat missions and his DFC to attest to that simple, beautiful fact. Now, as for being a family man . . . aw, hell . . . he meant to be a family man, but somehow his career, or something, always got in the way. He and his wife, Betty, had gotten married as soon as they finished high school in Mitchell. Right from the very first he found himself in situations where he became separated from her. He didn't plan it that way, but it kept happening. Right after they got married, he was due to start his freshman year at Purdue, and so he went off to Purdue to find some place for them to live. Well . . . somehow the only place he could find was a basement room. She said that was okay, she didn't mind, they'd live in a basement room. He said, well, the problem was, he was going to have to share the room with another guy—it was the only way he could afford it—so he would get back to Mitchell as often as he could, on weekends. So that was the way they started out, with him on the campus at Purdue and Betty living with her parents in Mitchell. Military flying was hard on home life, too. Gus graduated and started his Air Force training at Randolph Field in Texas. Betty was pregnant, and he was in training and was making only $100 a month anyway, and so why didn't she stay in Seymour, Indiana, with her sister Mary Lou, and Gus would visit her when he could. The only trouble was, he couldn't visit her very often, it being so expensive to travel from Texas to Indiana. When Betty gave birth to their first child, Scott, Gus was at an important part in his flight training and couldn't get to Indiana for that, either. He gets on the phone and says to her: "Well, you tell me what you really want me to do" . . . and she says, "Well, I guess you ought not to interrupt your training." In fact, he didn't quite manage to see his first child until six months later. Now, that sort of thing could happen in the service, because a fellow could get sent overseas at a moment's notice. But he, Gus, hadn't been sent anywhere except down the road to Arizona, to Williams Air Force Base, for advanced training. At the time . . . well, it just seemed damned hard to get all the way from the Southwestern U.S.A. to Indiana when you were in the thick of flight training. Then the Air Force *did* send him overseas, to Korea, and Betty went back to Indiana again. Korea! He *loved* it! He liked combat missions so much that when he completed a hundred missions, he volunteered for twenty-five more. He wanted to *stay* there! But the bastards made him come back. Somehow he and Betty managed to get by through all this. He gruffed a lot of Hoosier gus gruffisms at her and she gruffed some back at him. They didn't get in many fights. Most weekends he could manage it, he would fly cross-country, piling up flight time. But how different was he from the other pilots at this table, if the truth were known—except for this unbelievable Marine, Glenn, who was sitting here next to him painting some goddamned amazing picture of the Perfect Pilot wrapped up in a cocoon of Home & Hearth and God & Flag!

Neither he nor any of the others set about altering that picture, however. At first it was hard to figure out what was happening. Glenn could have never gone off on these fantastic outside loops of his if it were not for the fact that practically every question had to do with families and faith and motivation and patriotism and so on. There had not been a single question about their achievements or experience as pilots. Then one of the reporters gets up and says:

"Could I ask for a show of hands of how many are confident that they will come back from outer space?"

Gus and the others started looking down the table at each other, and then they all started hoisting their hands in the air. It really made you feel like an idiot, raising your hand this way. If you didn't think you were "coming back," then you would really have to be a fool or a nut to have volunteered at all. As the seven of them looked at each other sitting there with their hands up in the air like schoolchildren, they began grinning in embarrassment, and then the heart of the matter dawned on them. This question about "coming back" was nothing other than an euphemistic way of asking: Aren't you afraid you're going to die? That was the question these people had been circling around the whole time. That was what they really wanted to know, all these wide-eyed reporters and their grunting crawling beggar photographers. They didn't care whether the seven Mercury astronauts were pilots or not. Infantrymen or acrobats would have done just as well. The main thing was: they had volunteered to sit on top of the rockets—which *always blew up!* They were brave lads who had volunteered for a suicide mission! They were kamikazes going forth to vie with the Russians! And all the questions about wives and children and faith and God and motivation and the Flag . . . they were really questions about widows and orphans . . . and how a warrior talks himself into going on a mission in which he is bound to die.

And this man, John Glenn, had given them an answer as sentimental as the question itself, and Gus and the rest had gone along with it. Henceforth, they would be served up inside the biggest slice of Mom's Pie you could imagine. And it had all happened in just about an hour. The seven of them sat there like fools with their hands hung up in the air, grinning with embarrassment. But that was all right; they would get over the embarrassment soon enough. Glenn, one couldn't help noticing, had *both* hands up in the air.

By the next morning the seven Mercury astronauts were national heroes. It happened just like that. Even though so far they had done nothing more than show up for a press conference, they were known as the seven bravest men in America. They woke up to find astonishing acclaim all over the press. There it was, in the more sophisticated columns as well as in the tabloids and on television. Even James Reston of *The New York Times* had been so profoundly moved by the press conference and the sight of the seven brave men that his heart, he confessed, now beat a little faster. "What made them so exciting," he wrote, "was not that they said anything new but that they said all the old things with such fierce convictions . . . They spoke of 'duty' and 'faith' and 'country' like Walt Whitman's pioneers . . . This is a pretty cynical town, but nobody went away from these young men scoffing at their courage and idealism." Manly courage, the right stuff—the Halo Effect, with Deacon Glenn leading the hallelujah chorus, had practically wiped the man out. If Gus and some of the others had been worried that they weren't being regarded as hot pilots, their worries were over when they saw the press coverage. Without exception, the newspapers and wire services picked out the highlights of their careers and carefully massed them to create a single blaze of glory. This took true journalistic skill. It meant citing a great deal from John Glenn's career, his combat flying in two wars, his five Distinguished Flying Crosses with eighteen clusters, and his recent speed record, plus the combat that Gus and Wally Schirra had seen in Korea and the medals they had won, one DFC apiece, and the bombing missions Slayton had flown in the Second World War and a bit about the jet fighters he had helped test at Edwards and the ones Shepard had tested at Pax River—and going easy on the subject of Scott Carpenter and Gordon Cooper, who had not flown in combat (Shepard had not, either) or done any extraordinary testing. John Glenn came out of it as tops among seven very fair-haired boys. He had the hottest record as a pilot, he was the most quotable, the most photogenic, and the lone Marine. But all seven, collectively, emerged in a golden haze as the seven finest pilots and bravest men in the United States. A blazing aura was upon them all.

It was as if the press in America, for all its vaunted independence, were a great colonial animal, an animal made up of countless clustered organisms responding to a single nervous system. In the late 1950's (as in the late 1970's) the animal seemed determined that in all matters of national importance the *proper emotion,* the *seemly sentiment,* the *fitting moral tone* should be established and should prevail; and all information that muddied the tone and weakened the feeling should simply be thrown down the memory hole. In a later period this impulse of the animal would take the form of blazing indignation about corruption, abuses of power, and even minor ethical lapses, among public officials; here, in April of 1959, it took the form of a blazing patriotic passion for the seven test pilots who had volunteered to go into space. In either case, the animal's fundamental concern remained the same: the public, the populace, the citizenry, must be provided with *the correct feelings!* One might regard this animal as the consummate hypocritical Victorian gent. Sentiments that one scarcely gives a second thought to in one's private life are nevertheless insisted upon in all public utterances. (And this grave gent lives on in excellent health.)

Even so, why was the press aroused to create *instant* heroes out of these seven men? This was a question that not James Reston or the pilots themselves or anyone at NASA could have answered at the time, because the very language of the proposition had long since been abandoned and forgotten. The forgotten term, left behind in the superstitious past, was *single combat.*

Just as the Soviet success in putting Sputniks into orbit around the earth revived long-buried superstitions about the power of heavenly bodies and the fear of hostile control of the heavens, so did the creation of astronauts and a "manned space program" bring back to life one of the ancient superstitions of warfare. Single combat had been common throughout the world in the pre-Christian era and endured in some places through the Middle Ages. In single combat the mightiest soldier of one army would fight the mightiest soldier of the other army as a substitute for a pitched battle between the entire forces. In some cases the combat would pit small teams of warriors against one another. Single combat was not seen as a humanitarian substitute for wholesale slaughter until late in its history. That was a Christian reinterpretation of the practice. Originally it had a magical meaning. In ancient China, first the champion warriors would fight to the death as a "testing of fate," and then the entire armies would fight, emboldened or demoralized by the outcome of the single combat. Before Mohammed's first battle as the warrior-prophet, the Battle of Badr, three of Mohammed's men challenged the Meccans to pick out any three of their soldiers to fight in single combat, proceeded to destroy them with all due ceremony, whereupon Mohammed's entire force routed the entire Meccan force. In other cases, however, the single combat settled the affair, and there was no full-scale battle, as when the Vandal and Aleman Armies confronted each other in Spain in the fifth century A.D. They believed that the gods determined the outcome of single combat; therefore, it was useless for the losing side to engage in a full-scale battle. The Old Testament story of David and Goliath is precisely that: a story of single combat that demoralizes the losing side. The gigantic Goliath, with his brass helmet, coat of mail, and ornate greaves, is described as the Philistine "champion" who comes forth to challenge the Israelites to send forth a man to fight him; the proposition being that whoever loses, his people will become the slaves of the other side. Before going out to meet Goliath, David—an unknown volunteer commoner—is given King Saul's own decorative armor, although he declines to wear it. When he kills Goliath, the Philistines regard this as such a terrible sign that they flee and are pursued and slaughtered.

Naturally the brave lads chosen for single combat enjoyed a very special status in the army and among their people (David was installed in the royal household and eventually superseded Saul's own sons and became king). They were revered and extolled, songs and poems were written about them, every reasonable comfort and honor was given them, and women and children and even grown men were moved to tears in their presence. Part of this outpouring of emotion and attention was the simple response of a grateful people to men who were willing to risk their lives to protect them. But there was also a certain calculation behind it. The steady pressure of fame and honor tended to embolden the lads still further by constantly reminding them that the fate of the entire people was involved in their performance in battle. At the same time—and this was no small thing in such a high-risk occupation—the honor and glory were in many cases rewards *before the fact;* on account, as it were. Archaic cultures were quite willing to elevate their single-combat fighters to heroic status even *before* their blood was let, because it was such an effective incentive. Any young man who entered the corps would get his rewards here on earth, "up front," to use the current phrase, come what may.

With the decline of archaic magic, the belief in single combat began to die out. The development of the modern, highly organized army and the concept of "total war" seemed to bury it forever. But then an extraordinary thing happened: the atomic bomb was invented, with the result that the concept of total war was nullified. The incalculable power of the A-bomb and the bombs that followed also encouraged the growth of a new form of superstition founded upon awe not of nature, as archaic magic had been, but of technology. During the Cold War period small-scale competitions once again took on the magical aura of a "testing of fate," of a fateful prediction of what would inevitably happen if total nuclear war did take place. This, of course, was precisely the impact of Sputnik I, launched around the earth by the Soviets' mighty and mysterious Integral in October of 1957. The "space race" became a fateful test and presage of the entire Cold War conflict between the "superpowers," the Soviet Union and the United States. Surveys showed that people throughout the world looked upon the competition in launching space vehicles in that fashion, i.e., as a preliminary contest proving final and irresistible power to destroy. The ability to launch Sputniks dramatized the ability to launch nuclear warheads on ICBMs. But in these neo-superstitious times it came to dramatize much more than that. It dramatized the entire technological and intellectual capability of the two nations and the strength of the national wills and spirits. Hence . . . John McCormack's rising in the House of Representatives to say that the United States faced "national extinction" if she did not overtake the Soviet Union in the space race.

The next great achievement would be the successful launching of the first man into space. In the United States—no one could say what was taking place in the land of the mighty Integral—the men chosen for this historic mission took on the archaic mantles of the single-combat warriors of a long-since-forgotten time. They would not be going into space to do actual combat; or not immediately, although it was assumed that something of the sort might take place in a few years. But they were entering into a deadly duel in the heavens, in any event. *(Our rockets always blow up.)* The space war was on. They were risking their lives for their country, for their people, in "the fateful testing" versus the powerful Soviet Integral. And even though the archaic term itself had disappeared from memory, they would receive all the homage, all the fame, all the honor and heroic status . . . *before the fact* . . . of the single-combat warrior.

Thus beat the mighty drum of martial superstition in the mid-twentieth century.

It was glorious! It was crazy! The following month, May, the House Committee on Science and Astronautics, of which McCormack was a member, called the seven astronauts before them in closed session. So the brave lads went to the Capitol for the secret proceedings. It was a strange and marvelous time. It became immediately apparent that from the committee chairman, Representative Overton Brooks, on down, no one had anything pertinent to ask them and nothing to tell them. The congressmen kept saying things like "As you men know, you are an outstanding group in our country and in our country's history." Or they would ask them questions that elicited choral responses. Brooks himself said, "All of you gentlemen have been in this type of work—that is, handling experimental planes—in the past, haven't you? You know what this is. You know that in handling any new experimental flying machine there is a certain type of risk. You understand that, isn't that right?" Then he peered beseechingly at the seven of them until they began chiming in, all at once: "Yes, sir!" "Right, sir!" "Certainly, sir!" "That's correct, sir!" There was something gloriously goofy about it. The congressmen in the room just wanted to see them, to use their position to arrange a personal audience, to gaze upon them with their own eyes across the committee table, no more than four feet away, to shake hands with them, occupy the same space on this earth with them for an hour or so, fawn over them, pay homage to them, bathe in their magical aura, feel the radiation of their righteous stuff, salute them, wish upon them the smile of God . . . and do their bit in bestowing honor upon them *before the fact* . . . upon our little Davids . . . before they got up on top of the rockets to face the Russians, death, flames, and fragmentation. *(Ours all blow up!)*

Chuck Yeager was in Phoenix to make one of his many public appearances on behalf of the Air Force. By now the Air Force couldn't publicize Yeager, breaker of the sound barrier, enough. Like the other branches of the service, the Air Force now saw that there was nothing like heroes and record holders for getting good press and winning appropriations. The only problem was that, in terms of publicity, every other form of flier was now overshadowed by the Mercury astronauts. As a matter of fact, today, in Phoenix, what was it the local reporters wanted to ask Chuck Yeager about? Correct: the astronauts. One of them got the bright idea of asking Yeager if he had any regrets about not being selected as an astronaut.

Yeager smiled and said, "No, they gave me the opportunity of a lifetime, to fly the X–1 and the X–1A, and that's more than a man could ask for, right there. They gave this new opportunity to some new fellows coming along, and that's what they ought to do.

"Besides," he added, "I've been a pilot all my life, and there won't be any flying to do in Project Mercury."

No flying?—

That was all it took. The reporters looked stunned. In some way they couldn't comprehend immediately, Yeager was casting doubt on two undisputable facts: one, that the seven Mercury astronauts were chosen because they were the seven finest pilots in America, and two, that they would be pilots on the most daring flights in American history.

The thing was, he said, the Mercury system was completely automated. Once they put you in the capsule, that was the last you got to say about the subject.

Whuh!—

"Well," said Yeager, "a monkey's gonna make the first flight."

A *monkey?*—

The reporters were shocked. It happened to be true that the plans called for sending up chimpanzees in both suborbital and orbital flights, identical to the flights the astronauts would make, before risking the men. But to just *say* it like that! . . . Was this national heresy? What the hell was it?

Fortunately for Yeager, the story didn't blow up into anything. The press, the eternal Victorian Gent, just couldn't deal with what he had said. The wire services wouldn't touch the remark. It ran in one of the local newspapers, and that was that.

But f'r chrissake . . . Yeager was only saying what was obvious to all the rocket pilots who had flown at Edwards. Here was everybody talking as if the Mercury astronauts would be the first men to ride rockets. Yeager had done precisely that more than forty times. Fifteen other pilots had done it also, and they had reached speeds greater than three times the speed of sound and an altitude of 126,000 feet, nearly twenty-five miles, and that was just the beginning. This very next month, June 1959, Scott Crossfield would begin the first testing of the X-15, designed for a pilot (a *pilot,* not a passenger) to take up to more than fifty miles, into space, at speeds approaching Mach 7.

All of this should have been absolutely obvious to anyone, even people who knew nothing about flying—and surely it would become clear that anybody in Project Mercury was more of a test subject than a pilot.

Two of the people they chose weren't even in Fighter Ops. They had one excellent test pilot from Edwards, Deke Slayton, but he had never been high on the list of those considered for something like the X series. The other Air Force pilot, Grissom, was assigned to Wright-Pat and was doing more secondary testing than prime work. Two of the Navy guys, Shepard and Schirra, were good experienced test pilots, solid men, even though neither had done anything to make anybody's mouth fall open at Edwards. Glenn had made a name for himself by setting the speed record in the F8U, but he had not done much major flight test work, at least not by Edwards standards. Well, hell, what did anybody expect? *Naturally* they hadn't picked the seven hottest pilots they could find. They wouldn't be doing any flying!

Surely all this would become obvious in time . . . and yet it wasn't becoming obvious. Here at mighty Edwards itself the boys could feel the earth trembling. A great sliding of the templates was taking place inside the invisible pyramid. You could feel the old terrain crumbling, and . . . seven rookies were somehow being installed as the hottest numbers in flying—and they hadn't done a goddamned thing yet but turn up at a press conference!

VI. On the Balcony

From the very beginning this "astronaut" business was just an unbelievable good deal. It was such a good deal that it seemed like tempting fate for an astronaut to call himself an astronaut, even though that was the official job description. You didn't even refer to the others as astronauts. You'd never say something such as "I'll take that up with the other astronauts." You'd say, "I'll take that up with the other fellows" or "the other pilots." Somehow calling yourself an "astronaut" was like a combat ace going around describing his occupation as "combat ace." This thing was such an unbelievable good deal, it was as if "astronaut" were an honorific, like "champion" or "superstar," as if the word itself were one of the infinite variety of *goodies* that Project Mercury was bringing your way.

And not just *goodies* in the crass sense, either. It had *all* the things that made you feel good, including the things that were good for the soul. For long stretches you'd bury yourself in training, in blissful isolation, good rugged bare-boned isolation, in Low Rent surroundings, in settings that even resembled hallowed Edwards in the old X–I days, and with that same pioneer spirit, which money cannot buy, and with everybody pitching in and working endless hours, so that rank meant nothing, and people didn't even have the inclination, much less the time, to sit around and make the usual complaints about government work.

And then, just about the time you were entering a good healthy state of exhaustion from the work, they would take you out of your isolation and lead you up to that balcony that all fighter jocks secretly dreamed of, the one where you walked out before the multitudes like the Pope, and . . . it actually happened! The people of America cheered their brains out for thirty minutes or so, and then you went back into your noble isolation for more work . . . or for a few proficiency runs at nailing down the holy coordinates of the fighter jock's life, which were, of course, Flying & Drinking and Drinking & Driving and the rest of it. These things you could plot on the great graph of Project Mercury in the most spectacular way, with the exception of the first: Flying. The lack of flying time was troubling, but the other items existed in such extraordinary dimensions that it was hard to concentrate on it at first. Any man who wasn't above a little regrouping now and then, to keep the highly trained mechanism from being wound up too tight, to "maintain an even strain," in the Schirra parlance, found himself in absolute Fighter Jock Heaven. But even the rare pilot who was aloof from such cheap thrills, such as the deacon, John Glenn, found plenty of *goodies* to even out the strain of hard work and mass adoration.

Each of them had an eye on Glenn, all right. Glenn's own personal conduct was a constant reminder of what the game was really all about. To all but Scott Carpenter, and perhaps one other, the way Glenn was going about this thing was irritating.

The seven of them were stationed at Langley Air Force Base in the Tidewater section of Virginia on the James River, about 150 miles due south of Washington. Langley had been the experimental facility of the old National Advisory Committee for Aeronautics and was now the headquarters of NASA's Space Task Group for Project Mercury. Every morning they could count on seeing John Glenn up early, out on the grounds, in the middle of everything, where nobody could miss him, doing his roadwork. He'd be out there in full view, on the circular driveway of the Bachelor Officers Quarters, togged out in his sweatsuit, his great freckled face flaming red and shining with sweat, going around and around, running a mile, two miles, three miles, there was no end to it, in front of everybody. It was irritating, because it was so unnecessary. There had been a vague medical directive to the effect that each of them would engage in at least four hours of "unsupervised exercise" per week, but that was the last that was heard of. The medical staff assigned to Project Mercury were mainly young military doctors, a bit dazzled by the mission, some of them, and they were not about to call an astronaut on the carpet and demand an accounting of his four hours. Fighter jocks, as a breed, put physical exercise very low on the list of things that made up the right stuff. They enjoyed the rude animal health of youth. They put their bodies through dreadful abuses, often in the form of drinking bouts followed by lack of sleep and mortal hangovers, and they still performed like champions. ("I don't advise it, you understand, but it *can* be done"—provided you have the right stuff, you miserable pud-knocker.) Most agreed with Wally Schirra, who felt that any form of exercise that wasn't fun, such as waterskiing or handball, was bad for your nervous system. But here was Glenn, pounding through everybody's field of vision with his morning roadwork, as if he were preparing for the championship fight.

The good Marine didn't just do his roadwork and leave it at that, either. Oh, no. The rest of them had their families installed at Langley Air Force Base or at least in the Langley vicinity. Gordon Cooper and Scott Carpenter and their families were packed into apartments on the base, the usual sort of worn-out base housing that junior officers rated. Wally Schirra, Gus Grissom, and Deke Slayton lived in a rather sad-looking housing development on the other side of the Newport News airport. Around the development was a stucco wall of the color known as glum ocher. Alan Shepard and his family lived a little farther away in Virginia Beach, where they happened to be living when he was chosen for Project Mercury. But Glenn . . . Glenn has his family housed 120 miles away in Arlington, Virginia, outside of Washington, and at Langley he stays in the Bachelor Officers Quarters, the BOQ, and does his running out front in the driveway. If this had been some devilishly clever scheme for him to get away from home and hearth and indulge in Drinking & Driving & so forth, that would have been one thing. But he wasn't the type. He was living in a bare room with nothing but a narrow bed and an upholstered chair and a little desk and a lamp and a lineup of books on astronomy, physics, and engineering, plus a Bible. On the weekends he would faithfully make his way home to his wife, Annie, and the children in an ancient Prinz, a real beat-up junker that was about four feet long and had perhaps forty horsepower, the sorriest-looking and most underpowered automobile still legally registered to any fighter pilot in America. A jock with any natural instincts at all, with any true devotion to the holy coordinates, either possessed or was eating his heart out for the sort of car that Alan Shepard had, which was a Corvette, or that Wally Schirra had, which was a Triumph, i.e., a sports car, or some kind of hot car, anyway, something that would enable you to hang your hide out over the edge with a little class when you reached the Driving juncture on the coordinates several times a week, as was inevitable for everyone but someone like John Glenn. This guy was putting on an incredible show! He was praying in public. He was presenting himself in their very midst as the flying monk or whatever the Presbyterian version of a monk was. A saint, maybe; or an ascetic; or maybe just the village scone crusher.

Being a good Presbyterian, John Glenn knew that praying in public was no violation of the faith. The faith even encouraged it; it set a salubrious example for the public. Nor did John Glenn feel the slightest discomfort because now, in post–World War II America, virtue was out of style. Sometimes he seemed to enjoy shocking people with his clean living. Even when he was no more than nine years old, he had been the kind of boy who would halt a football game to read the riot act to some other nine-year-old who said "Goddamn it" or "Aw shit" when a play didn't go right. This was an unusual gesture even where he grew up, which was New Concord, Ohio, but not so extraordinary as it might have been a lot of other places. New Concord was a sort of town, once common in America, whose peculiar origins have tended to disappear in the collective amnesia as *tout le monde* strives to be urbane. Which is to say, it began as a religious community. A hundred years ago any man in New Concord with ambitions that reached as high as feed-store proprietor or better joined the Presbyterian Church, and some of the awesome voltage of live Presbyterianism still existed when Glenn was growing up in the 1920's and 1930's. His father was a fireman for the B&O Railroad and a good churchgoing man and his mother was a hardworking churchgoing woman, and Glenn went to Sunday school and church and sat through hundreds of interminable Presbyterian prayers, and the church and the faith and the clean living served him well. There was no contradiction whatsoever between the Presbyterian faith and ambition, even soaring ambition, even ambition grand enough to suit the invisible ego of the fighter jock. A good Presbyterian demonstrated his *election* by the Lord and the heavenly hosts through his success in this life. In a way, Presbyterianism was tailor-made for people who intended to make it in this world, as well as on the Plains of Heaven; which was a good thing, because John Glenn, with his sunny round freckled country-boy face, was as ambitious as any pilot who had ever hauled his happy burden of self-esteem up the pyramid.

So Glenn went pounding around the driveway of the BOQ of Langley Air Force Base in his sweatsuit, doing his roadwork, and he frankly didn't care if most of the others didn't like it. The running was good for him on several levels. At thirty-seven he was the oldest of the fellows, and there was a little more pressure on him to demonstrate that he was in good condition. Besides that, he had a tendency to put on weight. From the waist up he was of only average size and musculature and, in fact, had surprisingly small hands. But his legs were huge, real kegs, muscular and fleshy at the same time, and he tended to pack on weight in the thighs. He was pushing 185 when he was selected for this thing, and he could well afford to get down to 170 or even less. As for living in the BOQ . . . why not? He and his wife, Annie, had bought their house in Arlington because the children would be in excellent public schools there. Why transplant them again when he would be on the road half the time and probably wouldn't see them except on weekends, anyway?

If it looked to the others as if he were living a monastic life . . . that wouldn't hurt too much . . . Competition was competition, and there was no use pretending it didn't exist. He already had an advantage over the other six because of his Marine flying record and the way he tended to dominate the publicity. He was ready to give a 110 percent on all fronts. If they wanted four hours of unsupervised exercise per week—well, give them eight or twelve. Other people could think what they wanted; he happened to be completely sincere in the way he was going about this thing.

The goal in Project Mercury, as in every important new flight project, was to be the pilot assigned to make the first flight. In flight test that meant your superiors looked upon you as *the* man who had the right stuff to challenge the unknowns. In Project Mercury the first flight would also be the most historic flight. They had been told that the first flight would be suborbital. There might be as many as ten or eleven suborbital flights, going to an altitude of about one hundred miles, fifty miles above the commonly accepted boundary line between the earth's atmosphere and space. These flights would not go into orbit, because the rocket they would be using, the Redstone, could not generate enough power to take a capsule to orbital speed, which would be about 18,000 miles per hour. The capsule would go up and come down in a big arc, like an artillery shell's. As it came over the top of the arc, the astronaut would experience about five minutes of weightlessness. These suborbital flights were scheduled to begin in mid-1960, and all seven pilots would get a crack at them.

Other men would no doubt go farther into space, into earth orbit and beyond. But they, in turn, would be chosen from the first men to fly suborbitally; so the first astronaut would be the one the world remembered. When a man realized something like that, there was no use being shy about the opportunity he had. Glenn had not gotten this far in his career by standing still in a saintly fashion and waiting for his halo to be noticed. When he reached Korea, flying strafing and bombing missions in support of Marine ground troops, he realized that the biggest accolade was being assigned to Air Force fighter squadrons, on loan (like Schirra), for air-to-air combat up at the Yalu River. So he had gone after that assignment and had gotten it and had shot down three MiGs during the last few days of the war. As soon as the war ended, he realized that flight test was the hot new arena and had gone straight to his superiors and asked to be assigned to the Navy's Patuxent River Test Pilot School, and they sent him there. He had been in flight test barely three years when he dreamed up the F8U transcontinental run. He dreamed it up himself, as a major in the Marines! Although everyone knew it was possible, no one had ever made a sustained coast-to-coast flight at an average speed of greater than Mach 1. Glenn developed the whole scheme, the aerial rendezvous with three different AJ1 refueling tankers, the way he would dive down to 22,000 feet to meet them, the whole thing. He pulled it off on July 16, 1957, flying from Los Angeles to Floyd Bennett Field in New York, in three hours and twenty-three minutes. The word was that there were some test pilots who were put out because he got the assignment. They seemed to think they had done the major test work on the F8U, and so forth and so on. But it was his idea! He got it launched! If he hadn't put himself forward, it wouldn't have happened at all. Last year, 1958, it was obvious to him that all the services were working on the problems of manned space flight. There was no Project Mercury yet, and no one knew who would be running the show when a manned program began. All he knew was that it was not likely to be the Marines but he wanted to play a part in it. So he had himself assigned to the Navy Bureau of Aeronautics. He volunteered for runs on the Navy's human centrifuge machine at Johnsville, Pennsylvania, exploring the high g-forces associated with rocket flight. By March of this year, 1959, just a month before the seven of them were selected as astronauts, he had been at the McDonnell Aircraft plant in St. Louis as a representative of the Bureau of Aeronautics on a NASA Mockup Review Board, reviewing progress of the manufacture of the Mercury capsule. He didn't know just how the seven of them were selected . . . but obviously all that hadn't hurt his chances.

And now the ante had been raised once more, and he was after nothing short of being the first man to go into space. NASA would have to beat the Russians to it, of course, for him or any other American to be first. But that was one of the things that made it exhilarating, exhilarating enough even to endure this sweaty pounding over a salty pine-tag circular driveway in Tidewater, Virginia. There was the same sort of *esprit*—usually called patriotism but better described as *joie de combat*—that had existed during the Second World War and, among pilots (and practically no one else), during the Korean War. Project Mercury was officially a civilian undertaking. But it struck Glenn as being like a new branch of the armed services. All seven of them were still in the military, drawing military pay, even though they wore civilian clothes. There was a warlike urgency and priority about the whole enterprise. And in this new branch of the military *no one outranked you.* It was almost too good to be true.

On the organization chart the seven of them had superiors. They reported to Robert Gilruth, the head of the new Space Task Group, who was a subordinate of Hugh Dryden, the deputy administrator of NASA. Gilruth was a superb engineer and a fine man; he had literally *written the book* on the handling characteristics of aircraft, the first scientific treatise on the subject, "Requirements for Satisfactory Flying Qualities of Airplanes," NACA Report No. 755, 1937, which had become a classic. He was a big, bald, shy man with a reedy voice. Most recently he had been head of the NACA Pilotless Aircraft Research Division, which had experimented with unmanned rockets. Gilruth was not used to marching the troops and certainly not a group of ambitious pilots. He was no Vince Lombardi. He was a genius among engineers, but he was not the type to take seven colossal stars who were suddenly the most famous pilots in America and mold them into Bob Gilruth's Astroteam.

They were so famous, so revered, so lavishly fussed and worried over at all times that they were without peers in this new branch of the military. Everywhere they went in their travels people stopped what they were doing and gave them a certain look of awe and sympathy. Sympathy . . . because *our rockets all blow up.* It was a nice, friendly, warm look, all right, and yet it was strange. It was a sort of glistening smile with tears and joy suffusing it; both tears and joy. In fact, it was an ancient look, from the primordial past, never seen in America before. It was the smile of homage and astonishment—at such bravery!—that had been given to single-combat warriors, in advance, on account, before the fact, since time was.

Well . . . Glenn was ready; he was ready for *election*; he was ready to be the first to go into the heavens when that debt of homage and honor and glistening faces came due.

One of the people who beamed that look at them with a sincere devotion was a Washington lawyer named Leo DeOrsey. Walter Bonney, the NASA public affairs officer who had run the press conference, had seen the frenzy of publicity building up around the seven men and concluded that they needed some expert help in their new role as celebrities. He approached DeOrsey. DeOrsey was a tax lawyer. Harry Truman had once considered making him head of the Internal Revenue Service. He had represented many show-business celebrities, including President Eisenhower's friend Arthur Godfrey. So the seven of them wound up having dinner with DeOrsey in a private room at the Columbia Country Club outside of Washington. DeOrsey was an affable gentleman with a little round pot belly. He had terrific clothes. He put on a long face and related how he had been approached by Bonney. He said he was willing to represent them.

"I insist on only two conditions," he said.

Glenn thought to himself, "Well, here it comes."

"One," said DeOrsey, "I will accept no fee. Two, I will not be reimbursed for expenses."

He kept the grave look on his face for a moment. And then he smiled. There were no catches and no angles. He was obviously sincere. He thought they were terrific and felt tickled pink to be involved with them at all. He couldn't do enough for them. And that was the way it went with Leo DeOrsey from that evening onward. He couldn't have been straighter or more generous.

DeOrsey proposed that the book and magazine rights to their personal stories be put up for sale to the highest bidder. Bonney was sure the President and NASA would allow it, because several military men had made such an arrangement since the Second World War, most notably Eisenhower himself. The selling point for NASA would be that if the seven of them sold exclusive rights to one organization, then they would have a natural shield against the endless requests and intrusions by the rest of the press and would be better able to concentrate on their training.

Sure enough, NASA approved the idea, the White House approved it, and DeOrsey started getting in touch with magazines, setting $500,000 as the floor for bids. The one solid offer—$500,000—came from *Life*, and DeOrsey closed the deal. *Life* had an excellent precedent for the decision. Few people remembered, but *The New York Times* had bought the rights to Charles Lindbergh's personal story before his famous transatlantic flight in 1927. It worked out splendidly for both parties. Having bought an exclusive, the *Times* devoted its first five pages to Lindbergh the day after his flight and the first *sixteen* the day after he returned from Paris, and all other major newspapers tried their best to keep up. In return for *Life*'s exclusive rights to their personal stories and their wives', the astronauts would share the $500,000 evenly; the sum amounted to just under $24,000 a year for each man over the three years Project Mercury was scheduled to run, about $70,000 in all.

For junior officers with wives and children, used to struggling along on $5,500 to $8,000 a year in base pay, plus another $2,000 in housing and subsistence allowances and perhaps $1,750 in extra flight pay, the sum was barely even imaginable at first. It wasn't real. They wouldn't see any of it for months, in any case . . . Nevertheless, the *goodies* were the *goodies*. A career military officer denied himself and his family many things . . . with the understanding that when the *goodies* came along, they would be accepted and shared. It was part of the unwritten contract. The *Life* deal even provided them with foolproof protection against the possibility that their personal stories might become all-too-personal. Although written by *Life*, the stories would appear in the first person under their own by-lines . . . "by Gus Grissom" . . . "by Betty Grissom" . . . and they would have the right to eliminate any material they objected to. NASA, moreover, would have the same right. So there was nothing to keep the boys from continuing to come across as what they had looked like at the first press conference: seven patriotic God-fearing small-town Protestant family men with excellent backing on the home front.

Here in the summer of 1959 that was fine with *Life* and with the rest of the press, for that matter. Americans seemed to be deriving profound satisfaction from the fact that the astronauts turned the conventional notions of Glamour upside down. It was assumed—and the Genteel Beast kept underlining the point—that the seven astronauts were the greatest pilots and bravest men in America *precisely because of* the wholesome circumstances of their backgrounds: small towns, Protestant values, strong families, the simple life. Henry Luce, *Life*'s founder and boss of bosses, had not played a major role, other than parting with the money, in making the astronaut deal, but eventually he came to look upon them as *his boys*. Luce was a great Presbyterian, and the Mercury astronauts looked like seven incarnations of Presbyterianism. This was no rural-American miracle, however. It was John Glenn who had set the moral tone of the Astronaut at the first press conference. The others had diplomatically kept their mouths shut ever since. From the Luces and Restons on down, the Press, that ever-seemly Victorian Gent, saw the astronauts as seven slices of the same pie, and it was mom's pie, John Glenn's mom's pie, from the sturdy villages of the American heartland. The Gent thought he was looking at seven John Glenns.

———

Among the seven instant heroes John Glenn's light shone brightest. Probably the least conspicuous, using that same measure, was Gordon Cooper. Cooper was a thin, apparently guileless soul, handsome in a down-home manner. He was from Shawnee, Oklahoma. He had a real Oklahoma drawl. He was also the youngest of the seven, being thirty-two years old. He had never flown in combat, nor had his test work at Edwards been of the sort that attracted much notice. Scott Carpenter was no further up the great ziggurat of flying, of course, but Carpenter was not at all reluctant to talk about his own relative lack of experience in jets, and so on. What seemed to annoy some of the boys was that none of the foregoing, obvious as it surely was, fazed Gordon Cooper in the slightest.

Two people who sometimes seemed to get impatient with Cooper were his Air Force comrades Gus Grissom and Deke Slayton. Grissom and Slayton had become great pals practically from the day they were selected as astronauts. They were from out of the same grim clay. Slayton was raised on a farm in western Wisconsin, up near the town of Sparta and the Elroy Sparta State Park Trail. He was taller than Grissom, more rugged, rather handsome, in fact, and quite intelligent, once you penetrated the tundra. When the subject was flying, his expression lit up, and he radiated confidence and had all the wit and charm and insights you could ask for. In other situations, however, he had Grissom's lack of patience for party manners and small talk and Grissom's way of lapsing into impenetrable blank stares, as if some grim wintertime north-country Lutheran cloud of Original Sin were passing in front of his face. Deke had started flying in the Second World War, when the Air Force was still part of the Army. In the Army one was continually around people who spoke Army Creole, a language in which there were about ten nouns, five verbs, and one adjective, or participle, or whatever it was called. There always seemed to be a couple of good buddies from Valdosta or Oilville or some place sitting around saying:

"I tol'im iffie tried to fuck me over, I was gonna kick'is fuckin' ass, iddnat right?"

"Fuckin' A."

"Soey kep'on fuckin' me over and I kicked 'is fuckin' ass in fo'im, iddnat right?"

"Fuckin' A."

"An' so now they tellin' me they gon' th'ow my fuckin' ass inna fuckin' *stoc*-kade! You know what? They some kinda fuckin' me over!"

"Fuckin' A well tol', Bubba."

Now that Deke was all of a sudden a celebrity, there were people who knew him who cringed every time he got near a microphone. They were afraid he was going to Army Creole the nationwide TV and scorch the brains of half the people of the U.S.A. The truth was, Deke was far too sharp for that. He was okay in Gus's book. They lived just two doors away from each other at Langley, and if they were both at home on the weekend, they usually did something together, such as go hunting or wangle a T–33 from Langley Air Force Base and fly cross-country, taking turns at the controls. Sometimes they would fly all the way to California and back, and it was likely that if they exchanged a total of forty sentences, transcontinental, they would come back feeling like they'd had a hell of an animated conversation and a deep talk.

Just a couple of years ago, at Wright-Patterson, it was Gus and Gordo—as Gordon Cooper was known—who had been the great weekend flying buddies. Then Gordo had been transferred to Edwards, where Deke Slayton happened to be. And now that all three of them were in the same corps, this extraordinary new corps of astronauts, there were nights when the others would hear Cooper's Oklahoma drawl getting cranked up . . . and the gorge would rise . . . They would all be knocking back a few at somebody's house, some Saturday night, and they would hear Cooper starting to talk about something extraordinary that happened when he was testing the F–106B or whatever at Edwards . . . and the blood would come into somebody's baleful eyes, and he'd say, "I'll tell you what Gordo did at Edwards. He was in *engineering*." The way *engineering* was pronounced, you would have thought Gordo had been a quartermaster or a drum major or a chaplain.

Deke Slayton took pride in the fact that he was in the hot branch of flight test at Edwards, which was Fighter Operations. The pilots in Fighter Ops at Edwards pushed the outsides of the envelopes of the hottest new airplanes made, the most recent examples being the Century series, of which Gordo's F–106B was one. But to be in engineering was to be an also-ran. Gus and Gordo remained friends and even did some rat-racing in their cars together and, later on, in speedboats. He was so friendly and easygoing, it was hard not to like the man. But sometimes Gus would cluck and fume over Gordo's yarns, too.

And none of this fazed Gordo in the slightest! He seemed to be oblivious of it all! He just went on drawling and lollygagging along as if he were sitting in the catbird seat the whole time! He was also given to sounding off now and then in ways the rest of them just couldn't comprehend. Like that business of the flight pay!

The truth was that none of them, not even Gus, who knew him fairly well, understood Cooper's particular makeup. Cooper may have had his blind spots, but if so, it was the blindness of the fighter jock resolutely making his way up the mighty ziggurat. So what if, by outward standards, he had not had the most brilliant career of all the seven astronauts? The day was young! He was only thirty-two! Cooper's fighter jock self-esteem seemed to be like a PAR lamp. It was as if wherever he landed, the light shone 'round about him, and that was the place to be. Cooper knew as well as anyone else that it was more prestigious to be in Fighter Ops than in engineering at Edwards. But once he was in engineering, the light shone 'round about him, and the picture of him in that place was good. As a pilot in engineering you saw the project from both sides, from the design and administrative side as well as from the test pilot's side. It was like being a project manager who also flew . . . that was what it was like . . . Much of Cooper's fireproof confidence was based on the fact that he was "a natural-born stick-and-rudder man," as the phrase went. When it came to sheer aplomb in controlling a winged aircraft, there was probably no other astronaut who could outdo him. His father had been a colonel in the old Army Air Force, a career officer, and Cooper had started flying before he was sixteen. He had met his wife, Trudy, at Hickam Field after he had enrolled in the University of Hawaii. She was also a pilot. Flying was like breathing to Cooper. He seemed to feel absolutely immune to the ordinary dangers of flight; in any event, he was absolutely cool when it came to dealing with them. As far as his career went, he was never troubled by doubts. It was only a matter of time before everything would go his way. Of that he seemed to be convinced.

When the tests for selection of astronauts began at Lovelace and Wright-Patterson . . . well, it was obvious, wasn't it? Everything was now going his way. He never had the slightest doubt that he would be chosen. His relative lack of credentials didn't trouble him at all. He would be chosen! He could tell! When it came to things like the rigors of the physical tests at Lovelace, he went through it all with a knowing wink. Things like scuttling down a corridor with barium exploding out of your tail—he figured it was intentionally set up as part of the stress testing. Nothing to it, once you understood the drill. Stress? He was so relaxed, the psychologists giving the stress tests at Wright-Patterson could hardly believe it. As soon as the tests at Wright-Pat were completed, Cooper told his commanding officer at Edwards that he had better look for a replacement for him. He was going to be chosen as an astronaut. This was more than a month before the choices were actually made.

Cooper turned out to be neither so naïve nor so guileless as some thought. From the beginning, in the interview sessions, the NASA psychologists had asked the candidates for astronaut many questions about their family lives. Quite aside from any possible public-relations considerations, there was a well-known theory in the psychology of flight to the effect that marital discord was a major cause of erratic behavior among pilots and often led to fatal accidents. The sound instincts of the career officer led Cooper to respond that his family life, with Trudy and the children, was real fine, terrific; regulation issue. This wasn't likely to bear much checking into, however, inasmuch as Cooper and Trudy were not living in the same house or even in the same latitude. They had separated; Trudy and the children were living down near San Diego, while he remained at Edwards. Clearly it was time for a reconciliation. Cooper took a quick trip to San Diego . . . he talked a whole rope . . . a veritable lasso . . . the separation, his prospects with NASA, and so on . . . In any case, Trudy and their two daughters returned to Edwards, and Cooper had the American dream back intact, under one roof, before the final round of the selection process, and nobody at NASA was the wiser.

After Cooper was selected, one of the *Life* writers brought up the point that he had less experience than most of the other astronauts. Cooper was not fazed for a moment. He said that he was also younger than the others and would probably be the only one of them to fly to Mars.

The only side of the whole deal that appeared to shake Cooper's confidence was the p.r. side of it, the publicity routine, the trips here and there, where various local worthies put you at the head table and whacked you on the back and asked you to get up and "just say a few words." Most of the trips were to cities where components of the Mercury system were being manufactured, such as St. Louis, where the capsule was being built at the McDonnell factory, or San Diego, where the Atlas rocket was being built at Convair. St. Louis, San Diego, Akron, Dayton, Los Angeles—somebody was always suggesting that you "just say a few words."

It was on such occasions that a man realized most acutely that America's seven astronauts were not by any means identical. Glenn seemed to eat this stuff up. He couldn't get enough grins or handshakes, and he had a few words filed away in every pocket. He would even come back to Langley and write cards to workers he had met on the assembly line, giving them little "attaboys," as if they were all in this thing together, partners in the great adventure, and he, the astronaut, would never forget his, the welding inspector's, beaming mug. The idea, much encouraged by NASA, was that the personal interest of the astronaut would infuse everyone working for the contractors with a greater concern for safety, reliability, efficiency.

Oddly enough, it seemed to work. Gus Grissom was out in San Diego in the Convair plant, where they were working on the Atlas rocket, and Gus was as uneasy at this stuff as Cooper was. Asking Gus to "just say a few words" was like handing him a knife and asking him to open a main vein. But hundreds of workers are gathered in the main auditorium of the Convair plant to see Gus and the other six, and they're beaming at them, and the Convair brass say a few words and then the astronauts are supposed to say a few words, and all at once Gus realizes it's his turn to say something, and he is petrified. He opens his mouth and out come the words: "Well . . . do good work!" It's an ironic remark, implying: ". . . because it's my ass that'll be sitting on your freaking rocket." But the workers started cheering like mad. They started cheering as if they had just heard the most moving and inspiring message of their lives: *Do good work!* After all, it's Little Gus's ass on top of our rocket!

They stood there for an eternity and cheered their brains out while Gus gazed blankly upon them from the Pope's balcony. Not only that, the workers—the workers, not the management but the workers!—had a flag company make up a huge banner, and they strung it up high in the main work bay, and it said: DO GOOD WORK.

All these people with their smiles of sympathy didn't ask for much. A few words here and there would do fine. *Do good work.* Nevertheless, that didn't make these public appearances any better for Cooper. He was in the same boat with Gus and Deke, who was also no Franklin D. Roosevelt when it came to public appearances. Everybody latched on to you during these trips, congressmen and businessmen and directors and presidents of this and that. Every hotshot in town wanted to be next to *the astronaut.* For the first ten or fifteen minutes it was enough for them to breathe the same air you breathed and occupy the same space as your famous body. But then they began looking at you . . . and waiting . . . Waiting for what? Well, dummy!—waiting for you to say a few words! They wanted something hot! If you were one of the seven greatest pilots and seven bravest men in America, then obviously you must be fascinating to listen to. *Riveting*—that was what you were supposed to be. A few war stories, man! And you would sit there with the clutch in, furiously trying to think of something, anything, and it would make you gloomier and gloomier. Your light no longer shone 'round about you.

It was on such occasions that the three Air Force men, Cooper and Gus and Deke, wouldn't have minded being like Alan Shepard. Shepard was all right. He didn't go for these public appearance stints any more than they did. But Al could shift gears anytime he had to. Al was a Naval Academy man, and if he had to glad-hand and shoot the breeze and trade the small talk with all these congressmen and realty board chairmen and rye distillers and get up and make extemporaneous remarks when called upon, then he could do it. Wally Schirra was another Naval Academy man, and he could play it any way he wanted to, also. Wally was a regular guy, a fighter jock through and through, but he also had the knack of turning on the old Academy charm around strangers. As for the other Navy guy, Carpenter, he wasn't an Academy man, but he was Mr. Charm himself, all the same. He knew how to turn on the party manners.

There was damned little social crap in the Air Force, and that was probably one reason why Cooper liked the blue suit. The "officers and gentlemen" business was kept at a minimum. At most bases the only well-to-do locals who invited Air Force officers to parties were the automobile dealers. They just loved the way those crazy blue-suit sombitches bought those cars and racked them up and then came back and bought some more. In the Air Force there was a nice piece of built-in democracy. Until an officer had reached the level of lieutenant colonel, there was only one way for him to make his mark and advance, and that was by proving himself as a pilot. If he could demonstrate that he had the right stuff in the air, there was nothing, short of gross character defects, that could keep him from rising through the middle ranks. In the Navy Air Force an officer also had to prove himself in the air, but at the test-pilot level the Navy began to insist on "leadership qualities" as well, meaning polish and the rest of it.

There you had a man like Al Shepard, who had come from out of what was sometimes called "the service aristocracy." Which was to say, Al was the son of a career officer. By now you ran into these fellows, the second-generation officers, all over the service. They seemed to make up half the graduates of West Point or Annapolis, like Al and Wally Schirra. Al's father was a retired Army colonel. Wally Schirra's was in effect a service family. His father had been a pilot in the First World War, had then left the service, but then became a civil engineer for the Air Force after the Second World War, helping to rebuild Japanese airfields. You very seldom ran into career officers who were the sons of businessmen, doctors, or lawyers. They steered their sons away from the service. They looked down on it. So what you found were the second-generation officers on the one hand, like Shepard and Schirra, and the sons of workingmen and farmers on the other, fellows like Gus and Deke and John Glenn. Fellows like Shepard and Schirra (and Carpenter) might come from small towns, strictly speaking, but it was a mistake to call them "small-town boys," the way you could apply that term to Gus or Deke, and it showed in the way they could handle themselves in public.

It wasn't long before Cooper began to miss flying, the stick 'n' rudder life, in the worst kind of way. He began to miss it the way another man might have missed food. The daily business of taking a high-performance aircraft aloft and hanging it out over the edge—this was at the heart of the fighter jock's life, even though its importance was never expressed except in the term "proficiency." Pilots devoutly believed that it was necessary to fly out to the edge with regularity in order to maintain proficiency or "decision-making ability." On one level it was a logical enough equivalent to an athlete's concern with staying in shape; but on another it had to do with the mysteries of the right stuff and the ineffable joys of showing the world, and yourself, that you had it. It was damned strange to be in flight training, as America's first astronauts, and yet to be doing no flying themselves, except as passengers.

There was no flying whatsoever on their training agenda! As the weeks went by, all seven men began to be bothered by this, but it was Cooper who voiced the complaint publicly. The early months included a heavy schedule of lectures, on astronomy, rocket propulsion, flight operations, capsule systems, and the trips to the contractors, and to the subcontractors, and to Cape Canaveral, where the rockets would be launched, to Huntsville, Alabama, where Wernher von Braun and his Germans were developing booster rockets, to Johnsville, Pennsylvania, where the human centrifuge was located. There was no end to it. On all these trips Cooper, like the others, had to travel by commercial airline. It seemed as if he spent half of every day standing around airports waiting for luggage or going through his pockets to see how much money he had. Here he was, flying half the month—as a passenger! On top of everything else, he was losing flight pay! It was no laughing matter! DeOrsey was negotiating the *Life* deal but had not yet closed it. If an Air Force captain kept up his proficiency flying, he stood to receive an extra $145 a month in flight pay, and there was not a sane blue-suiter alive who did not go out and get that flight pay each month unless bedridden or grounded. The extras—my God, it was impossible to explain to an outsider, but these things were built into the psyche of the career officer like first principles! Besides, your family always needed the money. Cooper, like the other six, was being paid by the military, and so he was losing a significant percentage of his income, which hadn't been much to begin with. Not only that, an officer in the military received a mere nine dollars a day in expenses for day trips and twelve dollars a day for overnight trips. To stay in hotels, to eat in restaurants—it was a losing proposition. Especially when they were supposed to be some sort of celebrities. They all felt like the biggest deadbeat celebrities in America. Say you were having lunch with five or six hotshots in Akron, where you went for pressure-suit fittings at B. F. Goodrich. You didn't dare reach for the check. Suppose through delayed psychomotor response or some other dreadful accident they let you *have* it! The damned thing might be for thirty-five dollars—and there went your family's food money for two weeks . . . And yet the flight pay itself was the least of it. It was more evidence of the curious *non-pilot* status of the astronaut. Cooper figured he was spending forty hours a month on commercial airliners in order to go through all this. What he wouldn't have given to have access to a supersonic fighter plane like the F–104B . . . Gus and Deke were managing to cadge rides on the weekends in T–33's at Langley. But the T–33 was pretty tame stuff, a subsonic trainer. The F–104B was something you could cut loose with. Langley Air Force Base wasn't even equipped to maintain such an aircraft, however. So Cooper was going all the way to McGhee-Tyson in Knoxville, where he had a buddy who could get him signed up for the occasional workout in the F–104B. With a ship like that he could *live and breathe* . . . and maintain proficiency and keep in touch with that righteous stuff . . .

Such thoughts were once more running through his mind as he sat down to lunch one day at Langley, when a reporter for the *Washington Star* named William Hines joined him and said hello. Well, they talked a little bit and one thing led to another, and pretty soon Cooper was painting the entire picture. When the story appeared in the *Star*—depicting Cooper's complaint, accurately, as a complaint common to practically all the astronauts—NASA officials were dumbfounded. Overton Brooks's House Committee on Science and Astronautics was dumbfounded. Gordo's fellow deservers of perks and goodies were dumbfounded, even though most of them agreed with him completely. They all looked a trifle petty. Here they were, seven heroes, warriors of the heavens, patriots, and they're all over the press complaining about flight pay and airplane rides . . .

Overton Brooks sent a committee investigator to Langley to see what the hell was going on. The report he brought back was a masterpiece, a veritable model performance, in the tactful handling of the grousing of his country's first single-combat warriors. "The astronauts," he wrote, "are fully aware of their responsibilities to the project and the American public, particularly with regard to the heroic role they are beginning to assume with the young people of the country. They have imposed upon themselves strict rules of conduct and behavior, which credits them with constructive and mature evaluation of their position as a cynosure of all eyes." The only thing is, they still want their goddamned flight pay and some hot airplanes.

Like most of the other wives, Betty Grissom was stuck at Langley with small children to take care of. At first she had thought she and Gus were at last going to be able to settle in for some ordinary home life, but somehow Gus was away as much as ever. Even when he had the weekends off, he would somehow wander over to Deke's house, and before she knew it, the two of them would be heading off to the base for some "proficiency" flying, and there went another weekend.

If Gus was home for the weekend, he was apt to get in some fast flurries of fatherhood for the benefit of their two boys, Mark and Scott. This might take the form of some good gruff-gus obedience lectures about obeying their mother when he wasn't there. Or it might take the form of something like the floating dock. The development they lived in backed up on a little lake. One weekend Gus set about building a floating dock so that the boys could use the lake as a proper swimming hole. The problem was that the older of the boys, Scott, was only eight, and Betty was afraid they were going to drown back there. She had nothing to worry about, as it turned out. The boys never took to the old swimming hole. They much preferred the swimming pool across the street at the community club. It had a diving board and a concrete apron and clear water and other children to play with. The floating dock remained out back moldering in the lake like a reminder of the kind of fatherhood that the astronaut life began imposing on all seven families.

Betty was not as upset about her husband's protracted absences as a lot of other wives would have been. When they had been stationed at Williams Air Force base, other wives had even put pressure on her not to let Gus have so many weekends off, because it was giving their husbands ideas. But few wives seemed to believe as firmly as Betty did in the unofficial Military Wife's Compact. It was a compact not so much between husband and wife as between the two of them and the military. It was because of the compact that a military wife was likely to say "*We* were reassigned to Langley" . . . *we*, as if both of them were in the military. Under the terms of the unwritten compact, they were. The wife began her marriage—to her husband and to the military—by making certain heavy sacrifices. She knew the pay would be miserably low. They would have to move frequently and live in depressing, exhausted houses. Her husband might be gone for long stretches, especially in the event of war. And on top of all that, if her husband happened to be a fighter pilot, she would have to live with the fact that any day, in peace or war, there was an astonishingly good chance that her husband might be killed, *just like that*. In which case, the code added: *Please omit tears, for the sake of those still living*. In return for these concessions, the wife was guaranteed the following: a place in the military community's big family, a welfare state in the best sense, which would see to it that all basic needs, from health care to babysitting, were taken care of. And a flying squadron tended to be the most tightly knit of all military families. She was also guaranteed a permanent marriage, if she wanted it, at least for as long as they were in the service. Divorce—still, as of 1960—was a fatal step for a career military officer; it led to damaging efficiency reports by one's superiors, reports that could ruin chances of advancement. And she was guaranteed one thing more, something that was seldom talked about except in comical terms. Underneath, however, it was no joke. In the service, when the husband moved up, the wife moved up. If he advanced from lieutenant to captain, then she became Mrs. Captain and now outranked all the Mrs. Lieutenants and received all the social homage the military protocol provided. And if her husband received a military honor, then she became the Honorable Mrs. Captain—all this regardless of her own social adeptness. Of course, it was well known that a gracious, well-spoken, small-talking, competent, sophisticated wife was a great asset to her husband's career, precisely because they were a team and *both* were in the service. At all the teas and socials and ceremonies and obligatory parties at the C.O.'s and all the horrible Officers Wives Club functions, Betty always felt at a loss, despite her good looks and intelligence. She always wondered if she was holding Gus back in his career because she couldn't be the Smilin' & Small-Talkin' Whiz that was required.

Now that Gus had been elevated to this extraordinary new rank—astronaut—Betty was not loath to receive her share, per the compact. It was as if . . . well, precisely because she had endured and felt out of place at so many teas and other small-talk tests, precisely because she had sat at home near the telephone throughout the Korean War and God knew how many hundreds of test flights wondering if the fluttering angels would be ringing up, precisely because her houses all that time had been typical of the sacrificial lot of the junior officer's wife, precisely because her husband had been away so much—it was as if precisely because that was the way things were, she fully intended to be the honorable Mrs. Captain Astronaut and to accept all the honors and privileges attendant thereupon.

Betty thought the *Life* deal was terrific. She didn't have to wrestle with the angels over that one for a second. They would be getting just under $25,000 a year from it, a sum almost beyond her imagining after all these glum ocher years. But that was only part of the beauty of this goodie. On the day it had been announced that Gus had been chosen as an astronaut, Betty had been even more terrified than Gus. Gus had only a NASA-controlled press conference to deal with. Betty, with practically no warning, had been mobbed, overrun, at their house in Dayton by the press. They came crawling in through the windows like ravenous termites, like fruit flies, taking pictures and yelling questions. She felt as if she had been engulfed in the monster Small-Talk Tea of all times, and merely the entire country would see her as an unsophisticated Hoosier grit. To her great relief, whatever answers she had come up with emerged as coherent whole sentences, and not at all foolish, in the newspapers the next day, and she looked splendid in the pictures. (Naturally she did not know that the press was an anachronistic colonial animal, a Victorian Gent who was determined to give to all important moments the proper tone.) Still, she wouldn't want to have to go through that sort of thing again. And now she wouldn't! She would only have to talk to *Life* reporters, and they turned out to be marvelous. They were polite, well-educated, well-dressed, friendly, kind, real ladies and gentlemen. They had no desire whatsoever to make her look bad. Betty and the other wives came bursting forth like great blossoms before the ten million readers of *Life* in a cover story in the September 21, 1959, issue. Their faces, smooth round white things with coronas of hair, were arranged on the cover like a corsage of flowers with Rene Carpenter's face in the middle—no doubt because the editors regarded her as prettiest. But who is that? Oh, that's Trudy Cooper. And who is *that?* Oh, that's Jo Schirra. And who is *that?* Oh, that's . . . They hardly recognized each other! Then they saw why. *Life* had retouched the faces of all of them practically down to the bone. Every suggestion of a wen, a hickie, an electrolysis line, a furze of mustache, a bag, a bump, a crack in the lipstick, a rogue cilia of hair, an uneven set of the lips . . . had disappeared in the magic of photo retouching. Their pictures all looked like the pictures girls can remember from their high-school yearbooks in which so many zits, hickies, whiteheads, blackheads, goopheads, goobers, pips, acne trenches, boil volcanoes, candy-bar pustules, rash marks, tooth-brace lumps, and other blemishes have been scraped off by the photography studio, you looked like you had just healed over from plastic surgery. The headline said: SEVEN BRAVE WOMEN BEHIND THE ASTRONAUTS.

Whether by design or not, *Life* had seized upon the idea that Luce's fellow Presbyterian John Glenn had put forth at the first press conference: "I don't think any of us could really go on with something like this if we didn't have pretty good backing at home." Pretty good backing? Perfect backing they were going to have: seven flawless cameo-faced dolls sitting in the family room with their pageboy bobs in place, ready to offer any and all aid to the brave lads. There was something crazy about it, but it was marvelous. The week before, in the September 14, 1959, issue, *Life* had ushered Gus and the other fellows out onto the Pope's balcony with a cover story headlined READY TO MAKE HISTORY that left no doubt whatsoever that these were the seven bravest men and the seven greatest pilots in American history, even if it was necessary to go easy on the details. Now *Life* was leading Betty and the other wives out onto that balcony.

Betty, for one, did not object to that at all.

They had to let the *Life* writers and photographers come into their houses and follow them around pretty much anywhere they wanted to, but that turned out to be no particular problem. Pretty soon they all realized they didn't even have to keep their guard up. The *Life* people were very sympathetic. The men among them obviously had a kind of male awe of Gus and the others; you could even detect a tinge of envy every now and then, because the *Life* reporters and the fellows were about the same age. But they were loyal. In any case, they were hamstrung, since Gus and Betty and the rest of the men and their wives had the right to censor anything that was going to appear under their names. And don't think they were bashful about it, either! Not for a minute! You'd hear one of the fellows on the telephone going over a manuscript with a *Life* writer line by line, telling him, in just so many words, what could stay in and what was coming out. Oh, the *Life* writers sometimes had their own notions of what was candid and colorful and "good copy." They liked to get on such subjects as the rivalries between the boys and such "colorful" matters as Driving & Drinking and the unspoken intrafraternal business of fear and courage . . . Well, the hell with that! It was not so much that the men wanted to come out sounding like the Hardy Boys in Outer Space—it was just that you'd have to be an idiot to let your personal story actually get personal. Every career military officer, and especially every junior officer, knew that when it came to publicity, there was only one way to play it: with a salute stapled to your forehead. To let yourself be turned into a *personality*, to become *colorful*, to be portrayed as an egotist or a rake-hell, was only asking for grief, as many people, including General George Patton, had learned. Scott Carpenter was a case in point. He was open and forthright by nature, and he happened to tell one of the *Life* writers how his teenage years had been anything but standard-issue astronaut-corps mom's-pie material, especially after his grandfather had died and he had drifted around Boulder raising hell when he felt like it—and some of this stuff came out in *Life,* without NASA being sent a draft of it, and Scott caught flak for weeks . . . on the grounds that he had put the program in a bad light.

As far as the wives were concerned, their outlook was the same as that of officers' wives generally, only more so. The main thing was not to say or do anything that reflected badly upon your husband. There wasn't much to worry about with *Life* on this score. If Betty or any of the others did happen to say anything wrong, she could always remove it before it saw print. As time went by, the *Life* writers must have despaired of getting any personal stuff at all into their personal stories.

Deke Slayton's wife, Marge, had been divorced, which was a matter of record, but that wasn't about to be printed in *Life* magazine. A *once-divorced astronaut's wife* was by now an unthinkable concatenation of words. When the selection process for astronaut had begun, Trudy Cooper, Gordon Cooper's wife, had been living by herself down at San Diego. The writers from *Life* may have known about it and they may not have. It was a moot point, because in any event there were not going to be any astronauts with washed-up marriages in the pages of *Life* magazine on the eve of the battle in the heavens with the Russians. The exclusive rights to the "personal stories" of the astronauts and their families that *Life* had purchased did not encompass any such tangled terrain as that.

And it didn't have to be that personal for them to wave the wand and make it disappear. Look at what they did with John Glenn's wife, Annie. Annie was a good-looking and highly capable woman, but she also had what was referred to as a "slight speech impediment" or "a hesitation in her speech." The truth was that she had a terrific stutter, the classic kind, the kind in which you get hung up on a syllable until you either force it out or run out of breath. Annie was game about it, and she would hang in there until she said what she wanted to say, but it was a real disability—everywhere except in *Life* magazine. In *Life* magazine there were going to be no ferocious stammering jackhammer stutters on the home front.

As for Betty, she came out in *Life* as the thoughtful, articulate, competent, much respected Honorable Mrs. Captain Astronaut. She didn't ask for much more than that. If it pleased them, the people at *Life* could sit around removing dour grim grit and zits until they earned a place beside the angels in Retouch Heaven.

VII. The Cape

Cape Canaveral was in Florida, but not any part of Florida you would write home about, except on one of those old Tichnor Brothers postcards on which there is a drawing of two grinning dogs positioned in front of a lamp post, each with a hind leg hoisted, and a caption that says: THIS IS A WONDERFUL PLACE . . . JUST BETWEEN YOU AND ME AND THE LAMP POST! No, Cape Canaveral was not Miami Beach or Palm Beach or even Key West. Cape Canaveral was Cocoa Beach. That was the resort town at the Cape. Cocoa Beach was the resort town for all the Low Rent folk who couldn't afford the beach towns farther south. Cocoa Beach was so Low Rent that nothing on this earth could ever change it. The vacation houses at Cocoa Beach were little boxes with front porches or "verandas" nailed onto them and a 1952 De Soto coupe with venetian blinds in the rear window rusting in the salt air out back by the septic tank.

Even the beach at Cocoa Beach was Low Rent. It was about three hundred feet wide at high tide and hard as a brick. It was so hard that the youth of postwar Florida used to go to the stock-car races at Daytona Beach, and then, their brains inflamed with dreams of racing glory, they would head for Cocoa Beach and drive their cars right out on that hard-tack strand and race their gourds off, while the poor sods who were vacationing there gathered up their children and their Scotch-plaid picnic coolers, and ran for cover. At night some sort of prehistoric chiggers or fire ants—it was hard to say, since you could never see them—rose up from out of the sand and the palmetto grass and went for the ankles with a bite more vicious than a mink's. There was no such thing as "first-class accommodations" or "red-carpet treatment" in Cocoa Beach. The red carpet, had anyone ever tried to lay one down, would have been devoured in midair by the No See'um bugs, as they were called, before it ever touched the implacable hardcracker ground.

And that was one reason why the boys loved it! Even Glenn—even Glenn, who did not partake of all of its Low Rent glories.

The place reminded them of what they had heard Edwards, or Muroc, was like in the legendary days of the late 1940's and early 1950's. It was one of those bleached, sandy, bare-boned stretches where the land that any sane man wants runs out . . . and the government takes it over for the testing of hot and dangerous machines, and the kings of the resulting rat-shack kingdom are those who test them. Just south of Cocoa Beach was Patrick Air Force Base, site for the headquarters of the Atlantic Missile Range, for the testing of the weaponry of the Cold War: guided missiles, intermediate-range ballistic missiles, and intercontinental ballistic missiles. North of Cocoa Beach, on the very tip of the Cape itself, was the huge new secret launching facility from which all these rockets and pilotless aircraft were fired, a stone boondock dune plain with the Atlantic Ocean on one side and the Banana River on the other, with soil so sandy that the scrub pines had trouble growing fifteen feet high, and yet malarial and so marshy that the cottonmouth moccasins stood their ground and stared you down, the sort of hopeless stone boondock spit where the vertebrates give up and the slugs and the No See'um bugs take over. The few buildings on the base were of the World War II Beaverboard Temporary variety. And like Edwards of old, the Cape, this poor godforsaken afterthought in the march of terrestrial evolution, turned out to be a paradise of Flying & Drinking and Drinking & Driving and Driving & the rest, for those who cared about such things. Or of Drinking & Driving & the rest, in any event. There was still no flying.

Langley remained the astronauts' headquarters, but the Cape would be their launching site, eventually, and they went there increasingly for training. They would fly in by commercial airliner, landing in Melbourne or Orlando. Most of them would rent convertibles and head for the Holiday Inn on Route A1A just north of the old part of Cocoa Beach, a motel run by a man named Henry Landiwirth, who soon found himself becoming innkeeper to the astronauts. Quite a little 1960's-style American Rat-Shack Strip was beginning to develop on Route A1A near the Holiday Inn: hamburger restaurants with plate-glass walls and hot magenta lights, night spots with Kontiki roofs, and little shopping slabs by the side of the highway, slabs of concrete with one-story cinderblock sheds on them broken up into store fronts with SPACE AVAILABLE signs posted.

Military units had always been great ones for creating "traditions" instantly, on the spot, and this unofficial corps of astronauts was no exception. The tradition was: the Cape is off limits to wives. This came about rather naturally. The Cape was not a good place for wives and children, because you couldn't count on finding kitchen facilities in the motels and there were none of the usual beach resort amenities, and none of them could afford the plane fares for family trips to Florida in the first place. Besides that, the boys' training hours were very long, sometimes ten or twelve hours a day. They did nothing at the Cape but work their butts off all day and then fall into bed, albeit this was a matter open to interpretation.

The boys' training at the Cape was not so much arduous as tedious. It was sedentary, even. It involved no flying. Some days they would be briefed on launch procedures. Or they would drive out to the launching base and go inside an old converted rat-shack hangar, Hangar S, and sit all day in a simulator known as the "procedures trainer," which on the inside was a replica of the capsule they would ride in during flight. Or technically they sat in there all day; in fact, they were lying down. It was as if you took a chair and pushed it over backward, so that its back was on the floor, and then sat in it. That was the position the astronaut would be in during his launch atop the rocket and the position he would be in as he came down toward the water inside the capsule at the end of the flight.

It was hard for Glenn or anyone else to explain exactly what you did for ten or twelve hours inside this thing. But clearly, once a man had had a day full of this tedious regimen, he was ready to limber up a little, get the blood flowing again, wiggle his fanny a bit. For Glenn it was enough to go out to that hardtack strand at Cocoa Beach and run two or three miles. It was the greatest long-distance running track you could possibly ask for, with pure ocean air to help your pump get going efficiently. And there would be John Glenn, the very picture of astronaut dedication, pounding along the same shore from which he would one day be hurled into the heavens. John Glenn Running for the Big One at Cocoa Beach was an even better picture than the one he had put on display at Langley. Glenn noticed that some of his confreres were loosening up in quite another way, however. Which is to say, they were checking in at the holy coordinates. After a long day of make-believe flying in the simulator . . . a little Drinking & Driving & the rest of the real pilot's life.

The *driving* eventually took on an extraordinary dimension here at the Cape. Gus Grissom and Gordon Cooper, and then Al Shepard and Wally Schirra, would discover Jim Rathmann. Rathmann was a big rugged character who had one of the largest automobile dealerships in the area, a General Motors agency about twenty miles south of Cocoa Beach near Melbourne. It was typical Air Force stuff that Gus and some of the others should become great pals of his. Rathmann was no ordinary auto dealer, however. He turned out to be a racing driver; the best, in fact. In 1960 he won the Indianapolis 500 after having finished second three times. Rathmann was a great friend of Ed Cole, the president of Chevrolet. Cole had helped Rathmann set up his agency. When he found out that Rathmann knew the Mercury astronauts, he became the astrobuff of all astrobuffs. America seemed to be full of businessmen like Cole who exercised considerable power and were strong leaders but who had never exercised power and leadership in its primal form: manly courage in the face of physical danger. When they met someone who *had it,* they wanted to establish a relationship with that righteous stuff. After meeting the astronauts, Cole, who had just turned fifty, was determined to learn to fly. Meantime, Rathmann set up a leasing arrangement whereby the boys could lease any type of Chevrolet they wanted for practically nothing per year. Eventually, Gus and Gordo had Corvettes like Al Shepard's; Wally moved up from an Austin-Healy to a Maserati; and Scott Carpenter got a Shelby Cobra, a true racing vehicle. Al was continually coming by Rathmann's to have his gear ratios changed. Gus wanted flared fenders and magnesium wheels. The fever gripped them all, but Gus and Gordo especially. They were determined to show the champ, Rathmann, and each other that they could handle these things.

Gus would go out rat-racing at night at the Cape, racing full-bore for the next curve, dealing with the oncoming headlights by psychokinesis, spinning off the shoulders and then scrambling back up on the highway for more of it. It made you cover up your eyes and chuckle at the same time. The boys were fearless in an automobile, they were determined to hang their hides right out over the edge—and they had no idea what mediocre drivers they actually were, at least by the standards of professional racing. Which is to say they were like every group of pilot trainees at every base in America who ever reached that crazed hour of the night when it came time to prove that the right stuff works in all areas of life.

Cocoa Beach had begun to take on the raw excitement of a boom town and the manic and motley cast of characters that goes with it. In boom towns of the oil-gush or gold-rush variety the excitement had always come from simple greed. But Cocoa Beach was more like a Second World War boom town. There was enough greed in the air to make things spicy, but the true fervor was the *joie de combat.* People coming to work at the Cape, for NASA, private contractors, or whomever, felt like part of the mad rush to battle the Soviets for dominion over the heavens. At Edwards, or Muroc, in the old days, the worthy warriors used to repair in the evening to Pancho's, which, though theoretically a public place, was like a club for the adventurers over the high desert. At the Cape, by 1960, the warriors had the motels on the rat-shack strip along Route A1A. At night the pool areas of the motels became like the roaring fraternity house lounge of Project Mercury. Very few people, no matter what their rank in the project, had a place big enough, much less attractive enough, to entertain in. But every night the fraternal lounge was open, under the skies, in the salt air, out near the beach, and the party was on, and one and all braved the palmetto bugs and the No See'um bugs and celebrated the fact that they were on the scene where this great Cold War adventure was taking place. Naturally nothing gave the party quite so much magic as the presence of an astronaut.

And Glenn could see that after eight, ten, twelve hours of lying cooped up in the procedures trainer out in Hangar S, most of his brethren were ready to provide the magic. No matter what time it was, it was beer-call time, as they said in the Air Force, and they would get in their cars and go barreling into Cocoa Beach for the endless, seamless party. And what lively cries and laughter would be rising up on all sides as the silvery moon reflected drunkenly on the chlorine blue of the motel pools! And what animated revelers were to be found! There were NASA people and the contractors and their people, and there were the Germans. Although they scrupulously avoided publicity, many of Wernher von Braun's team of V–2 experts had important jobs at the Cape and were happy to find a fraternal atmosphere in which they could take off their official long faces and let the funny bone out for a tap dance or two. And many were the midsummer nights in Cocoa Beach, nights so hot and salty that the No See'um bugs were sluggish, when sizzling *glühwein* materialized as if from out of a time warp and drunken Germans could be heard pummeling the piano in the cocktail lounge and singing the "Horst Wessel Song"! It was like some improbable echo of Pancho's along the hardtack Florida littoral. Oh, yes, it was! As at Pancho's, the most marvelous lively young cookies were materializing also, and they were just *there,* waiting beside the motel pools, when one arrived, young juicy girls with stand-up jugs and full-sprung thighs and conformations so taut and silky that the very sight of them practically pulled a man into the delta of priapic delirium. Some of them had come to work for the contractors, some to work for NASA, some to work for this or that business that was starting up in the little boom town—and some simply *got there, materialized.* And when an astronaut arrived, it was as if they dropped out of the sky or rose up from out of the Bermuda grass. In any event, they were always there and ready.

As even Glenn could tell, it was enough just to *be* an astronaut, whether a handsome devil like Scott Carpenter or a gruff little fellow like Gus Grissom. As soon as Gus arrived at the Cape, he would put on clothes that were Low Rent even by Cocoa Beach standards. Gus and Deke both wore these outfits. You could see them tooling around the Strip in Cocoa Beach in their Ban-Lon shirts and baggy pants. The atmosphere was casual at Cocoa Beach, but Gus and Deke knew how to squeeze casual until it screamed for mercy. They reminded you, in a way, of those fellows whom everyone growing up in America had seen at one time or another, those fellows from the neighborhood who wear sport shirts designed in weird blooms and streaks of tubercular blue and runny-egg yellow hanging out over pants the color of a fifteen-cent cigar, with balloon seats and pleats and narrow cuffs that stop three or four inches above the ground, the better to reveal their olive-green GI socks and black bulb-toed bluchers, as they head off to the Republic Auto Parts store for a set of shock-absorber pads so they can prop up the 1953 Hudson Hornet on some cinderblocks and spend Saturday and Sunday underneath it beefing up the suspension. Gus and Deke made a perfect pair, even down to their names. Not even the sight of the boys in their Mechanics & Tradesmen's Ban-Lon could turn off the girls to the presence of the astronauts.

There were juicy little girls going around saying, "Well, four down, three to go!" or whatever—the figures varied—and laughing like mad. Everybody knew what they meant but only halfway believed them. There was no question but that the temptations for the Fighter Jock Away from Home were enormous. It was all so easy and casual on these midsummer nights. Before the missiles came to the Cape, Cocoa Beach was a hard-shelled Baptist stronghold with more churches than gasoline stations, and practically all of them were of the pietistic or Dissenting Protestant variety. But the new Cocoa Beach, the Project Mercury boom town, was part of the new face of the 1960's: the little town whose life was completely keyed to the automobile. Naturally, nobody built hotels in Cocoa Beach, only motels; and when they built apartment houses, they built them like motels, so that you could drive up to your own door. At neither the motels nor the apartment houses did you have to go through a public lobby to get to your room. A minor architectural note, one might say—and yet in Cocoa Beach, like so many towns of the new era, this one fact did more than *the pill* to encourage what would later be rather primly named "the sexual revolution."

There had always been a part of the Military Wife's Compact that tacitly granted an officer a little latitude in this area. Naturally, there would be times when a military man would be sent far from home, perhaps for extended periods, and he might find it necessary to satisfy his healthy manly urges on these far-off terrains. There was even the implication that such urges were a good sign of a fighting man's virility. So the wife and the military itself would avert their eyes and stand mute—so long as the officer caused no scandal and did nothing to shake the solidity of his marriage and his family. This tradition had originated, of course, long before the airplane made it possible for an officer to reach the distant terrain in two or three hours for a long weekend or an overnight stand. Traditions often began on a moment's notice in the military; but they took a long time to die, and this one was in no danger of dying at Cocoa Beach.

That much John Glenn could discern also . . . and such was the background of the Konakai Séance.

Every now and then the seven pilots would shut the door of their office at Langley, and not even the secretary could come in. If anybody wanted to know what was going on in there, they were told that the astronauts were having a séance. A *séance?* Oh, it's just a name they thought up for a meeting in which they try to come up with a common position, a consensus, concerning certain problems. The implication was that the problems were mostly technical in nature. Wally Schirra would mention that they had had a séance before going to the engineers and insisting on changes in the design of the instrument panel of the Mercury capsule. The idea was to give the corps of astronauts some of the solidity of a squadron. The seven of them might have their rivalries, their differences in backgrounds and temperaments and approaches to the job at hand, but they should be able to arrive at firm decisions as a group, no matter how acrimonious the debate might be, and then close ranks and pull together, one for all, all for one. Whether or not the session at the Konakai qualified as a séance by the usual standards was hard to say. But God knows it dealt with a recurrent problem . . . and the debate was acrimonious . . .

One day all seven of them were out in San Diego for a tour of the Convair plant and a look at the latest progress on the Atlas rocket. Convair wanted to do it up right and had treated them all to their own rooms at the Konakai, a rather high-toned hotel built in a Polynesian motif on Shelter Island, overlooking the Pacific. It so happened that Scott Carpenter had drawn a room with a double bed. That evening one of the boys approached him in a comradely fashion and said that his room had two twin beds, whereas in fact he was going to require a double bed for the evening. Would Scott mind switching rooms? It was all the same to Scott, and so they switched rooms. Scott mentioned it to his buddy John Glenn with a smile, as an amusing local note, and thought no more about it.

The next day the seven of them were in the living room of a suite that had been set aside for their use, when Glenn launched into a lecture, along the following lines: the playing around with the girls, the cookies, had gotten out of hand. He knew, and they knew, that it could blow up into something very unfortunate. They were all squarely in the public eye. They had the opportunity of a lifetime, and he was sorry but he just wasn't going to stand by and let other people compromise the whole thing because they couldn't keep their pants zipped.

There was no doubt whatsoever that Glenn meant every word of it. When he got his back up, he was formidable. He was not to be trifled with. In his eyes burned four centuries of Dissenting Protestant fervor, nailed down by two million laps that his legs had pounded around the BOQ driveway.

But there was more than one hard customer in the room. Staring straight back at Glenn, volt for volt, was Al Shepard. The others, Glenn included, understood Shepard least of all, because there seemed to be two Al Shepards, and no one ever knew for sure which one he was dealing with. Back home at Langley you saw one Alan Shepard, the utterly, and if necessary, icily correct career Navy officer. Shepard's father, Colonel Alan Shepard, Sr., was an impressive figure whom few people cared to challenge. Shepard was always a good son. The colonel sent him to private schools, and in due course he followed the colonel's model of a military career, graduated from the Naval Academy, and became a pilot; and although he had never served in combat, he was considered one of the Navy's best test pilots, drawing important assignments in testing the F3H, the F8U, the F4D Skyray, the F11F Tigercat, the F2H3 Banshee, and the F5D Skylancer, including the tricky business of proving out some of these monsters in their first landings on the then-new angled carrier decks. He was regarded as a topnotch Navy aviator, tough, quick-witted, and a leader. He had married a good-looking woman of great charm and poise—"a real lady," people always said—named Louise Brewer. She was a Christian Scientist. Shepard was from

New Hampshire, and in New England the Christian Scientists had considerable social cachet, since they were on the average the wealthiest church members in the United States and had an intellectual tradition somewhat similar to the Unitarians'. Although this side of Christian Scientist life was not generally known in America, it was not lost upon the Navy, where the brass traditionally kept tabs on religious affiliations. Being an Academy man was the most important thing, but belonging to a socially correct Protestant denomination was the next best thing. The Episcopal Church ranked first, unofficially, throughout the military (both Schirra and Carpenter were Episcopalians). Well, the Christian Scientists, although smaller in numbers, were even tonier. Such were the general contours of the correct life of Commander Shepard, the icy career officer. But inside his locker he kept . . . Smilin' Al of the Cape! In point of fact, Shepard himself had never joined the Christian Scientist Church or even come close to it. In his secret heart he was probably stone atheist. At the original press conference he had rather adroitly finessed the point by saying that he belonged to no church but attended the Christian Science Church regularly. Somehow the impression was left that Shepard was a Christian Scientist who had done everything but sign on the dotted line. (The press, the ever-proper Gent, was happy enough to see it that way.) As long as he was at home, however, Shepard could have passed for a model Christian Scientist husband, had he chosen to. He did go to church with Louise regularly. He did not drink, smoke, swear, or let his lips—his eyes and lips were his most pronounced features—spread into a warm and winning fighter jock grin when a pretty girl came by.

No, he didn't flash that famous Smilin' Al Shepard look until he stepped out of his airplane Away from Home—and most especially at the Cape. Then Al looked like a different human being, as if he had removed his ice mask. He would come out of the airplane with his eyes dancing. A great goomba-goomba grin would take over his face. You halfway expected to see him start snapping his fingers, because everything about him seemed to be asking the question: "Where's the action?" If he then stepped into his Corvette—well, then, there you had it: the picture of the perfect Fighter Jock Away from Home.

But now, in this room at the Konakai Hotel, it was the Icy Commander who stared back at Glenn. Commander Al, the colonel's son, knew how to put on all the armor of military correctness, in the stern old-fashioned way. He informed Glenn that he was way out of line. He told him not to try to foist his view of morality on anybody else in the group. In the succeeding weeks the Glenn position and the Fighter Jock position began to form, with various hands adding their own amendments. As for the Fighter Jock position: The seven of them had volunteered to do a job and they were devoting long hours of training to prepare for it and were doing many things above and beyond the strict call of duty, such as the morale tours of the factories, and forgoing flight pay and vacations and any semblance of an orderly family life—and that therefore what one did with what little time he had to himself was his own business, so long as he used good sense.

Shepard had struck the tone of the by-the-book commander. He sounded as utterly convinced and correct in his own fashion as Glenn did in his. Commander Al was capable of a rather formal rhetoric in discussions of this sort, complete with a little litotes.

There was no reason why one should have an aversion to the company of women, so long as one's acquaintanceships did not impair one's performance in the program or reflect adversely upon it.

John Glenn, however, was buying none of that. He stared back at Smilin' Al of the Cape and the Icy Commander, both of them, with John Calvin's own eyes. As time went by, the Glenn position became: Look, whether we like it or not, we're public figures. Whether we deserve it or not, people look up to us. So we have a terrific responsibility. It's not enough not to get caught. It's not even enough to know to your own satisfaction that you've done nothing wrong. We've got to be like Caesar's wife. We've got to be above even the appearance of doing wrong.

It went on like that, with neither giving an inch. That line about "Caesar's wife" would not be forgotten. Everybody knew there was something to what the fellow was saying . . . Nevertheless . . . Could you believe it? Could you believe that the day would come when you would actually see a pilot, an equal among equals, give his comrades a little sermon about keeping their hands clean and their peckers stowed? Where did he get off setting himself above them this way, and what was his real game?

Glenn knew he was making no friends with this approach. Yet there were key moments in a military career when a man had to assume leadership. That was the essence of leadership caliber, and surely that fact would be appreciated—if not by the pilots themselves, then surely by . . . others who would hear about it. The competition for the first flight was not a popularity contest among the troops, after all. Bob Gilruth and his deputies in the Space Task Group would make the choice. Glenn had never been afraid to alienate his peers when he knew he was right; perhaps this, too, had always impressed his superiors—and he had *never* been left behind. His faith in what was right was part of his righteous stuff.

Glenn had one great ally among the other six, and that was Scott Carpenter. Carpenter looked up to him and backed him in the debate. Wally Schirra and Gordon Cooper tended to back Shepard, arguing that when you were on duty you should be a model of correctness, but that when you were off duty your personal life was your own lookout. Schirra was finding Glenn more and more irritating. Who the hell did he think he was? After a while, they barely spoke to each other unless the job forced them to.

Grissom and Slayton somewhat dourly sided with Glenn on this particular point. Since he was making such a federal case out of it, they would acknowledge the soundness of his logic. But this didn't mean they idolized him any more than Schirra or Shepard did. A basic division was building up in the group. It was the other five against the pious fair-haired boy and his sidekick, Carpenter. Some of them seemed to derive some satisfaction from lumping Carpenter with Glenn. What was Carpenter even doing here! They couldn't get over the fact that Scott and his wife, Rene, had flamboyant cushions on the floor of their living room and they actually sat there while Scott played the guitar and Rene sang. The fact that she had a trained voice made no difference. It was beatnik stuff. Not only that, Carpenter was a great pal of the doctors. He and Glenn were both like that. They went out of their way to cooperate with the Life Sciences people, too.

Glenn and Carpenter were even willing guinea pigs for the two psychiatrists who had just come on board, Sheldon Korchin from the University of California and George Ruff, who had been in charge of the psychiatric testing at Wright-Pat. Both men were likable enough as individuals, but this thing of submitting to psychiatric study seemed dispensable to some of the boys, particularly Schirra and Cooper. The two psychiatrists were continually having you urinate into bottles so that they could analyze your urine for corticosteroid levels, which supposedly were an index of stress. But Carpenter thought this was terrific stuff, too. He even *talked* to them about it!

What some of the other five found eccentric in Carpenter was what Glenn (and the doctors) found interesting. Scott was about the only one you could sit down with and talk about the broader and more philosophical sides of Project Mercury and space exploration. Scott was the only one with a touch of the poet about him, in the sense that the idea of going into space stirred his imagination. He would even go out at night and prop a telescope up on top of his car on a tripod and just stargaze and let himself drift into the most profound speculation of astronomy: *What is my place in the cosmos?*

Just try to imagine Grissom doing that! If Gus had a telescope, he might use the small end of it to try to whack a turkey joint out of the maw of the Disposall if the thing was stuck, but that would be the end of that. Gus and Deke were the duo at the other end of the spectrum. The main thing was to ride the bird up there into space and get the job done and get back, and let's hold the Mickey Mouse down to a minimum.

Shepard and Wally Schirra were paired somewhere in between. It was not that they were inseparable pals or even buddies, however. Shepard had no intimates, so far as anyone knew, and Schirra probably spent as much time wondering what made Shepard tick as the rest of them did. It was just that they came into the astronaut corps with similar backgrounds. There was no particular advantage to forming a clique in this seven-man corps, because only one man could win the competition, i.e., get the first flight, and it wasn't a voting situation, in any case. Nevertheless, if any such situation came up, Al and Wally would probably tend to side with Deke and Gus . . . As for Gordon Cooper, he seemed to be regarded as the odd man out. One got the impression that he was not in the running. As far as Gordo himself was concerned, however, he sided with Gus and Deke and Al and Wally on most of the pertinent questions, from the business of medical experiments to life after hours.

They were all beginning to realize that the stakes were tremendous. With the first flight into space, the holy first flight, one of them would become not only the pre-eminent *astronaut* . . . but also the True Brother at the top of the entire pyramid. The first American into space—who might very well be the first human being to go into space—would have an eminence that not even Chuck Yeager had ever enjoyed, because he would belong not just to the history of aviation but to world history.

And who would this one man be? Well . . . who else would it be but John Glenn! Glenn was just pouring it on. He was even assuming the role of natural group leader—giving them a little moral lecture at the séances!

The begrudging consensus that had developed from the Konakai séance was that, yeah, Glenn was right; they ought to watch themselves a little more carefully. But Al Shepard, for one, was not the type to let Glenn get away with it unmarked. Al kept putting the needle in. If there was anyone else around to enjoy it, Al would say to Glenn: "John, I think you need to loosen up a little bit, boy. What you need is a sports car. Why don't you get rid of that junk heap you're driving and do a little rat-racing. It'll do you good, John."

Al never missed an opportunity to stick it to Glenn about his terrible-looking underpowered Prinz and his need to get a car with a little more juice and loosen up. It became a refrain. Glenn knew how to roll with this kind of ribbing and grin through it. At the same time, you could tell it was getting under his skin. You couldn't help but get the feeling that the piece of equipment Al was really saying Glenn should loosen up and turn on the juice with was not an automobile.

One morning, when they came into the Astronaut Office, there was a big inscription up on the blackboard:

DEFINITION OF A SPORTS CAR: A HEDGE
AGAINST THE MALE MENOPAUSE.

VIII. The Thrones

In the eyes of the engineers assigned to project mercury the training of the astronauts would be the easy task on the list. Naturally you needed a man with the courage to ride on top of a rocket, and you were grateful that such men existed. Nevertheless, their training was not a very complicated business. The astronaut would have little to do in a Mercury flight except stand the strain, and the engineers had devised what psychologists referred to as "a graded series of exposures" to take care of that. No, the difficult, the challenging, the dramatic, the pioneering part of space flight, as the engineers saw it, was the technology.

It was only thanks to a recent invention, the high-speed electronic computer, that Project Mercury was feasible at all. There was an analogy here with the great Admiral of the Seas himself, Columbus. It was only thanks to a recent invention of his day, the magnetic compass, that Columbus had dared to sail across the Atlantic. Until then ships had stayed close to the great land masses for even the longest voyages. Likewise, putting a man into space the quick and dirty way without high-speed computers was unthinkable. Such computers had not been in production before 1951, and yet here it was, 1960 and engineers were already devising systems for guiding rockets into space, through the use of computers built into the engines and connected to accelerometers, for monitoring the temperature, pressure, oxygen supply, and other vital conditions of the Mercury capsule and for triggering safety procedures automatically—meaning they were creating, with computers, systems in which machines could communicate with one another, make decisions, take action, all with tremendous speed and accuracy . . .

Oh, genius-engineers!

Ah, yes, there was such a thing as self-esteem among engineers. It may not have been as grandiose as that of fighter jocks . . . nevertheless, many was the steaming enchephalitic summertime Saturday night at Langley when some NASA engineer would start knocking back that good sweet Virginia A.B.C. store bourbon on the patio and letting his ego out for a little romp, like a growling red dog.

The glorification of the astronauts had really gotten out of control! In the world of science—and Project Mercury was supposed to be a scientific enterprise—pure scientists ranked first and engineers ranked second and the test subjects of experiments ranked so low that one seldom thought about them. But here the test subjects . . . were national heroes! They created a zone of awe and reverence wherever they set foot! Everyone else, whether physicist, biologist, doctor, psychiatrist, or engineer, was a mere attendant.

At the outset it had been understood—it didn't even require comment—that the astronauts would be just that: test subjects in an experiment. Mercury was an adaptation of the Air Force's Man in Space Soonest concept, in which you would attach biosensors to your human subject, seal him up in a capsule, propel him into space ballistically—i.e., like an artillery shell—and bring him back to earth with completely automatic guidance and see how he made out. In November 1959, six months after the seven astronauts were chosen, Randy Lovelace and Scott Crossfield presented a paper at an aerospace medical symposium in which they said that biomedical research was "the sole purpose of the ride," so far as having an astronaut on board was concerned. They added that an aerodynamic space vehicle, such as the proposed X–15B or X–20, would require "a much more highly trained pilot." Since he was involved in the X–15 project, Crossfield had his own ax to grind, but what he and Lovelace were saying was perfectly obvious to any engineer who knew the difference between ballistic and aerodynamic space vehicles. In short, the astronaut in Project Mercury would not be a pilot under any conventional definition.

Even as late as the summer of 1960, at an Armed Forces–National

Research Council conference at Woods Hole, Massachusetts, on "the training of astronauts," various engineers and scientists from outside NASA thought nothing of describing the Mercury rocket-capsule vehicle as a fully automated system in which "the astronaut does not need to turn a hand." They would say, "The astronaut has been added to the system as a redundant component." (A *redundant component!)* If the automatic system broke down, he might step in as a repairman or manual conductor. Above all, of course, he would be wired with biosensors and a microphone to see how a human being responded to the stress of the flight. That would be his main function. There were psychologists who advised against using pilots at all—and this was more than a year after the famous Mercury Seven had been chosen. The pilot's, particularly the hot pilot's, main psychological bulwark under stress was his knowledge that he controlled the ship and could always *do something* ("I've tried A! I've tried B! I've tried C!" . . .). This obsession with active control, it was argued, would only tend to cause problems on Mercury flights. What was required was a man whose main talent was for *doing nothing* under stress. Some suggested using a new breed of military flier, the radar man, the Air Force Strategic Air Command radar observer or the Navy radar intercept officer, a man who had experience riding in the rear in high-performance aircraft under combat conditions and doing nothing but reading the radar, come what may, abandoning all control of the craft (and protection of his own life) to someone else, the pilot ("I looked over at Robinson—and he was staring at the radar like a *zombie!"*). An experienced zombie would do fine. In fact, considerable attention had been given to a plan to anesthetize or tranquillize the astronauts, not to keep them from panicking, but just to make sure they would lie there peacefully with their sensors on and not *do something* that would ruin the flight.

The scientists and engineers took it for granted that the training of the astronauts would be unlike anything ordinarily thought of as flight training. Flight training consisted of teaching a man how to take certain actions. He was taught how to control an unfamiliar craft or how to put a familiar craft through unfamiliar maneuvers, such as bombing runs or carrier landings. On the other hand, the only actions the astronauts would have to learn how to take would be to initiate the emergency procedures in the case of a bad rocket launch or a bad landing and to step in as a backup (redundant component) if the automatic control system failed to hold the heatshield in the correct position prior to re-entry through the earth's atmosphere. The astronaut would not be able to control the path or the speed of the capsule at all. A considerable part of his training would be what was known as de-conditioning, de-sensitizing, or adapting out fears. There was a principle in psychology that maintained that "bad habits, including overstrong emotionality, can be eliminated by a graded series of exposures to the anxiety-arousing stimulus." That was what much of astronaut training was to be. The rocket launch was regarded as a novel and possibly disorienting event, in part *because* the astronaut would have no control over it whatsoever. So they had devised "a graded series of exposures." They took the seven men to the Navy's human centrifuge facility in Johnsville, Pennsylvania. The centrifuge looked like a Wild Bolo ride; it had a fifty-foot arm with a cockpit, or gondola, on the end of it, and the arm could be whirled around at astonishing speeds, great enough to put up to 40 g's of pressure on the rider inside the gondola, one g being equal to the force of gravity. The high g-forces generated by combat aircraft in dives and turns during the Second World War had sometimes caused blackouts, red-outs, gray-outs, or made it impossible for pilots to lift their hands to the controls; the giant centrifuge at Johnsville had been built to explore this new problem of high-speed flight. By 1959 the machine had been computerized and turned into a simulator capable of duplicating the g-forces and accelerations of any form of flight, even rocket flight. The astronaut was helped into his full pressure suit, with his biosensors attached and his rectal thermometer inserted, and then placed into the gondola, in a contoured seat molded for his body, whereupon all the wires, hoses, and microphones he would have in actual flight were hooked up, and the gondola was depressurized to five pounds per square inch, as it would be in space flight. The interior of the gondola had been converted into a replica of the Mercury capsule's interior, with all the switches and console displays. The taped noise of an actual Redstone rocket firing was played over the astronaut's headset, and the ride began. Using the computers, the engineers would put the man through an entire Mercury flight profile. The centrifuge built up the g-forces at precisely the same rate they would build up in flight, up to six or seven g's, whereupon the g-forces would suddenly drop off, as they would in flight as the capsule went over the top of its arc, and the astronaut experienced a tumbling sensation, as he would, presumably, in flight. All the while the astronaut would be required to push a few switches, as he would in actual flight, and talk to a mock flight controller, forcing his words out into the microphone, no matter how great the pressure of the g-forces on his chest. The centrifuge could also duplicate the pressures of deceleration a man would experience during the return through the earth's atmosphere.

To get the seven men used to weightlessness, they took them on parabolic rides in the cargo holds of C–131 transports and backseat in F–100Fs. When the jet came up over the top of the arc of the parabola, the subject would experience from fifteen to forty-five seconds of weightlessness. This was the only flying scheduled for the astronauts' entire training program; and they were, of course, merely passengers on board, as they would be in the Mercury flights.

The only way the astronaut would be able to move the capsule in the slightest would be to fire hydrogen-peroxide thrusters during the interval of weightlessness, tipping or swinging the capsule this way or that, in order to get a particular view out the portholes, for example. NASA built a machine, the ALFA trainer, to accustom each trainee to the sensation. He sat in a seat resting on air bearings and used a hand controller to make it pitch up and down or yaw back and forth. On a screen in front of him, where the capsule's periscope screen would be, aerial photographs and films of the Cape, the Atlantic Ocean, Cuba, Grand Bahama Island, Abaco, all the landmarks, rolled by . . . and veered off as the astronaut pitched or yawed, just as they would in actual flight. The ALFA even made a whooshing sound like that of hydrogen-peroxide thrusters when the astronaut pushed the stick.

By mid-1960 the engineers had developed the "procedures trainer," which was in fact a simulator. There were identical procedures trainers at the Cape and at Langley. At the Cape the trainer was in Hangar S. It was there that the astronaut spent his long day's training. He climbed into a cubicle and sat in a seat that was aimed straight up at the roof. The back of the seat was flat on the floor of the cubicle, so that the astronaut rested on his back. He looked up at a replica of a console that would be used in the Mercury capsule. It was as if he were on top of the rocket, with his face aimed at the sky. The console was wired to a bank of computers. About twenty feet behind the astronaut, on the floor of Hangar S, sat a technician at another console, feeding simulated problems into the system.

The technician would start off saying, "Count is at T minus fifty seconds and counting."

From inside the trainer, over his microphone, the astronaut would answer: "Roger."

"Check your periscope—fully retracted?"

"Periscope retracted."

"Ready switch on?"

"Ready switch on."

"T minus ten seconds. Minus eight . . . seven . . . six . . . five . . . four . . . three . . . two . . . one . . . Fire!"

Inside the trainer the dials in front of the astronaut would start indicating that he was on his way, and he was supposed to start reading the gauges and reporting to the ground. He would say, "Clock is operating . . . okay, twenty seconds . . . one thousand feet [altitude] . . . one-point-five g's . . . Trajectory is good . . . Twelve thousand feet, one-point-nine g's . . . Inner cabin pressure is five p.s.i. . . . Altitude forty-four thousand, g-level two-point-seven . . . one hundred thousand feet at two minutes and five seconds . . ." The instructor might pick this point to hit a button on his console marked "oxygen." A red warning light marked O_2EMERG would light up, and the astronaut would say: "Cabin pressure decreasing! . . . Oxygen is apparently leaking! . . . It's still leaking . . . Switching to emergency reserve . . ." The astronaut could throw a switch that brought more oxygen into the simulator system—i.e., into its computer calculations—but the instructor could hit his "oxygen" button again, and that meant that the leak was continuing, and the astronaut would say: "Still leaking . . . It's approaching zero-flow rate . . . Abort because of oxygen leak! Abort! Abort!" Then the astronaut would hit a button, and a button marked MAYDAY would light up red on the instructor's console. In actual flight the escape tower was supposed to fire at this point, pulling the capsule free of the rocket and bringing it down by parachute.

The astronauts spent so much time hitting the abort handle in the procedures trainer that it got to the point where it seemed as if they were training for an abort rather than for a launch. There was very little action that an astronaut could take in a Mercury capsule, other than to abort the flight and save his own life. So he was not being trained to *fly* the capsule. He was being trained to ride in it. In a "graded series of exposures" he was being introduced to all sights, sounds, and sensations he might conceivably experience. Then he was reintroduced to them, day after day, until the Mercury capsule and all its hums, g-forces, window views, panel displays, lights, buttons, switches, and peroxide squirts became as familiar, as routine, as workaday as an office. All flight training had a certain amount of desensitizing built into it. When a Navy pilot practiced carrier landings on the outline of a flight deck painted on an airfield, it was hoped that the maneuver might also desensitize his normal fear of landing a hurtling machine in such a small space. Nevertheless, he was there chiefly in order to learn to land the machine. Not until Project Mercury had there been a flight training program so long and detailed, so sophisticated, and yet so heavily devoted to desensitizing the trainee, to adapting out man's ordinary fears, and enabling one to think and use his hands normally in a novel environment.

Oh, all of this had been well known at the outset! . . . so much so that the original NASA selection committee had been afraid that the military test pilots they were interviewing would regard the job as boring or distasteful. Since they figured they needed six astronauts for Mercury, they had considered training twelve—on the assumption that half of them would resign once they fully understood how passive their role would be. And now, in 1960, they began to realize that they had been correct; or halfway, in any case. The boys were, indeed, finding the role of biomedical passenger in an automated pod, i.e., the role of human guinea pig, distasteful. That much had proved to be true. The boys' response, however, had not been resignation or anything close to it. No, the engineers now looked on, eyebrows arched, as the guinea pigs set about . . . *altering the experiment.*

The difference between pilot and passenger in any flying craft came down to one point: control. The boys were able to present some practical, workmanlike arguments on this score. Even if an astronaut were to be a redundant component, an observer and repairman, he should be able to *override* any of the Mercury vehicle's automatic systems *manually*, if only to correct malfunctions. So went the argument. But there was another argument that could not be put into so many words, since one was forbidden to state the premise itself: the right stuff.

After all, the right stuff was not bravery in the simple sense of being willing to risk your life (by riding on top of a Redstone or Atlas rocket). Any fool could do that (and many fools would no doubt volunteer, given the opportunity), just as any fool could throw his life away in the process. No, the idea (as all *pilots* understood) was that a man should have the ability to go up in a hurtling piece of machinery and put his hide on the line and have the moxie, the reflexes, the experience, the coolness, to pull it back at the last yawning moment—but how in the name of God could you either hang it out or haul it back if you were a lab animal sealed in a pod?

Every signal they received told the boys that the true brethren at Edwards looked upon them as glorified "klutzes," to use Wally Schirra's phrase. Schirra knew the Edwards outlook in such matters well enough. He had done some major testing of the F–4H at Edwards for the Navy in 1956. But it was Deke Slayton who felt the condescension of the brethren most of all. He had come into Project Mercury straight from Fighter Ops at Edwards, and his pals there kidded him unmercifully. "A monkey's gonna make the first flight." That was the typical refrain. When the boys went to Edwards for their briefings on the X–15 program and their weightless parabolas—riding backseat with Edwards pilots—they picked up a whiff of . . . contempt . . . It hadn't helped any that Scott Carpenter and a couple of the others had taken over the controls on the F–100Fs and tried to fly the weightless parabolas from the front seat . . . and had failed. They hadn't been able to fly the correct profile and produce the weightless interval. Of course, with a little practice they could have no doubt mastered it . . . Nevertheless! . . . Rightly or wrongly, some of the boys felt that rocket pilots like Crossfield were high-hatting them. And what about the Society of Experimental Test Pilots? The SETP was the main organization within the fraternity. Several of the boys didn't even qualify for membership. The SETP required that a member have at least twelve months' experience in the first flights of new aircraft, probing the outer limits of the envelope. The SETP was not about to accept astronauts until they had done a hell of a lot more than volunteer for Mercury and sign a contract with *Life*. On the upper elevations of the pyramid the brave lads—they could sense it—were viewed as seven green rookies; and all the while there was the infuriating question: "Are astronauts even *pilots?*"

Deke Slayton, who was a member of the Society of Experimental Test Pilots, had been invited to address the annual conference in Los Angeles in September 1959 on that very subject: the role of the astronaut in Project Mercury. The meeting happened to come just two weeks after *Life* had started its sunburst of stories categorizing the seven astronauts as the best and bravest pilots in American history. No reader of *Life* would have recognized the Deke Slayton who went to the podium in a hotel convention hall to speak to the brotherhood. From the start his tone was defensive. He said he had some "stubborn, frank" comments on the role of *the pilot* in Project Mercury. There were people in the military, he said, who wondered "whether a college-trained chimpanzee or the village idiot might not do as well in space as an experienced test pilot." *(A monkey's gonna make the first flight!)* He knew there was that kind of talk going around, and it annoyed him. These people were confusing Mercury "with the Air Force Man in Space Soonest or Army Adam programs, which were essentially man-in-a-barrel approaches." His audience looked at him blankly, since such had been precisely the origin of the Mercury program. "I hate to hear anyone contend that present-day pilots have no place in the space age and that non-pilots can perform the space mission effectively," he said. "If this were true, the aircraft driver could count himself among the dinosaurs not too many years hence." That was hardly likely, he went on. A non-pilot might be able to do part of the job. But in those critical moments when it was necessary to keep your head and make observations and record data while cantilevered out over the bottomless Gulp . . . who else could cope with it but someone made of the stuff of the professional test pilot?

Slayton possessed a forcefulness that people often failed to detect at first. His remarks may not have convinced many skeptics within the

Society of Experimental Test Pilots. Nevertheless, they became, in effect, the keynote address of the campaign that began inside NASA.

By now, September 1959, Slayton and the rest of them realized that, as Glenn had first divined, the astronaut corps was like a new branch of the service and that in this new branch no one outranked them. Certainly Robert Voas didn't outrank them. Voas was a Navy lieutenant who had been designated as the astronauts' training officer. Voas was neither a flight instructor nor an aeronautical engineer but an industrial psychologist who had been chosen precisely because the training of astronauts was regarded not as a form of pilot training but as a form of psychological adaptation. Voas was no older than they and ranked below them even in the regular military; so one of the boys' first moves was to see that Voas, as training officer, functioned more like a trainer on a sports team and, in any case, not like the coach. *They* began telling *him* what their training schedule was going to be. Voas became a coordinator and spokesman for the astronauts in matters of training.

Gordon Cooper had been frowned upon a few months before when he complained about the lack of supersonic fighter planes for "proficiency" flying, but now the boys took up his complaint within the corridors of NASA, with Slayton and Schirra leading the way and Voas arguing their case for them. Soon they had two F–102s on loan from the Air Force. The ships were somewhat the worse for wear, however—absolute junkers, in fact, in the eyes of the seven pilots. The Air Force had sloughed these wrecks off on them like hand-me-downs. The poor condition of the F–102s wasn't the worst of it, however. The galling thing was that the F–102, which had been one of the first in the Century series, was by now a back number. It would go supersonic but just barely, Mach 1.25 being about top speed. Wally Schirra knew how to formulate the argument on this score. Wally was not merely an expert prankster; he could also turn stern and bang the table and conjure up the aura of the right stuff and its privileges and prerequisites without once uttering the unspoken things. Wally would say to the brass: You're presenting us to the American people as the seven best test pilots in America, and we are *among* the best, all the p.r. aside, and yet you're not even giving us the opportunity to keep up our proficiency! Before I joined this program I was flying fighter aircraft capable of Mach 2 or better. And now we're supposed to keep up our proficiency with a couple of old clunkers that will hardly go Mach 1 even when they're in half-decent shape. It doesn't make any sense! It's as if you decided to prepare a major-league ball club for World Series competition by having them take a year off to play against a bunch of old crocks in a Parks & Recreation league in south Jersey. Wally was terrific in moments like this; and by and by, the boys would get a couple of F–106s, which were second-generation F–102s and capable of Mach 2.3. In the meantime, they tried to make do with the F–102s. But, hell, even flying F–102s was a big step beyond the original training agenda—which assumed that proficiency flights of whatever sort would be of no use for the astronaut in Project Mercury. Nor had this assumption yet died, Wally and Deke or no Wally and Deke.

At the Woods Hole conference Voas described the advantages of the F–102 flights in sustaining the astronauts' "decision-making abilities," and an aviation psychologist from the University of Illinois, Jack A. Adams, could scarcely believe what he was hearing.

"Frankly," he said, "I cannot see how decision making, or any other type of response, for that matter, in the F–102 can transfer significantly to the comparatively unique responding required of the astronaut in the Mercury vehicle." Then he added: "The astronaut's task is actually more like a radar observer's job than a pilot's." Another aviation psychologist, Judson Brown of the University of Florida, was just as baffled: "It has been frequently mentioned that skilled pilots must be used for Mercury, for the X–15, and the Dyna-Soar projects. Clearly, the use of skilled pilots seems to be of much less importance for Mercury than for the other two. There is a serious question whether positive transfer will occur from pilot training to Mercury capsule operation."

Inside NASA, however, this position was no longer tenable. From a sheerly political or public relations standpoint, *the astronaut* was NASA's prize possession, and the seven Mercury astronauts had been presented to the public and the Congress as great pilots, not as test subjects. If they now insisted on *being* pilots, great or otherwise—who was going to step in and say no? The boys sensed this; or as Wally Schirra put it, they realized they had "a fair amount of prestige around the country." So next they began whittling down the number of medical and scientific experiments they were expected to take part in—the guinea-pig stuff—simply by characterizing them as useless or stupid and cutting them out of their schedules. Here they tended to have the support of Gilruth's chief of operations, Walt Williams. Williams was a big hearty powerful-looking engineer who had been one of the true geniuses of the X series at Edwards, the man who had turned supersonic flight test into a precise and rational science. Williams was a flight-line engineer; he didn't have much patience with matters of flight test that were not *operational.* The one engineer who didn't mind letting it be known that Astropower, as it came to be called, was getting out of hand was one of Williams's lieutenants, Christopher Columbus Kraft, Jr. Chris Kraft was a hard-driving young man, thirty-six years old, urbane and sharp-witted, as aeronautical engineers went, and he was scheduled to be flight director for Mercury; but he didn't yet have the clout to do much of anything where the astronauts were concerned. The seven men pressed on. They were tired of the designation of "capsule" for the Mercury vehicle. The term as much as declared that the man inside was not a pilot but an experimental animal in a pod. Gradually, everybody began trying to work the term "spacecraft" into NASA publications and syllabuses. Next the men raised the question of a cockpit window for the spacecraft. As it was now designed, the Mercury capsule had no window, just a small porthole on either side of the astronaut's head. His main way of seeing the outside world would be through a periscope. A window had been regarded as an unnecessary way of inviting rupture due to changes in pressure. Now the astronauts insisted on a window. So the engineers went to work designing a window. Next the men insisted on a hatch they could open by themselves. The hatch, as currently designed, would be bolted shut by the ground crew. In order to leave the capsule after splashing down, the astronaut would either have to slither out through the neck, as if he were coming out of a bottle, or wait for another crew to unbolt the hatch from the outside. So the engineers went to work designing a hatch with explosive bolts so that the astronaut could blow it off by hitting a detonator. It was too late to incorporate the new items into the capsule—the *spacecraft*—that would be used for the first Mercury flight. That vehicle was already badly behind schedule as it was. But they would be in each craft thereafter . . .

And why? Because *pilots* had windows in their cockpits and hatches they could open on their own. That was what it was all about: being a *pilot* as opposed to a guinea pig. The men hadn't stopped with the window and the hatch, either. Not for a moment. Now they wanted . . . *manual control of the rocket.* They weren't kidding! This was to take the form of an override system: if the astronaut believed, in his judgment, as captain of the ship (not *capsule),* that the booster rocket engine was malfunctioning, he could take over and guide it himself—like any proper pilot.

How could they be serious!—the engineers would say. Any chance of a man being able to guide a rocket from inside a ballistic vehicle, a projectile, was so remote as to be laughable. This proposal was so radical the engineers knew they would be able to block it. It was no laughing matter to the seven pilots, however. They also wanted complete control of the re-entry procedure. They wanted to establish the capsule's angle of attack manually and fire the retro-rockets themselves without any help from the automatic control system. This suggestion made the engineers wince. Slayton even wanted to redesign the hand controller that would activate the hydrogen peroxide thrusters to make the capsule pitch, roll, or yaw. He wanted the hand controller to operate the pitch and roll thrusters only; yaw would be controlled by pedals which the astronaut would operate with his feet. That was the conventional setup on aircraft: a two-axis stick plus pedals. That was the way *pilots* established attitude control.

Life magazine and the worshipful public and the worshipful politicians and all the others who had already exalted the seven astronauts did not care in the slightest whether they functioned like pilots or not. It was enough that they were willing to climb atop the rocket at all, in the name of the battle with the Soviets for the high ground, and be exploded into space or to the harp farm. It was not enough for the men themselves, however. All of them were veteran military pilots, and five of them had already reached the higher elevations of the invisible ziggurat when Project Mercury began, and they were determined to go into space as pilots and as nothing else.

Control—in the form of overrides at the very least—was the one thing that would neutralize the recurring taunt within the fraternity: *A monkey's gonna make the first flight.* In his speech before the brotherhood Slayton had brought that out into the open with his crack about the "college-trained chimpanzee." That had seemed like a Slayton sortie into sarcasm and hyperbole. He made no reference to the fact that such a college actually existed.

But in the deserts of New Mexico, about eighty miles north of El Paso and the Mexican Border, at Holloman Air Force Base, which was part of the White Sands missile-range complex, NASA had set up a Project Mercury chimpanzee colony. There was nothing secret about it, but it attracted little notice. The chimpanzee program had been devised mainly to satisfy "the medical Cassandras." From the moment a Joint Armed Forces–National Research Council Committee on Bioastronautics had visited the new NASA facility at Langley in January 1959, there had been doctors warning that the weightless state or high g-forces, or both, could be devastating and that animal flights should be mandatory. So NASA had twenty veterinarians training forty chimpanzees in a compound at the Aeromedical Research Laboratory at Holloman. Eventually one of the beasts would be chosen for what amounted to a dress rehearsal of the first manned flight. The idea would be not only to see if the chimpanzee could stand the strain but also to see if he could use his brain and his hands normally throughout the ride.

Chimpanzees were chosen as much for their intelligence as their physiological similarity to humans. They could be trained to do fairly complex manual tasks, on cue, particularly if one got hold of them when they were young. Once they were trained to do the tasks on the ground, it would be possible to give them the cues to do the same tasks during a space flight and see if the weightless condition impeded them. Early in the game the vets decided that rewards, mere positive reinforcement, would not be sufficient for the job at hand. The only sure-fire training technique was *operant conditioning*. The principle here was the avoidance of pain. Or, to put it another way, if the ape didn't do the job right, he was punished with electric shocks in the soles of his feet.

The Holloman veterinarians, like most veterinarians, were compassionate men who were interested in relieving pain in animals and not in inflicting it upon them. But this was war! The chimpanzee program was an essential part of the battle for the high ground! It was no time for halfway measures! As congressmen told you every day, national survival was at stake! The veterinarians' brief was to get the job done with the utmost speed and efficacy. There were several ways of training animals, but only *operant conditioning*, based on concepts developed by B. F. Skinner, seemed anywhere near foolproof. In any case, the "psychomotor stimulus plates" were attached to the beasts' feet and they were strapped into chairs, and the process began . . . And when the apes did well, you gave them hugs and nuzzles, to be sure, first taking the precaution of making certain they weren't in a mood to bite your goddamned nose off.

Oh, the apes knew a thing or two! Their intelligence was only just below that of man. They had memories; they could figure out the situation. At an early age they had been seized in West Africa and separated from their mothers by this new species, the humans, and removed from all familiar surroundings and put in cages and shipped to this godforsaken alien landscape, the New Mexico desert, where they remained in cages . . . when they weren't in the hands of a bunch of human ballbreakers in white smocks who strapped them down and zapped them and put them through insane exercises and routines. The beasts tried everything they could think of to escape. They snapped, snarled, spit, bit, thrashed at the straps, and made runs for it. Or they bided their time and used their heads. They would go along with a training task, seeming to cooperate, until the white smocker seemed to let his guard down—then they'd make a break for it. But the resistance and the wiles were of no avail. All they got for their struggles was more zaps and blue bolts. Some of the brightest apes were also the most intractable and resourceful; they would take the electric shocks and then seem to give up and submit to fate—and *then* try to tear the white smockers a new asshole or two and make a run for it. With these implacable little bastards it was sometimes necessary to give them a tap or two with a rubber hose, or whatever.

Then, at last, the sophisticated part of the training could begin. It took two main forms: the desensitizing or adapting out of the fears that a rocket flight would ordinarily hold for the animal (just confining an untrained chimpanzee in a Mercury capsule would have driven him berserk with fear); and placing the animal in a procedures trainer, a replica of the capsule he would be inside in flight, and teaching him to respond to lights and buzzers and throw the proper switches on cue—and having him do this day after day until it became a thoroughly familiar environment, as familiar, routine, and workaday as an office.

The vets took the chimpanzees by airplane to Wright-Patterson for rides on the centrifuge the Air Force had there. They would strap each ape into the gondola, close the hatch, and pipe in the sound of a Redstone rocket launch and start him spinning, gradually introducing him to higher g-forces. They took them for parabolic rides backseat in fighter planes to familiarize them with the feeling of weightlessness. They put them in the simulator for endless hours and endless days of on-cue manual-task training. Since the chimpanzee would not be wearing a pressure suit in flight, he was put inside a pressurized cubicle, which in turn would be placed in the Mercury capsule. The monkey's instrument panel was inside the inner cubicle. Therein, day by day, month by month, the monkey learned to operate certain switches in different sequences when cued by flashing lights. If he did the job incorrectly, he received an electric shock. If he did it correctly, he received banana-flavored pellets, plus some attaboys and nuzzles from the vets. Gradually the beasts were worn down. They were tractable now. The *operant conditioning* was taking place. A life of avoiding the blue bolts and gratefully accepting the attaboys and pellets had become the better part of valor. Rebellion had proved to be a dead end.

The apes had begun their training at the same time as the astronauts, i.e., in the late spring of 1959. By now, 1960, they had been through almost every phase of astronaut training, other than abort and re-entry emergency sequences and attitude control.

Some of the apes could operate their procedures trainer like a breeze, almost as rapidly as a man. The vets had every reason to be proud of what they had accomplished. On the outside the animals were as mild, tractable, smart, and lovable as the best little boy on the block, although inside . . . something was building up like Code Blue in the boiler room.

About eight hundred miles west of Holloman Air Force Base, in the same latitude of the great American desert, was Edwards. The X–15 program had begun to pick up some momentum. There were even journalists coming to Edwards—in the midst of this, the Era of the Astronaut—and talking about the X–15 as "America's first spaceship." There were two men on hand at the base writing books about the project; one of them was Richard Tregaskis, who had written the best seller *Guadalcanal Diary*. The X–15, *America's first spaceship* . . . could it be? A year ago it would have seemed impossible. But now the Mercury program was beginning to lag. NASA had talked of making the first manned flight in mid-1960; well, it was now mid-1960, and they didn't even have the capsule ready for unmanned testing.

NASA's prime pilot for the X–15 project was Joe Walker. He looked like a young towheaded version of Chuck Yeager, the country boy who loved to fly. He talked like Yeager. Well, hell, who didn't around here? But with Walker it came naturally. Just as Yeager was from the coal country of West Virginia, Walker was from the coal country of Pennsylvania, and Walker liked to do that Yeager thing where you mixed up a lot of up-hollow talk—"The mother liked to blowed up on me"—with postwar engineerese about parameters, inputs, and extrapolations. As a matter of fact, Yeager had let it be known that he thought Walker was the pick of the litter at Edwards now.

Yes, Walker looked and sounded like a younger version of Yeager—but in fact he was two years older. Yeager was still only thirty-seven, and Walker was thirty-nine. Walker was seven months older than Scott Crossfield. So aside from everything else, Walker didn't have time to cool his heels. If the X–15 and X–20 programs at Edwards got stalled while all the money and attention went to Project Mercury, it would be bad news.

Edwards had grown until it was about twenty times the size it was during the heyday of Yeager. Pancho Barnes's Happy Bottom Riding Club was long gone. The Air Force had taken her property by eminent domain for the building of a new runway. There had been a bitter fight in court, during which a base commander had accused Pancho of running a whorehouse, and Pancho had told the court that she had it on good authority that the old peckerwood had instructed his pilots to accidentally-on-purpose napalm her ranch. Pancho had gone into retirement, with her fourth husband, her erstwhile ranch foreman, over in the town of Boron, to the northeast of the base.

There were now about three thousand Air Force personnel at Edwards and about seven thousand civilians, some with NASA, including Walker himself. Yet the high desert was so vast and so open that it swallowed up all ten thousand of them with no trouble at all, and the place didn't look terribly different except during the afternoon traffic jam, when all the civil servants got off work and sped toward the air conditioners that awaited them in their tract homes. Walker and his wife and two children lived in Lancaster, a desert town about a half hour's drive west of Edwards. Walker had built a house in a tract that some inspired developer—inspiration was the choicest item in the real-estate boom of the period—had named White Fence Farms. You had to build a white fence around your house in order to live there. That he did. As for the Farm part—here you had yourself a problem, unless you farmed Joshua trees. The developer's idea, in his sales pitch, was that you could build chicken coops at the rear end of your lot and have a second income.

At that, Walker's place looked like a little bit of heaven compared to Bob White's. But then, on the surface, Walker and White were different in every respect. White, who was a major, was the Air Force's prime pilot for the X–15 project. He was the eternally correct and reserved Air Force blue-suiter. He didn't drink. He exercised like a college athlete in training. He was religious. He was an usher in the Roman Catholic chapel of the base and never, but never, missed Mass. He was slender, black-haired, handsome, intelligent—even cultivated, if the truth were known. And he was terribly serious. He was not a beer-call fighter jock. Not many people picked out Bob White to just shoot the breeze with. White and his family lived on the base itself at 116 Thirteenth Street in a miserable grid of military housing plots known as the Wherry housing section. Or it had been known as Wherry at the outset. By 1960 it was usually referred to as Weary housing. Children grew up there thinking that Weary housing was the real name. Parked out front of White's place was an unpainted Model A Ford. The Air Force, being the newest branch of the service, was strong on instant tradition. This old junker, the Ford, was bestowed, as an ironic sculpture of the Right Stuff, upon whomever was the number-one Air Force test pilot at Edwards. Scott Crossfield, the prime pilot for the manufacturer, North American, had completed the first phase of testing the X–15, checking out the power system and basic aerodynamics. White and Walker had been chosen to push the rocket plane to its outer limits, which were envisioned as speeds in excess of Mach 6, or about 4,000 miles per hour, and, more important, an altitude of 280,000 feet. Just where "space" began was a matter of definition that had never been fully resolved. But fifty miles up was generally accepted as the boundary line. There was very little atmosphere left at that altitude; in fact, once a ship reached 100,000 feet, there was not enough air remaining to provide aerodynamics. The X–15's target of 280,000 feet was 53 miles up.

White and Walker had begun to fly the X–15 with the so-called Little Engine. This was, in fact, two X–I engines built into a single fuselage. They provided 16,000 pounds of thrust. The X–15 was the most evil-looking beast ever put into the air. It was a 7.5-ton black chimney with little fins on it and an enormous blocky tail. The black paint had been created to withstand the heat generated by friction when the ship went up above 100,000 feet and re-entered the denser atmosphere below. Everyone was waiting for the delivery of the Big Engine, the XLR–99. This was a rocket with 57,000 pounds of thrust, or four times the base weight of the ship. Once the XLR–99 was installed . . . well, Walker just might become the first man to cross the boundary into space. The engine's 57,000 pounds of thrust were only 21,000 pounds less than that of the Redstone rocket, which—eventually—was supposed to take the astronauts on their first flights. As a matter of fact, it was the development of the Redstone as a missile that had first given NASA engineers like Walt Williams the idea for the X–15, back in the early 1950's.

How, then, could there be so much excitement over Project Mercury and so little over the X–15? Here was the thing that got to the boys after a while, no matter how nonchalant they tried to appear: the Mercury astronauts were national heroes without ever having left the ground—all because they had volunteered to ride on top of rockets. Well . . . Walker and White and Crossfield, like Yeager before them, had *already* ridden rockets, from the X–I to the X–15. And they had ridden them as *pilots*. Your own brain was the guidance system for the X–15, and your own hand maneuvered the ship. In the Mercury-Redstone system, a bank of computers was the pilot, and the astronaut was a passenger. Why couldn't everyone comprehend such a simple fact? Was it because the astronauts were seen as America's front runners in the race with the Russians? Well, if so, that was pretty ironic. By now, mid-1960, the astronauts were supposed to have gone up in their first ballistic flights. That was the whole point of choosing the Mercury system. It was dirty—but it was quick; supposedly. But the Mercury capsule wasn't even ready yet. There had been one delay after another. It was beginning to look unlikely that there would be a manned launch before 1961. The X–15 project was now actually *ahead* of Project Mercury in the attempt to reach space.

On May 7 Walker had cut loose the X–15 on its first real speed run with the Little Engine and reached Mach 3.19 or 2,111 miles an hour, just a shade faster than Mel Apt's world record of 2,094 miles an hour in the X–2. On May 19 Bob White took the X–15 on its first bid for maximum altitude with the Little Engine and reached 109,000 feet, which was 17,000 feet under Iven Kincheloe's record in the X–2. And that was another point that everybody should have known about . . . and didn't. Kinch and Mel were now dead. Mel Apt died just a few minutes after he set his world speed record, the victim of a demon that was waiting especially for rocket ships reaching speeds of Mach 2 or more in the thin air up around 70,000 feet: instability in the yaw or roll axis . . . followed by an uncontrollable tumble. Sometimes it took the form of "inertia coupling," which usually occurred when a pilot tried to bank a rocket ship and it snapped into a full roll and then began pitching and yawing—*and* rolling violently. This would throw it end over end. Some pilots felt that the formal term "inertia coupling" added damned little to your understanding of the phenomenon. The ship simply "uncorked" (as Crossfield liked to put it) and lost all semblance of aerodynamics and fell out of the sky like a bottle or a length of pipe. There was no way to maneuver out of a rocket-plane tumble. The pilot took a furious beating from the g-forces and from being thrown about the cockpit. The more he experimented with the controls, the worse fix he was in. Yeager had been the first rocket pilot to go through this particular hole in the supersonic envelope, and it was during the flight in the X–1A in which he set a speed record of Mach 2.42. He was battered unconscious and fell seven miles before hitting the denser atmosphere at 25,000 feet and coming to and managing to put the ship into a spin. That was good; a mere spin he knew how to get out of, and he survived. Kinch went into a tumble during his record flight and came out of it at low altitude, as Yeager had done. That was just twenty days before Mel Apt augered in. Mel went into the wild tumble and tried to eject, but wasn't able to complete the sequence in the X–2. Yeager had always figured it was useless to try to punch out of a rocket plane. Crossfield called it "committing suicide to keep from getting killed." Inertia coupling nearly killed Kit Murray in 1954, when he set an altitude record of 94,000 feet in the X–1A, and it had hit Joe Walker twice, once in the XF–102 and again in the X–3.

When he talked about it, Joe Walker would say he got out of it each time through "the J.C. maneuver." He'd say: "In the J.C. maneuver you take your hands off the controls and put the mother in the lap of a super-na-tu-ral power." And, in fact, that was the only choice you had.

The way Walker talked about it, with his big mountain-boy grin on, it was . . . just like talking about sports . . . But every prospective X-15 pilot had seen the on-board film from Mel Apt's flight, and it was not a droll experience to watch that film. The camera had been mounted just behind Apt in the cockpit. It was a stop-frame camera that took one picture per second. In one frame Apt and his white helmet would be upright in the cockpit. In the next you would see his head, body, and helmet keeled over, crashing into the wall of the cockpit. In the first you saw a mountain ridge framed in the cockpit window, as if he were headed down in a dive, and in the next you saw empty sky: he was going end over end like an extra-point kick. The film seemed to go on forever. It was eerie looking at it, because you knew that at the end that little figure bouncing around in the white helmet would be dead.

Life magazine was writing about how Deke Slayton had once been in an inverted spin in an F–105. No picnic, to be sure, and yet the rocket pilots looked at inverted spins as their friends on the way out of supersonic instability. People were impressed because the seven Mercury astronauts were willing to risk having Redstone rockets blow up under them. Christ! Rockets had already blown up under good men! Skip Ziegler's X–2 exploded while still attached to the mother ship, a B–29, killing Skip and a B–29 crewman. The same thing had very nearly happened to Pete Everest in the X–1D—and to Walker himself in the X–1A. Walker was strapped into the X–1A, under the bomb bay of a B–29, at 35,000 feet, seventy seconds from launch, when a fuel tank exploded in the rear of the rocket plane. Walker got out, climbed back up into the B–29, passed out from lack of oxygen, was revived by a "walkaround" oxygen bottle, went back down into the burning X–1A, and tried to jettison the rest of the fuel so as to prevent both ships, the X–1A and the B–29, from burning up. The rocket plane was finally dropped, like a bomb, over the desert. Walker received the Distinguished Service Medal for his trip back into the burning ship.

That was back in August of 1955, and the newspapers talked about it for a little while, but now no one remembered, or comprehended, that all of these things had been adventures in *manned rocket flight.* With the Big Engine already on the way, the XLR–99—well, it was likely that if NASA would just pour the money and personnel and emphasis into the X–15 project and the X–20 project, the United States could have orbiting spacecraft in reasonably short order. *Ships,* vehicles with a pilot who took them aloft and brought them back through the atmosphere with his own hand and then landed them . . . on the dome of the world, at Edwards. It wasn't merely that the Mercury plan of a man in a pod splashing down in the middle of the ocean under a parachute was "dirty," primitive, and an embarrassing way for a pilot to come down, as the Edwards pilots saw it. It was also needlessly dangerous. A slight error in trajectory or timing and he might hit the water scores or hundreds of miles off target; and any man who had ever flown a search plane knew how hopeless it could be to spot a small object in the open sea, particularly in bad weather.

It could even be argued that the X–15 pilots were a year or so ahead of the astronauts when it came to training for space flight. The Mercury training program had borrowed a lot of X–15 training—without flying. Each X–15 flight was so expensive—about $100,000, if you figured in the time and wages of all the support personnel—it was impractical to have a pilot use the X–15 itself for his basic training. Using the new piece of engineering technology, the computer, NASA built the first full-scale flight simulator. The realism of it was uncanny. Of course, they couldn't simulate the g-forces of rocket flight—so they had dreamed up the idea of using the Navy's human centrifuge at Johnsville.

Up above the centrifuge arm there was a balcony, and this balcony was known as the Throne Room, because arrayed upon it was a lineup of green plastic seats with high backs. Each had been custom-made, molded to the contours of the torsos and legs of a rocket pilot. Each had his name on it: "A. Crossfield" (Scott Crossfield's first name was Albert), "J. Walker," "R. White," "R. Rushworth," "F. Petersen," "N. Armstrong," and so forth. They looked like royal mummies when they were lined up like that, and they were already there in the Throne Room when the shells of "J. Glenn," "A. Shepard," "W. Schirra," and the four others joined the tableau. The astronauts took centrifuge training that had first been worked out for Walker and the X–15 pilots. The astronauts' procedures trainer was a modified version of the X–15 simulator. NASA even rigged up an inertia-coupling trainer for the astronauts, a device called the Wild Mastiff that spun you in all three axes, pitch, roll, and yaw, at once; but the ride was so horrendous it wasn't used much. Joe Walker & Co. had taken that ride in real time . . . at altitude . . . And where did the astronauts go for their parabolic rides in the F–100Fs, to experience weightlessness? To Edwards. Chuck Yeager himself had flown the first weightless parabolas for the Air Force, and then Crossfield had flown them for NASA. Edwards pilots took the astronauts up in the back seat.

For the most part, the men involved in the X–15 program were realistic about the situation. Technically there was no reason why the X–15 should not lead to the X–15B or the X–20 or some other aerodynamic spaceship. Politically, however, the chances were not good and hadn't been good since October 1957, when Sputnik I went up. The politics of the space race demanded a small manned vehicle that could be launched as soon as possible with existing rocket power. And as the Edwards brethren knew, there was no use trying to wish the politics of the situation away.

But now, in mid-1960, the political reality itself had begun to change. The first signs had come in May. This was the same month, it so happened, in which Walker and White had begun to unlimber the X–15 and the Little Engine. But the change was being caused by events quite outside of their control.

The starting point was the so-called U–2 incident. A Soviet surface-to-air missile—no one even knew the Soviets had created such a weapon—shot down an American CIA "spy plane," the U–2, flown by a former Air Force pilot named Francis Gary Powers. Khrushchev used the incident to humiliate President Eisenhower at a summit conference in Paris. This was an election year, of course, and both of the main Democratic contenders, Lyndon Johnson and John F. Kennedy, began citing the Soviets' superiority in rockets as a means of attacking the Eisenhower Administration. Meanwhile, the Soviets and their mighty Integral began pouring it on in earnest. They sent up a series of huge, five-ton Korabl ("Cosmic") Sputniks, carrying dummy cosmonauts or dogs or both; they obviously had a system powerful enough and sophisticated enough to put a man into orbit. NASA was not only unable to keep up with its original schedule of a manned ballistic flight in 1960, it couldn't even deliver a finished capsule—and its test rocket launches, all public events, went from bad to worse.

On July 29, NASA brought the seven astronauts and hundreds of VIPs to Cape Canaveral for a highly publicized first test of the Mercury-Atlas vehicle, a Mercury capsule atop an Atlas rocket. The Atlas, with its 367,000 pounds of thrust, would be used for manned orbital flights; the first Mercury flights, which would be suborbital, would use the smaller Redstone. July 29 was a dark rainy day, which only made the lift-off of the mighty rocket all the more spectacular. The earth rumbled underfoot, and the rocket rose slowly on three columns of flame. It was a terrific show. After sixty seconds it seemed to be directly overhead and gradually nosing over on its long arc toward the horizon, and the astronauts and everybody else had their necks up and their heads bent back, watching the Ahura-Mazda surge, when—*kaboom!*—it blew up. Just like that, right over their heads. For a moment it seemed as if it was going to come down in a few thousand enormous flaming pieces, right on everybody's bean. There was no danger, in fact; the rocket's momentum carried the debris away from the launch site. It was damned sobering, however, with your gullet stuck up in the air like a bird's . . . And it was very bad news for Project Mercury.

It was not the ultimate fiasco, however. The ultimate fiasco came later in the year when NASA put on a test at Cape Canaveral designed to show all the politicians that the Mercury capsule-and-rocket system was now almost ready for manned flight. They flew five hundred VIPs, including many congressmen and prominent Democrats, down to the

Cape for the big event. The rocket, the Redstone, was not powerful enough to place the capsule into orbit, but it was supposed to take it up more than one hundred miles, fifty miles above the earth's atmosphere, and then it would re-enter the atmosphere and splash down in the Atlantic by parachute about three hundred miles from the Cape, near Bermuda. Everything except an astronaut was on the launch pad. The dignitaries were all seated in grandstands, and the countdown was intoned over the public-address system: "Nine . . . eight . . . seven . . . six . . ." and so forth, and then: "We have ignition!" . . . and the mighty belch of flames bursts out of the rocket in a tremendous show of power . . . The mighty white shaft rumbles and seems to bestir itself—and then seems to change its mind, its computerized central nervous system, about the whole thing, because the flames suddenly cut off, and the rocket settles back down on the pad, and there's a little *pop*. A cap on the tip of the rocket comes off. It goes shooting up in the air, a tiny little thing with a needle nose. In fact, it's the capsule's escape tower. As the great crowd watches, stone silent and befuddled, it goes up to about 4,000 feet and descends under a parachute. It looks like a little party favor. It lands about four hundred yards away from the rocket on the torpid banks of the Banana River. Five hundred VIPs had come all the way to Florida, to this goddamned Low Rent sandspit, where bugs you couldn't even see invaded your motel room and bit your ankles until they ran red onto the acrylic shag carpet—all the way to this rock-beach boondock they had come, to see the fires of Armageddon and hear the earth shake with the thunder—and instead they get this . . . this *pop* . . . and a cork pops out of a bottle of Spumante. It was the original Project Vanguard fiasco all over again, except that it was worse in a way. At least with Vanguard, back in December of 1957, the folks got lots of flames and explosion. It at least looked halfway like a catastrophe. Besides that, it was very early in the game, in the contest for the heavens. But this—it was ridiculous! It was pathetic!

Kennedy had won the election, and during the campaign he had made such a point of attacking NASA's ineptness that it was a foregone conclusion that NASA's chief, T. Keith Glennan, who was a Republican in any case, would be replaced. The question now was how many other heads would roll. What about Bob Gilruth? After all, he was in charge of Project Mercury, which was going nowhere. Or von Braun, the alleged German rocket genius? Much sarcasm was creeping into the debate, and even von Braun was being attacked. For that matter, what about the seven brave lads . . .

As this sort of talk began to circulate, people at Edwards began to beam up the radar . . . For months the word within NASA had been that the X–15 project would be the last hurrah for "the flyboys." Now that was all changing. No one was saying it publicly yet, but the unthinkable was now possible: for the first time, Project Mercury itself was regarded as expendable. Kennedy's advisor in scientific areas was Jerome Wiesner of M.I.T. He had drawn up a report for Kennedy that said the following, in effect:

Project Mercury had been sold to the Eisenhower Administration during the original Sputnik panic as the "quick and dirty" solution to getting a man into space ahead of the Russians. It had merely proved to be dirty or, hopeless, as in the Popped Cork business, which demonstrated that NASA did not even have the primitive Mercury-Redstone system ready. Even if the system worked, the Redstone could put a man into only a suborbital trajectory, with just fifteen minutes in space. The mighty Soviet Integral had already launched a series of huge Korabls and was probably on the verge of putting a man not only into space but into earth orbit. But in one area, Wiesner was telling Kennedy, the United States was ahead of the Soviets, and this was in unmanned scientific satellites. Why not concentrate on that program for the time being and play down—in effect, forfeit—the losing race to put a man into space? Why not abandon all these frantic attempts to convert the underpowered Redstone and Atlas missiles into space rockets and instead develop a careful, solid, long-range program using bigger rockets, such as the Titan, which might be ready in eighteen months?

And there you had it! As Joe Walker and everyone at Edwards knew, the "solid, long-range program" using the Titan was the X–20 or Dyna-Soar program, which would begin at Edwards as soon as the X–15 project was completed. The Air Force, which was in charge of the X–20 project, had never abandoned its hopes of running the entire manned space program. All along it had seemed unjust that NASA had been able to appropriate all the research and planning that had gone into

Flickinger's Man-in-Space-Soonest program and convert it into Project Mercury. Perhaps with the change in administrations the situation could be corrected.

Joe Walker was feeling good. In August he had pushed the X–15 just about as fast as the Little Engine could take it, to a new world speed record of Mach 3.31, or 2,196 miles per hour. After he landed it on Rogers Lake, he cut loose with a cowboy yell that startled everyone on the radio circuit: "Yippeeeee!" That was Joe Walker. A week later Bob White went up in the X–15 and set a new altitude record of 136,500 feet, or slightly more than twenty-five miles. It had been a perfect flight. It was as much as you could expect from the Little Engine. The conditions had been almost precisely the conditions of space flight. He took the ship up in a ballistic arc, the same sort of arc the Mercury-Redstone vehicle was supposed to go on . . . someday . . . He experienced five g's during the rocket thrust on the way up. An astronaut in Mercury was supposed to experience six. He was weightless for two minutes as he came up over the top of the arc. An astronaut was supposed to be weightless for five minutes. At 136,500 feet the air was so thin, White had no aerodynamic control at all. It was absolutely silent up there. He could see for hundreds of miles, from Los Angeles to San Francisco.

It was much like the Mercury flights were supposed to be—except that Bob White was *a pilot* from beginning to end! He was in control! He took the ship up and he brought it back down through the heavy atmosphere and he landed it at Edwards! He didn't splash down in the water like a monkey in a bucket! Bob White's picture wound up on the cover of *Life*. There was Justice, there was Logic, in the universe, after all. Bob White on the cover of *Life!* For a solid year *Life* had been the fraternity bulletin for the Mercury astronauts. But now even Henry Luce and that bunch had woken up to the truth. Perhaps they had been betting on the wrong horses! Right? Walker and White and Crossfield could afford a little jealousy now . . . toward one another for a change. People from the TV show *This Is Your Life* had turned up at Edwards and were talking to everybody they could find who knew Joe Walker. This was one of the most popular shows on television, and it was run like a surprise party; the subject, in this case Walker, didn't learn about it until the moment of the show itself, after a biography of him had been put together on film. Scott Crossfield had a book contract, to write his autobiography, and Time-Life was talking to Bob White about a contract like the astronauts'.

Bob White was all right. You could read the cover story they had already written about him in *Life,* and you could see that White had not unbent so much as one inch for the occasion. You could see them straining to manufacture one of those "personality profiles" about White, and all he would give them was the Blue Suit and a straight arrow. That was Bob White.

A True Brother!

IX. The Vote

Even the scenery was depressing. There was nothing to see but the snow blowing over the road and the stunted countryside rolling by in slow motion. Between Langley and Arlington even the woods looked stunted. There had been a blizzard the day before, but the landscape was all so raggedy it didn't even look good in the snow. Over the car radio he could hear John F. Kennedy delivering his inaugural address. The reception was poor, and the broadcast kept fading in and out through the static. The announcer, who spoke in hushed tones as if he were describing a tennis match, had said that it was seventeen degrees in Washington and a wind was blowing on Capitol Hill and Kennedy was bareheaded and wore no overcoat. Kennedy had his voice set at a strangely high pitch. He seemed to be screaming to keep warm. He was screaming a great many sonorous rhetorical figures. The words merely drifted by John Glenn, as he drove, like the snow and the stunted scrub pines outside.

That was ironic, because at first Loudon Wainwright thought that Glenn was totally absorbed in the new President's inaugural address. He kept fiddling with the dial, battling the static, trying to make the broadcast come in better. When Wainwright made the occasional comment, Glenn had no response at all. Wainwright was one of the *Life* writers assigned to the personal stories of the astronauts, and he had come to know Glenn fairly well. At this very moment Glenn was giving him a lift to National Airport before heading home. If John had been determined to digest every single word and nuance of Kennedy's address, it would not have been terribly surprising. John was one of those rare celebrities who came pretty much as advertised. He really was serious about God, country, home, and hearth. He probably even had it in him to take a Presidential inaugural address seriously. But then Wainwright noticed that John not only wasn't reacting to what he said, he wasn't reacting to what Kennedy said, either. He was a thousand miles away, as the saying goes, and not particularly happy to be there.

The curious thing, in light of what happened yesterday, was that for about three months the intense competitive feeling among the seven of them had died down. The entire Mercury project, the astronauts included, had been in serious—no, *horrendous*—trouble. After the MA-1 fiasco and the Popped Cork fiasco, it had no longer been a question of which one of them was going to get the first flight, but whether any of them would go into space at all, or even continue to bear the title *astronaut.*

Naturally it would have been crushing to Bob Gilruth, Hugh Dryden, Walt Williams, Christopher Kraft, and all of the NASA brass to have Mercury canceled for delays or ineptness or whatever. But not so crushing as it would have been to the Mercury astronauts! Oh, no! To be pronounced the seven bravest lads in America, the fearless pioneers of space, to be on the cover of *Life,* to have factory workers in San Diego agonize over your hide and every sweet little Konakai cookie on both shores lust over it . . . and then to be told, "Thanks a lot, but we've called the whole thing off" . . . They would become factory seconds! They would be back in uniform, in the Air Force, the Navy, and the Marines, saluting and flapping in the breeze as the seven most laughable duds in the service!

One had only to imagine it . . . and it was easy to imagine by late 1960. All of them, astronauts, administrators, engineers, technicians, were suddenly in such trouble that a wagon-train phase began. Everyone, from top to bottom, began pulling together like pioneers besieged in the pass. It was now of supreme importance to push the Mercury-Redstone program forward before the new President and his science advisor, Wiesner, had time to go to work on NASA. The frantic hope was to complete some tests that would bring the program so close to the first manned flight that Kennedy could hardly afford to dismantle Project

Mercury without letting them have at least one try. So everyone rode hell-for-leather for the top of the next hill, and never mind the ordinary precautions. The number of unmanned tests was cut down drastically. Tests were scheduled one on top of the other, so that the first manned flight might be scheduled within three months. They were ready to try things they would never have thought of doing before. Rather than prepare a new rocket for the next test, they used the one that had been left on the pad after the Popped Cork fiasco. After all, it hadn't blown up; it had merely refused to leave the ground.

That was the spirit of the hour—ride! *más allâ!* over the next hill! don't look back!—when Bob Gilruth called in Glenn and the other six for a meeting at the office at Langley just before Christmas. Gilruth had always been a sympathetic soul around the seven of them; and now that the cowboy rush was on, his concern for them was written all over his face. The message seemed to be: "It's terrible, but I may have to send one of you boys up without all the precautions that I would like to take." When they assembled in his office, he told them he wanted them to take a little "peer vote," along the following lines: "If you can't make the first flight yourself, which man do you think should make it?" Peer votes were not unknown in the military. They had been used among seniors at West Point and Annapolis for some time. For that matter, during the selection process for astronaut, the groups of finalists at Lovelace and Wright-Patterson took peer votes. But peer votes had never amounted to anything more than what they were *prima facie:* an indication of how men at the same level regarded one another, whether for reasons of professionalism or friendship or jealousy or whatever. Pilots regarded peer votes as a waste of time, because a man either had the right stuff in the air or he didn't, and a military career, particularly among those with "the uncritical willingness to face danger," was not a personality contest. But there was something about Bob Gilruth's deep concern . . . They were to think the whole thing over and put their choices down on paper and drop them off at Gilruth's office. The look on Bob Gilruth's face set off a neural alarm.

Even so, the esprit throughout NASA was tremendous over that Christmas holiday. Everyone was working like a zealot. Christmas itself was a mere syncopation in the mad cowboy rush. Bureaucratic lines no longer meant anything. Anyone in Project Mercury could immediately get in to see anybody else about any problem that came up. At Langley, if some GS–14 wanted to get hold of Gilruth face to face, all he had to do was wait around in the cafeteria at lunchtime and walk up to him while he waited in line, edging his tray along the stainless-steel tubing. There weren't enough hours in the day to do all the things that had to be done.

On January 19, the day before Kennedy's inauguration, Gilruth again convened the seven of them in the astronaut office. He said that what he was about to tell them must be kept in strictest confidence. As they all knew, he said, the original plan had been to select the pilot for the first flight on the very eve of the flight itself. But he had thought better of that, because it now seemed clear that the prime pilot should have maximum access to the procedures trainer and other training facilities during the final weeks before the flight. So a decision had been made as to who was to be the prime pilot and as to which two men would be the backup pilots for the first flight. In due course, the press would be given the names of all three men, but the fact that the prime pilot had already been chosen would not be revealed. The press and the public would be told only that it would be one of these three men. All three men would go through the same training, and it would seem that the original plan was being adhered to, and the prime pilot would be spared the public pressure that would otherwise bear down on him.

This had been a very difficult decision, he said, because all seven of them had worked with such dedication, and he knew that any of them would make a capable pilot for the first flight. But it had been necessary to come to a decision. And the decision was that the prime pilot would be . . . Alan Shepard. The backup pilots would be John Glenn and Gus Grissom.

The words hit Glenn like a ray. The cause, effect, and outrageous results thereof came to him in a flash, and he was stunned. Al was staring at the floor. Then Al looked up at him and the rest of them with his eyes gleaming, resisting the temptation to break into a grin of triumph. Yet Al had triumphed! It was unbelievable, and yet it was really happening. Glenn knew what he had to do and he made himself do it. He made himself smile the earnest-looking smile of the runner-up and congratulate Al and shake his hand. Now the other five were doing the same thing, coming up to Al and smiling earnest smiles and shaking his hand. It was a shocking thing, and yet it had happened—Glenn was absolutely sure of it. To get around the agony of having to *designate* someone to sit up on top of the first rocket—*a peer vote!* After he, Glenn, had spent twenty-one months doing everything humanly possible to impress Gilruth and the rest of the brass, it had been turned into a popularity contest among the boys.

A peer vote!—it was unbelievable! Every move Glenn had made undoubtedly worked against him like a captured weapon in the peer vote. In the peer vote he was the prig who had risen at the séance like John Calvin himself and told them all to keep their pants zipped and their wicks dry. He was the Eddie Attaboy who had gotten up every morning at dawn and done all that ostentatious running and tried to make the rest of them look bad. He was the Harry Hairshirt who lived like an Early Christian martyr in the BOQ. He was the Willie Workadaddy who drove around in a broken-down Prinz, like a lonely beacon of restraint and self-sacrifice in a squall of car crazies.

But Smilin' Al Shepard—Smilin' Al was the fighter jock's fighter jock, if all you were doing was taking a peer vote. He was His Lordship of Langley and the King of the Cape. He gave off the *aura* of the hot pilot. Since they had done practically no flying for twenty-one months, there had been no way for Glenn or anyone else to impress the others in the air—and so it came down to a question of which one other than he, Glenn, the Offending Saint, *looked* the most like a hot pilot. It was not only a popularity contest, it was a *cosmetic* popularity contest. How else was he to regard it? It was as if the past twenty-one months of training had never happened.

And so now Glenn was driving back home, through the stunted Virginia countryside and the raggedy snowflakes, to tell Annie the secret bad news. And the new President was screaming over the car radio: "Together let us explore the stars . . ." Shepard would be first! It was incredible. Shepard would be . . . *the first man to go into space!* He would be famous throughout eternity! And here was something even more unbelievable: he, John Glenn, for the first time in his career, would be one of those who were *left behind.*

The astronaut headquarters on the base at the Cape were in the building known as Hangar S. The hangar had been rebuilt inside to house the procedures trainer, a pressure chamber, and most of the other facilities an astronaut would need in the final preparations for a flight. There was a suite of rooms for living quarters, a dining room, a medical examination room, a ready room in which the astronaut would put on his pressure suit, a special doorway where the astronaut would get into a van to be driven out to the launch pad, and so forth. The boys seldom stayed there overnight, however, much preferring the motels in Cocoa Beach—and of course they had yet to use Hangar S for an actual flight. In fact, the first creatures to use Hangar S fully, from procedures trainer to rocket launch, were the chimpanzees. The chimpanzees were already at Hangar S on the morning before the inauguration, when Gilruth told the seven men that Alan Shepard had won the competition for the first flight. The chimpanzees had been there, ready for the first flight for almost three weeks. The veterinarians at Holloman Air Force Base had cut the original field of forty chimpanzees down to eighteen and, finally, to six, two males and four females, and had flown them to the Cape and installed them out back of Hangar S in a fenced-in compound. In the middle were two long, narrow trailer units, each made up of two eight-foot-wide trailers hooked up end to end. Around them was an assortment of other trailers and vans, including a special transfer van for carrying a chimpanzee from Hangar S to the rocket launch pad. The press was not invited into the little trailer park, nor would the good Gent have been interested. The chimpanzee test seemed to be merely one more tedious preliminary to the main event. Not even people at the base had much of an idea what was going on out back of Hangar S. The apes spent most of the day inside the two double-trailer units. The trailers were home and office, cage and *operant conditioning* cubicle, for the beasts. Inside each unit were three cages, two procedures trainers, and a mockup of the Mercury capsule. The veterinarians in their white smocks and the attendants in their white T-shirts and white ducks had a trailer unit of their own; they were on hand, in shifts, twenty-four hours a day. Within those long skinny trailers the biggest countdown in the brief history of Project

Mercury was underway. Every day for twenty-nine days, at the heart of the American space facility, out back of a great moldering hangar, on a stone boondock scrub-pine spit on the point of Cape Canaveral, a tribe of six scrawny apes and twenty humans in white were up in the early morning, on the move, restless, earnest, driven, scratching, flapping, ricocheting through the trailers' innards, yammering at fate and each other. The humans were giving the apes physical examinations and wiring them from top to bottom, sticking eight-inch thermometers up their rectums, imbedding sensors in the thoracic cages, clamping the psychomotor stimulus plates on the soles of their feet, threading them into their restraint harnesses, strapping them into their procedures-trainer cubicles, closing the hatches, pressurizing the cubicles with pure oxygen, inserting the cubicles in the mock Mercury capsules, and then flashing the lights. The apes had to throw their switches on cue or get the dread dose of volts in the soles of the feet. How their scrawny fingers did fly! All six apes were stringy-looking, like those overtrained fleaweight college wrestlers who have run so many laps and taken so many B–12s and diuretics at the training table that they appear to be little dried-up lengths of gristle, nodes, and nerve ganglia. But out back of Hangar S the little bastards could play their Mercury consoles like a dream.

The white-smock humans zapped the apes through their workouts right up to the last. On January 30, on the eve of the flight, they made the final selection. Originally the first astronaut was to have been chosen at this same stage. They chose a male chimpanzee as the prime flier and a female as the backup. The Air Force had bought the male from a supplier in the Cameroons, West Africa, eighteen months ago, when he was about two years old. All this time the animals had been known by numbers. He was test subject Number 61. On the day of the flight, however, his name was announced to the press as Ham. Ham was an acronym for Holloman Aerospace Medical Center.

Before dawn on January 31 they woke Number 61 up and led him out of his cage and fed him and gave him a medical examination and attached his biosensors—and put the zap plates on his feet—and put him in his cubicle and closed the hatch and depressurized it. Another goddamned day with these earnest hard-zapping ballbreaking white-smock humans. The vets put the cubicle in the transfer van, and the chimpanzee was taken to the launch pad, out by the sea. The sun was up now, and a white rocket with a Mercury capsule and an escape tower on top of it stood gleaming, and they took Number 61 in his cubicle up an elevator on a gantry beside the rocket and then put the cubicle in the capsule. There were more than a hundred NASA engineers and technicians nearby, working on the flight, monitoring consoles, and an entire team of veterinarians was monitoring the dials that read the ape's heartbeat, respiration, and temperature. Hundreds more NASA and Navy personnel were strung out across the Atlantic, toward Bermuda, in a communications and recovery network. This was the most crucial test in the entire history of the space program and they were giving it everything they had.

It was four hours before they were able to fire the rocket. The biggest problem was an inverter, a device that was supposed to prevent rogue surges of power in the Mercury capsule's control system. The inverter kept overheating. All the while during the "hold," as they called the delay, Chris Kraft, director for the first ape flight, just as he would be for the first human flight, kept asking how the ape was doing, apparently on the presumption that the long confinement would make the beast anxious. The doctors checked their dials. The ape didn't seem to have a nerve in his body. He was lying up there in his cubicle as if he lived there. And why not? For the ape every hour of the delay was like a holiday. No lights! No zaps! Peace . . . bliss! They gave him two fifteen-minute workouts with the lights, just to keep him alert. Otherwise it was terrific. "Hold" for an eternity! Don't let anything stop you!

When they fired the rocket, shortly before noon, it climbed at a slightly higher angle than it was supposed to, driving Number 61 back into his couch with a force of seventeen g's, i.e., seventeen times his own weight, five g's more than anticipated. His heart rate shot up as he strained against the force, but he didn't panic for a moment. He had been through this same sensation many times on the centrifuge. As long as he just took it and didn't struggle, they wouldn't zap all those goddamned blue bolts into the soles of his feet. There were a lot worse things in this world than g-forces . . . He was weightless now, hurtling toward Bermuda, and they flashed the lights in his cubicle, and the ape's pulse returned to normal, to no more than what it was on the ground. The usual shit was flowing. The main thing was to keep ahead of those blue bolts in the feet! . . . He started pushing the buttons and throwing the switches like the greatest electric Wurlitzer organist who ever lived, never missed a signal . . . Then the Mercury retro-rockets were fired, automatically, and the capsule came down through the atmosphere at the same angle at which it had gone up. Another 14.6 g's hit Number 61 on the way down, making him feel as if his eyeballs were coming out of his head. He had been through so-called eyeballs-out g's, too, many times, on the centrifuge. It could get a lot worse. There were worse things than feeling as if his eyeballs were coming out . . . The goddamned zap plates on his feet, for a start . . . So far as mere space flight was concerned, Number 61 was fearless. The beast had been operantly conditioned, aerospatially desensitized.

The high angle of the launch also caused the capsule to overshoot the planned splashdown area by 132 miles. So it took two hours for a Navy helicopter crew to find the capsule out in the Atlantic and bring it aboard a recovery ship. The capsule and the ape were riding up and down in seven-foot swells. Water had begun to seep in where the landing bag had torn loose in the heavy seas. The capsule was wheezing and gurgling with water and pitching up and down like a ball in the waves. It wouldn't have stayed afloat too much longer. Eight hundred pounds of water had seeped in. For the ordinary prudent human being it would have been two hours of freaking terror. They brought the capsule to a recovery ship, the *Donner,* and opened it and brought out the ape's cubicle and opened the hatch. The ape was lying there with his arms crossed. They offered him an apple and he took it and ate it with considerable deliberation, as if gloriously bored. Those two hours of being slung up and down in the open sea in seven-foot gulps inside a closed coffin-like cubicle had been . . . perhaps the best time ever in this miserable land of the white smocks! No voices! No zaps! No bolts, no lengths of hose, no more breaking of his bloody balls . . .

There was great elation among the astronauts and almost everyone involved in Project Mercury. There seemed to be no way that Kennedy and Wiesner could intervene and keep them from trying at least one manned flight. The pall of the Day the Cork Popped had been removed.

Late the next day they flew Number 61 back to the Cape and Hangar S, where a great mob of reporters and photographers were now waiting out in the compound by the Mercury capsule that had been used for the training. The vets led the ape out of a van. As the mob closed in and the flashbulbs began exploding, the animal—brave little Ham, as he was now known—became furious. He bared his teeth. He began snapping at the bastards. It was all the vets could do to restrain him. This was immediately—on the spot!—interpreted by the press, the seemly Gent, as an understandable response to the grueling experience he had just been through. The vets took the ape back inside the van until he calmed down. Then they led him out again, trying to get him near a mockup of a Mercury capsule, where the television networks had set up cameras and tremendous lights. The reporters and photographers surged forward again, yammering, yelling, exploding more camera lights, shoving, groaning, cursing—the usual yahoo sprawl, in short—and the animal came unglued again, ready to twist the noodle off anybody he could get his hands on. This was interpreted by the Gent as a manifestation of Ham's natural fear upon laying eyes once again on the capsule, which looked precisely like the one that had propelled him into space and subjected him to such severe physical stresses.

The stresses the ape was reacting to were probably of quite another sort. Here he was, back in the compound where they had zapped him through his drills for a solid month. Just two years ago he had been captured in the jungles of Africa, separated from his mother, shipped in a cage to a goddamned desert in New Mexico, kept prisoner, prodded and shocked by a bunch of humans in white smocks, and here he was, back in a compound where they had been zapping him through their fucking drills for a solid month, and suddenly there was a whole new mob of humans on hand! Even worse than the white smocks! Louder! Crazier! Totally out of their gourds! Yammering, roaring, brawling, exploding lights beside their bug-eyed skulls! Suppose they threw him to these assholes! *Fuck this—*

At some point in the madhouse scene out back of Hangar S, a photograph was taken in which Ham was either grinning or had on a grimace that looked like a grin in the picture. Naturally, this was the picture that went out over the wire services and was printed in newspapers throughout America. Such was the response of the happy chimpanzee to being the first ape in outer space . . . A fat happy grin . . . Such was the perfection with which the Proper Gent observed the proprieties.

Well, there were some big grins, all right, up in the high desert, at Edwards, among the brethren. Here were some men who had something to smile about. Now the whole business of Project Mercury was no doubt clear to everyone. No one, not even in the general public, could possibly miss the point now. It was that obvious. The first flight—the coveted *first flight of the new bird*—that full-bore first flight that every test pilot strove for—had just been made in Project Mercury. *And the test pilot was an ape!* An ape made the first flight! "A college-trained chimpanzee"!—to use the very words that they had heard Astronaut Deke Slayton himself use before the Society of Experimental Test Pilots. And the ape had performed flawlessly, done as well as any man could possibly have done—for there was nothing for a man to do in the Mercury system except push a few perfunctory buttons and switches. This any college-trained chimpanzee could do also! He hadn't missed a beat! Give him a signal and he'll toggle you a switch! To see the point—and surely all the world now saw the point—you only had to imagine sending an ape up for the first flight of the X-15. You would have a twenty-million-dollar hole in the ground and a pulverized ape. But in Project Mercury an ape was fine! First-rate! In fact . . . the ape *was* an astronaut! He was the first one! Perhaps the female ape who backed him up deserved the next flight. Let *her* fly, goddamn it! She has earned it as much as the seven human ones—she's been through the same training! . . . and so forth and so on . . . The brethren let their beer-call brains soar. Perhaps the ape would go to the White House and get a medal. (Why not!) Perhaps the ape would address the September meeting of the Society of Experimental Test Pilots in Los Angeles. (Why not!—another *astronaut* had done it, Deke Slayton, without flying at all!) Oh, it was a laugh and a half, the whole thing. For now the truth was out—it was obvious in a way that no one in the world could miss.

And in the days that followed, the first days of February 1961, the True Brothers waited for this revelation to sweep across the press and the public and the Kennedy Administration and the military brass. But, strange to say, there was not even one such sign anywhere. In fact, they could begin to see signs of something quite the opposite. It was incredible, but the world was now full of people who were saying:

"My God, do you mean there are *men* brave enough to try what the ape has just gone through?"

John Glenn found himself in a ridiculous situation. It was nothing less than a charade. He had to pretend to be in the running for the first flight—and then he would read in the newspapers that he was the front runner. Since he had always been the fair-haired boy of the seven, that was the way it kept coming out. He and Gus Grissom had to tag along with Shepard through the training grind to keep up the fiction that the decision had not yet been made. In fact, Shepard was now the king—and Al knew how to *act* like the king, His Majesty the Prime Pilot—and Glenn was just a spear carrier.

Yet the charade, which Gilruth himself insisted on, also offered Glenn one last chance. No more than a handful of people knew that Shepard had been chosen for the first flight. Therefore, it was not too late to change the decision, to rectify what Glenn viewed as the incredible and outrageous business of the peer vote. But if this meant going over someone's head in one way or another . . . well, in the military it was a grave error, a serious breach of all that was holy, to go over the head of a superior unless (1) it was a critical situation and you were in the right and (2) your brash move worked (i.e., those at the top backed you up). On the other hand, there was nothing in the Presbyterian faith, another code Glenn knew very well, that said you had to stand around shy and obedient while the Pharisees waffled and oscillated, making imaginary snowballs. And was not NASA a civilian agency? (God knew it was not run like the Marines.) Glenn seemed to favor the Presbyterian course. He began talking to people in the hierarchy, asking what they thought of the decision.

He did not argue that he should be the one who was chosen, or not in so many words. He argued that the choice could not be made from a narrow perspective. America's first astronaut would not be merely a test pilot with a mission to carry out; he would be a historic representative of

America, and his character would be viewed in that light. If he did not measure up to that test, it would be unfortunate not only for the space program but for the nation.

The new administrator of NASA, appointed by Kennedy to replace T. Keith Glennan, was James E. Webb, a former oil company executive and a political grand master in the Democratic Party. Webb was of a valuable breed well known in Washington: the off-the-ballot politician. The off-the-ballot politician usually looked like a politician, talked like a politician, walked like a politician, loved to mix with politicians, moved and shook with politicians, winked with politicians, sighed ruefully with them. He was the sort of man of whom a congressman or a senator was likely to say: "He speaks my language." The ablest and most distinguished of the off-the-ballot politicians, like Webb, were likely to end up with high-level appointments. Webb had been director of the Bureau of the Budget and Undersecretary of State under Truman. He was also a great friend of Lyndon Johnson and Senator Robert Kerr of Oklahoma, who was chairman of the Senate Committee on Aeronautical and Space Sciences. For six years Webb had been head of a subsidiary of the Kerr family's oil empire. Webb was the sort of man whom corporations doing work for the government, such as McDonnell Aircraft and Sperry Gyroscope, liked to have on their boards of directors. He looked the part. He had big smooth jowls like Glennan's and even better hair, wavy, thick, as if every strand were nailed in, and dark, but graying smartly, and combed straight back in the manner favored by all serious men of the day. He had the sort of record that made him an ideal appointee to commissions like the Municipal Manpower Commission, which had occupied much of his time since 1959. He was known as a man who could make bureaucracies run. He was used to corner offices with terrific views. He was no fool. What would he have made of this business of Astronaut Glenn's dissatisfaction with the selection of Astronaut Shepard for the first Mercury flight? Gilruth said he had made the decision himself; and it was based on a wide range of criteria, many of them quite objective. Shepard had performed best on the procedures trainer, for example. When Gilruth considered all criteria, and not just the peer vote, Shepard ranked first and Glenn ranked second. So what was Glenn objecting to? It was a bit baffling. But one thing was sure: Webb was not likely to start his tenure as administrator of

NASA by taking a flying leap into some incomprehensible dogfight among the seven bravest lads in the history of the United States. Astronaut Glenn's objections—and his last chance to become the world's first man in space—simply sank without a bubble one day, and that was that.

By now, late February of 1961, Glenn was not the only supremely miffed astronaut. Gilruth had finally published the names of the men who would make the first three flights—Glenn, Grissom, and Shepard, always in alphabetical order—implying that no decision had been made as to which of them would make the first flight, due to take place within ninety days. So *Life* ran a big story with pictures of Glenn, Grissom, and Shepard on the cover with the headline: THE FIRST THREE. *Life* was really excited about the whole thing. They tried to get NASA to label the first three "the Gold Team" and the rest of them "the Red Team." The Gold Team and the Red Team. Jesus! The picture possibilities alone were fabulous.

This being the fraternal bulletin, *Life*, the notion of "the first three" struck Slayton, Wally Schirra, Scott Carpenter, and Gordon Cooper as a humiliation. In their minds they were now labeled "the Other Four." There were now the First Three and the Other Four. They had been . . . *left behind!* In some hard-to-define way, it was the equivalent of washing out.

Life really did it up in the best *Life* style. They flew the First Three and the First Three Wives and the First Three Children down to the Cape and took a lot of Inseparable Astronaut Family pictures on Cocoa Beach. The results were bizarre evidence of the determination of the Proper Gent to make everything come out in a seemly fashion. For a start, the travel schedules of the astronauts had made an absolute hash of ordinary home life. To show three astronauts having an outing with their families at the same time, even in different locations, would have been stretching the truth considerably. To present such a spectacle at the Cape—which was, in effect, off limits to wives—was an absolute howler. On top of that, if you were going to put astronaut families together for a frolic on the beach, you could scarcely come up with a less likely combination than the Glenns, the Grissoms, and the Shepards—

the clans of the Deacon, the Hossier Grit, and the Icy Commander. They would have passed like ships in the night in even the calmest of times, and these were not the calmest of times. Not even *Life* with all its powers of orchestration (and they were great) could make it come out right. They ran a big double-truck picture of the First Three and their wives and broods, the glorious First Three tribe, out on the hardtack sands of Cocoa Beach, engrossed (the caption would have one believe) in the sight of an exploratory rocket rising from the base several miles away. In fact, they looked like three families from warring parts of our restless globe who had never laid eyes on each other until they were washed up upon this godforsaken shore together after a shipwreck, shivering morosely in their leisure togs, staring off into the distance, desperately scanning the horizon for rescue vessels, preferably three of them, flying different flags.

As for the Other Four, they might as well have dropped through a crack in the earth.

Glenn worked at being backup astronaut and charade master as if these were the roles the Presbyterian God had elected him to play. He gave it "a hundred percent," to use one of his favorite phrases. Besides . . . if, in the Lord's mysterious workings . . . it so happened that Shepard was not able to make the first flight, for some reason or another, he would be a hundred percent ready to step in. As a matter of fact, by April it had become blessedly, healingly possible for a fighter jock like Glenn to swallow his personal ambitions and lose himself in the mission itself. A true *sense of mission* had taken over Project Mercury. The mighty Soviet Integral had just sent two more huge Korabls into orbit with dummy cosmonauts and dogs aboard, and both flights had been successful from beginning to end. The race was coming down to the wire. Gilruth had even considered sending Shepard up in March, but Wernher von Braun had insisted on one last test of the Redstone rocket. The test went perfectly and everyone now wondered, in hindsight, if valuable time had been lost. Shepard's flight was scheduled for May 2, although it was not referred to publicly as Shepard's flight. The charade continued in full force, with Glenn still reading about himself in the newspapers as the likely choice. Around Hangar S there were NASA people who were talking about bringing all three, Glenn, Grissom, and Shepard, out to the launch pad on May 2 in their pressure suits with hoods over their heads, so that no one would know who was taking the first flight until he was inside the capsule. The reason why it should even matter had been long since forgotten.

NASA engineers and technicians at the Cape were pushing themselves so hard in the final weeks people had to be ordered home to rest. It was a grueling time and yet the sort of interlude of adrenal exhilaration that men remember all their lives. It was an interlude of the dedication of body and soul to a cause such as men usually experience only during war. Well . . . this *was* war, even though no one had spelled it out in just that way. Without knowing it, they were caught up in the primordial spirit of single combat. Just days from now one of the lads would be up on top of the rocket for real. Everyone felt he had the life of the astronaut, whichever was chosen (only a few knew), in his hands. The MA–1 explosion here at the Cape nine months ago had been a chilling experience, even for veterans of flight test. The seven astronauts had been assembled for the event, partly to give them confidence in the new system. And their gullets had been stuck up toward the sky like everybody else's, when the whole assembly blew to bits over their heads. In a few days one of those very lads would be lying on top of a rocket (albeit a Redstone, not an Atlas) when the candle was lit. Just about everybody here in NASA had seen the boys close up. NASA was like a family that way. Ever since the end of the Second World War the phrase "government bureaucracy" had invariably provoked sniggers. But a bureaucracy was nothing more than a machine for communal work, after all, and in those grueling and gorgeous weeks of the spring of 1961 the men and women of NASA's Space Task Group for Project Mercury knew that bureaucracy, when coupled with a spiritual motivation, in this case true patriotism and profound concern for the life of the single-combat warrior himself—bureaucracy, poor gross hideously ridiculed twentieth-century bureaucracy, could take on the aura, even the ecstasy, of communion. The passion that now animated NASA spread out even into the surrounding community of Cocoa Beach. The grisliest down-home alligator-poaching crackers manning the gasoline pumps on Route A1A would say to the tourists, as the No-Knock flowed, "Well, that Atlas vehicle's given us more fits than a June bug on a porch bulb, but we got real confidence in that Redstone, and I think we're gonna make it." Everyone who felt the spirit of NASA at that time wanted to be part of it. It took on a religious dimension that engineers, no less than pilots, would resist putting into words. But all felt it.

Whoever had possessed any doubts about Gilruth's powers of leadership dismissed them now. He had all phases of Project Mercury coming together in a coda. His calmness was all at once like a seer's. Wiesner, who had become Kennedy's Cabinet-level science advisor, had ordered a full-scale review of the space program and its progress, meaning of course its lack of progress, and he and a special committee under his jurisdiction kept sending queries and memoranda to NASA about careless planning, disregard of precautions, and the need for an entire series of chimpanzee flights before risking the life of one of the astronauts. At Langley and at the Cape they treated Wiesner and all his minions as if they were aliens. They ignored their paperwork and didn't return their telephone calls. Finally, Gilruth told them that if they wanted that many more chimpanzee flights, they ought to move NASA to Africa. Gilruth seldom said anything cutting or even ironic. But when he did, it stopped people in their tracks.

The launch procedures were now rehearsed endlessly and with great fidelity. The three of them, Shepard, Glenn, and Grissom, were staying in motels in Cocoa Beach, but they would get up early in the morning, before dawn, drive to the base, to Hangar S, have breakfast in the same dining room where Shepard would eat on the morning of the flight, go to the same ready rooms he would use on that morning for the physical exams and for putting on the pressure suit, have the biosensors attached and the suit pressurized, get into the van at the door and ride out to the launch pad, go up in the gantry elevator, get into the capsule on top of the rocket, and go through the procedures training—"Abort! Abort!"—the whole thing—using the actual instrument panel that would be used in flight and the actual radio hookups. All of this was done over and over. They were now using the capsule itself for simulation—just as the chimpanzees had. The idea was to decondition the beast completely, so that there would not be a single novel sensation on the day of the flight itself.

All three of them were taking part, but naturally Shepard, as prime pilot (no other word was used now), had precedence. And he *took* precedence. The little group in Hangar S now saw Al in both his varieties . . . and both were king, both the Icy Commander and Smilin' Al. Usually he left the Icy Commander back in Langley and brought only Smilin' Al to the Cape. But now he had both installed at the Cape. As the pressure built up, Al set a standard of coolness and competence that would be hard to top. In the medical examinations, in the heat-chamber sessions, in the altitude-chamber runs, he was as calm as ever. By now the White House had become extremely jittery—fearing what the debacle of a Dead Astronaut would do to American prestige—and so some dress rehearsals were conducted on the centrifuge at Johnsville, with Al and his two charade hands, Glenn and Grissom, taking part, and Al was imperturbable. Likewise, in the eleventh-hour simulations atop the rocket at the Cape. Al showed only one sign of stress: the cycles—Smilin' Al/Icy Commander—now came one on top of the other, in the same place, and alternated so suddenly that the people around him couldn't keep track. They learned a little more about the mysterious Al Shepard here in the eleventh hour. Smilin' Al was a man who wanted very much to be liked, even loved, by those around him. He wanted not just their respect but also their affection. Now, in April, on the eve of the great adventure, Smilin' Al was more jovial and convivial than ever. He did his José Jiménez routine. His great grin spread wider and his great beer-call eyes beamed brighter than ever before. Smilin' Al was crazy about a comedy routine that had been developed by a comedian named Bill Dana. It concerned the Cowardly Astronaut and was a great hit. Dana portrayed the Cowardly Astronaut as a stupid immigrant Mexican named José Jiménez, whose tongue wrapped around the English language like a taco. The idea was to interview Astronaut Jiménez like a news broadcaster.

You'd say things like: "What has been the most difficult part of astronaut training, José?"

"Obtaining de maw-ney, señor."

"The money? What for?"

"For de bus back to Mejico, you betcha, reel queeck, señor."

"I see. Well, now, José, what do you plan to do once you're in space?"

"Gonna cry a lot, I theeeenk."

Smilin' Al used to crack up over this routine. He liked to do the José Jiménez part; and if he could get someone to feed him the straight lines, he was in Seventh Heaven, Smilin' Al version. Feed him the lines for his José Jiménez knock-off, and he'd treat you like the best beer-call good buddy you ever had. Of course, the Cowardly Astronaut routine was also a perfectly acceptable way for bringing up, on the oblique, as it were, the subject of the righteous stuff that the first flight into space would require. But that was probably unconscious on Al's part. The main thing seemed to be the good fun, the camaraderie, the closeness and blustery affection of the squadron on the eve of battle. In these moments you saw Smilin' Al supreme. And in the next moment—

—some poor Air Force lieutenant, thinking this was the same Smilin' Al he had been joking and carrying on with last night, would sing out, "Hey, Al! Somebody wants you on the phone!"—and all at once there would be Al, seething with an icy white fury, hissing out: "If you have something to tell me, Lieutenant . . . you will call me 'Sir'!" And the poor devil wouldn't know what hit him. Where the hell did that freaking arctic avalanche come from? And then he would realize that . . . all at once the Icy Commander was back in town.

Of course those few who knew he was the man who was going up first were in a mood to forgive him all . . . well, save for an astronaut or two . . . As for the NASA technicians and the military personnel assigned to the mission, they were in a mood of utter adoration of the *single-combat warriors,* all three of them, for one of them would be placing his hide on top of the rocket. *(And our rockets always blow up.)* Toward the end the three of them would enter a room for some sort of test . . . and the technicians and workmen would stop what they were doing and break into applause and beam at them that warm moist smile of sympathy. Without knowing it, they were bestowing homage and applause in the classic manner: before the fact. These little scenes pressed Glenn's powers as charade master to the limit. More of these warm beams would be aimed at him than at either of the other two. He was the one mentioned in the press as the most likely choice. Not only that, he was the warmest of the three, the most consistently friendly toward one and all when he ran into them. It was just about too much. He had to keep smiling and aw-shucking and playing Mr. Modest, just as if it might, in fact, be he who was going up on top of the rocket on May 2 as the first man in the world to risk the mighty shot into space.

And then the omnipotent Integral intervened . . . a practical joker to the end! Early on the morning of April 12, the fabulous but anonymous Builder of the Integral, Chief Designer of the Sputniks, struck another of his cruel but dramatic blows. Just twenty days before the first scheduled Mercury flight he sent a five-ton Sputnik called *Vostok I* into orbit around the earth with a man aboard, the first cosmonaut, a twenty-seven-year-old test pilot named Yuri Gagarin. *Vostok I* completed one orbit, then brought Gagarin down safely, on land, near the Soviet village of Smelovka.

The omnipotent Integral! NASA had really believed—and the astronauts had really believed—that somehow, in the religious surge of *the mission,* Shepard's flight would be the first. But there was no putting one over on the Integral, was there! It was as if the Soviets' Chief Designer, that invisible genius, was toying with them. Back in October 1957, just four months before the United States was supposed to launch the world's first artificial earth satellite, the Chief Designer had launched Sputnik I. In January 1959, just two months before NASA was scheduled to put the first artificial satellite into orbit around the sun, the Chief Designer launched *Mechta I* and did just that. But this one, *Vostok I,* in April 1961, had been his *pièce de résistance.* Given the huge booster rockets at his disposal, he seemed to be able to play these little games with his adversaries at will. There was the eerie feeling that he would continue to let NASA struggle furiously to catch up—and then launch some startling new demonstration of just how far ahead he really was.

The Soviets persisted in offering no information as to the Chief Designer's identity. For that matter, they identified no one involved in Gagarin's flight other than Gagarin himself. Nor did they offer any pictures of the rocket or even such elementary data as its length and its rocket thrust. Far from casting any doubt as to the capabilities of the Soviet program, this policy seemed only to inflame the imagination. The Integral! Secrecy was by now accepted as "the Russian way." Whatever the CIA might have been able to do in other parts of the world, in the Soviet Union they drew a blank. Intelligence about the Soviet space program remained very sketchy. Only two things were known: the Soviets were capable of launching a vehicle of tremendous weight, five tons; and whatever goal NASA set for itself, the Soviet Union reached it first. Using those two pieces of information, everyone in the government, from President Kennedy to Bob Gilruth, seemed to experience an involuntary leap of the imagination similar to that of the ancients . . . who used to look into the sky and see a clump of stars, sparks in the night, and deduce therefrom the contours of . . . an enormous bear! . . . the constellation Ursa Major! . . . On the evening of Gagarin's flight, April 12, 1961, President Kennedy summoned James E. Webb and Hugh Dryden, Webb's deputy administrator and NASA's highest-ranking engineer, to the White House; they met in the Cabinet room and they all stared into the polished walnut surface of the great conference table and saw . . . the mighty Integral! . . . and the Builder!—the Chief Designer! . . . who was laughing at them . . . and it was awesome!

In Washington, at Langley, and at the Cape, NASA was deluged with telephone calls from newspapers, wire services, magazines, radio stations—and most callers wanted to know what the astronauts' reaction was to Gagarin's flight. So the First Three, Glenn, Grissom, and Shepard, all prepared statements. Shepard cranked out something that said next to nothing at all; standard government issue. Privately he was put out with Gilruth and von Braun and everyone else for not sending him up in March, as it now appeared they could have.

As usual, it was Glenn whom the press quoted most. He as much as said: "Well, they just beat the pants off us, that's all, and there's no use kidding ourselves about that. But now that the space age has begun, there's going to be plenty of work for everybody." Glenn was considered especially forthright, gracious, and magnanimous. He was big about the thing, as the saying goes—and that seemed especially commendable, since he was still considered the American front runner for the flight that would have made him "the first man in space." He had swallowed his own disappointment like a man.

X. Righteous Prayer

Alan Shepard finally got his turn on May 5. He was inserted in the capsule, on top of a Redstone rocket, about an hour before dawn, with an eye toward a launch shortly after daybreak. But as in the case of the ape, there was a four-hour hold in the countdown, caused mainly by an overheating inverter. Now the sun was up, and all across the eastern half of the country people were doing the usual, turning on their radios and television sets, rolling the knobs in search of something to give the nerve endings a little tingle—and what suspense awaited them! An astronaut sat on the tip of a rocket, preparing to get himself blown to pieces.

Even in California, where it was very early, highway patrolmen reported a strange and troubling sight. For no apparent reason drivers, hordes of them, were pulling off the highways and stopping on the shoulders, as if controlled by Mars. The patrolmen were slow in figuring it out, because they did not have AM radios. But the citizenry did, and they had become so excited as the countdown progressed at Cape Canaveral, so ravenously curious as to what would happen to the mortal hide of Alan Shepard when they fired the rocket, it was too much. Even the simple act of driving overloaded the nervous system. They stopped; they turned up the volume; they were transfixed by the prospect of the lonely volunteer about to be exploded into hash.

This tiny lad, up on the tip of that enormous white bullet, appeared to have about one chance in ten of living through it. Over the three weeks since the great Soviet triumph of Gagarin's flight, one terrible event had followed another. The United States had sent in a puppet army of Cuban exiles to conquer the Soviets' puppet regime in Cuba, and instead suffered the humiliation that became known as the Bay of Pigs. This had nothing directly to do with the space flight, of course, but it heightened the feeling that this was not the time to be trying brave and desperate deeds in the contest with the Soviets. The sad truth was, *our boys always botch it.* Eight days after that, on April 25, NASA had another big test of an Atlas rocket. It was supposed to carry a dummy astronaut into orbit, but it went off course and had to be blown up by remote control after forty seconds. The explosion nearly wiped out Gus Grissom, who was following the rocket's ascent as chase pilot in an F–106. Three days after that, April 28, a so-called Little Joe rocket with a Mercury capsule on top of it went off on another crazy trajectory and had to be aborted after thirty-three seconds. Both of these were tests of the Mercury-Atlas system, which would be used for orbital flights, and they had nothing to do with the Mercury-Redstone system, which Shepard would be riding—but it was far too late to make fine points. *Our rockets always blow up and our boys always botch it.*

And so now, on the morning of May 5, thousands, millions, stopped by the side of the road, paralyzed by the drama. This was the greatest death-defying hell-driver stunt ever broadcast, a patriotic stunt, a hash-mad stunt bound up with the fate of the country. People were beside themselves.

What must be going through the man's mind? Him and his poor wife . . . Then the radio announcer would tell how Shepard's wife, Louise, was following the countdown over television inside their home in Virginia Beach, Virginia. What a state the poor woman must be in! And so forth and so on. Brave lad! He hasn't resigned yet!

As for Shepard, what was going through his mind at that moment, and through much of his body, from his brain to his pelvic saddle, was a steadily increasing desire to urinate. It was no joke. He had been through 120 complete simulations of his flight, simulations that included the smallest details anyone could imagine: the early morning wakeup by the official astronaut physician, Dr. William Douglas, the physical exam, the attaching of all the biosensors, the slipping of the thermometer tube up the rectum, the putting on of the suit, the hooking up of the oxygen tube and the communications lead, the ride out to the launch pad, the insertion, as it was known, into the capsule, the closing of the hatch, the works. They even went through the process of sucking the air out of the capsule with a hose and pressurizing the interior with pure oxygen. Then Shepard would go through yet more simulated rides and aborts, using the capsule itself as if it was a procedures trainer.

Three days ago, it turned out, even *the mental atmosphere* of the real thing was simulated. It was three days ago, May 2, that Shepard was originally scheduled to be launched. The weather made it a doubtful proposition, but they went ahead with the countdown, and Shepard had dinner in crew quarters the night before the flight, with much comradely banter, and Dr. Douglas tiptoed into his room the next morning and woke him up, and then he had the pre-launch breakfast, steak wrapped in bacon, plus eggs—in fact, Shepard went through everything, right up to the point where he would have climbed into the van and ridden out to the rocket, in the belief that this just might be it. Then the launch was postponed because of bad weather. Only at this point did NASA finally reveal that it was Shepard who had been assigned to the flight and was suited up and waiting behind the door in Hangar S. So Shepard had even been through the actual feeling of . . . *this is the day*. But no one had ever seriously envisioned the problem he now faced.

There was no easy way out when one's bladder kept getting larger and the capsule kept getting smaller. The dimensions of this little pod had been kept as tight as possible in order to hold down the weight. Once the various tanks, tubes, electrical circuits, instrument panels, radio hookups, and so on, were crammed in, along with the astronaut's emergency parachute, the space left over was not much more than a holster you could slip two legs and a torso into, with a tiny bit of room remaining for arms. The word they used, *insertion*, was not far off. The seat was literally a mold of Shepard's back and legs. They had packed the plaster right onto his body up at Langley Field. He was now in his seat, but resting on his back. It was as if a man were sitting in a very small sports car that had been upended so that it pointed straight up at the sky. In the rehearsals Shepard had reached the point where he could slip into his slot with one continuous series of moves. But this time, for the real thing, he had on a new pair of white boots, and the boot slipped on the armrest of the couch when he was snaking his right leg up into the capsule. That threw him off, and he wound up with everything inside except his left arm. The capsule was so small that getting his left arm inside became a terrific operation, with him wrenching this way and that and crewmen out on the gantry offering advice. Now he was so jammed in that the cuff on his right wrist, where his glove joined the sleeve of the pressure suit, kept catching on the parachute. He looked at the parachute and all of a sudden wondered what good it was. Technicians were craning in and fastening him to the couch with knee straps and a lap belt and a chest strap and screwing hoses to his pressure suit to maintain the pressure and control the temperature and wiring up leads for the biomedical sensors and the radio hookup and attaching and sealing a hose to the faceplate of his helmet, for oxygen. In all likelihood, if he ever needed the parachute, he'd be a hole in the ground before he could get all these rigs undone. Then they closed the hatch, and he could feel his heartbeat quicken. But it soon subsided, and he was stuffed into this little thimble, lying on his back, practically immobile, with his legs jackknifed.

It was like being a china Cossack packed in a box full of Styrofoam. His face was pointed straight up toward the sky, but he couldn't see it because he had no window. All he had were two little portholes, one on either side, above his head. The *true pilot's* window and hatch wouldn't be ready until the second Mercury flight. He might as well have been inside a box. A greenish fluorescent light filled the capsule. He could see outside only through the periscope window on the panel in front of him. The window was round, about a foot in diameter, in the middle of the panel. Outside, in the dark, the launch crewmen on the gantry could see the lens of the periscope if he pointed it their way. They kept walking in front of it and giving him big grins. Their faces filled the window. There was a wide-angle distortion, so that their noses protruded about eight feet out in front of their ears. When they grinned, they seemed to have more teeth than a perch. Once the dawn broke he could look out of the periscope and turn it this way and that, and see the Atlantic over here . . . and some people down on the ground . . . although the perspectives were a bit strange, because he was lying on his back and the periscope window was not terribly big and the angles were unusual. But then the sun grew brighter and brighter and he kept getting bursts of sunlight in the periscope window, lying on his back like this and looking up, and so he reached up with his left hand and clicked a gray filter into place. That helped a great deal, even though it neutralized most colors. Now that the hatch had been bolted shut, Shepard could hear practically nothing from the outside world except the voices that came over the headset inside his helmet. He spent part of the time, as before any test flight, going over the checklist. Over the headset came the voice of the leader of the launch pad team, saying:

"Auto retro jettison switch. Arm?"

And Shepard would say, "Roger. Auto retro jettison switch. Arm."

"Retro heater switch. Off?"

And Shepard would say, "Retro heater switch. Off."

"Landing bag switch. Auto?"

"Landing bag switch. Auto."

And on down the line. After a while, however, as the holds dragged on, people were coming on the circuit to keep him company and find out how he was bearing up. He could hear Gordon Cooper, who was serving as the "capcom," the capsule communicator, in the blockhouse near the launch pad, and Deke Slayton, who would be the capcom in the flight control center during the launch itself. Cooper had a telephone hookup with the capsule, and every now and then Bill Douglas or another of the doctors would come on the line, apparently as much to assess his spirits as anything else. Wernher von Braun talked to him at one point. The countdown progressed very slowly. Shepard asked Slayton to have someone call his wife to make sure she understood the reasons for the delay. And then he was back in the tight little world of the capsule. There was an irritating tone, very high up in the audible range, coming over the headset the whole time, some sort of feedback sound, apparently. He could hear the hum of the cabin fans and the pressure-suit fans, both used for cooling, and he could hear the inverters moaning up and down. So here he was, stuffed into this little blind thimble, spliced into the loop by every conceivable sort of wire and hose, leading from his body, his helmet, and his suit, listening to the hums, moans, overtones . . . and the minutes and the *hours* began to go by . . . and he'd swivel his knee and ankle joints a few centimeters this way and that to keep his circulation up . . . and two annoying little pressure points began to build up where his shoulders pressed back into the couch . . . and then the tide began to build up in his bladder.

The problem was, there was nothing to urinate into. Since the flight would last only fifteen minutes, it had never occurred to anybody to include a urine receptacle. Some of the simulations had dragged on for so long the astronauts had ended up urinating in their pressure suits. It had been the only thing to do, short of spending hours removing the man from all his hookups and the capsule and the suit itself. The chief danger in introducing liquid into a pure-oxygen environment, such as that of the capsule and the pressure suit, was of causing an electrical short circuit that might start a fire. Luckily, the only wires that urine was likely to come in contact with inside the pressure suit were the low-voltage leads to the biomedical sensors, and the procedure had seemed safe enough. There was even a sponge mechanism inside the suit for removing excess moisture, which ordinarily would come in the form of perspiration. Nevertheless, no one had seriously studied the possibility that on *the day* itself, the day of the first American manned space flight, the astronaut might end up on top of the rocket and stuffed into the capsule with his legs practically immobile for more than four hours . . . with his bladder to answer to. There was no way the astronaut could simply urinate into the lining of the pressure suit and have it go unnoticed. The suit had its own cooling system, and the temperature was monitored by interior thermometers, which led to consoles, and in front of these consoles were some by now highly keyed-up technicians whose sole mission was to stare at the dials and account for every fluctuation. If a nice steaming subdermal river of 98.6 degrees was introduced into the system with no warning, the Freon flow would suddenly increase—Freon was the gas used to cool the suit—and, well, God knew what would result. Would they hold up the whole production? Terrific. Then the No. 1 astronaut could explain, over a radio transmitter, while the nation waited, while the Russians girded themselves for round 2 of the battle of the heavens, that he had just peed in his pressure suit.

Compared to the prospect of such a flap, no matter how minor, in the final phase of the countdown, any possible danger of blowing up on the launch pad was far down an astronaut's list of worries. For a test pilot the right stuff in the prayer department was not "Please, God, don't let me blow up." No, the supplication at such a moment was "Please, dear God, don't let me fuck up."

To come this far . . . and *fuck up* . . .

All along the constant fear of the righteous pilot was not of death but of ending up where John Glenn was this morning: standing by as a mere spear carrier in the drama. You had to hand it to Glenn, however. He had really buckled down and worked like a Trojan as backup pilot during the last month of the training. He had really proved helpful. He had even come up with a little Dawn Patrol camaraderie this morning. Shepard was in a mood for that. Ever since he got up, he had been in his Smilin' Al of the Cape cycle. On the way out in the van he had gotten Gus Grissom to play straight man for the José Jiménez routine. Shepard liked to do the Mexican accent the way Dana did it. "Eef you osk me what make de good astronaut, I tail you dat you got to have de courage and de good blood pressure and de four legs." "Four legs?" Gus asked dutifully. "Weh-ayl, eet ees a dog dey want to send, bot dey theenk dot ees too croo-el." Yes, they would remember him as having been loose as a goose on the way to the top of the rocket. When he came up the gantry elevator and stepped out, to enter the capsule, Glenn was already up there, dressed in white like the technicians. He was smiling. When Shepard finally squeezed himself into the capsule and looked at the instrument panel, there was a little sign on it saying NO HANDBALL PLAYING IN THIS AREA. That was Glenn's little joke, and he grinned and reached in and took the sign out. Actually, it was funny enough . . .

Too late, John! Shepard wasn't going to get run over or break a leg or be struck down by an angry God. He was in the capsule and they were bolting the hatch, and all the others were . . . *left behind* . . . out there . . . beyond the portal . . . There was no way a righteous pilot could have explained to anyone but another pilot what this feeling was, and of course he would not have dared to try to explain it even to a pilot. The holy *first flight!*—and he would be up there at the apex of the entire pyramid if he survived it.

And if he didn't? This would have been even more difficult to explain: the evil odds were essential to the enterprise. That unmentionable *stuff*, after all, involved a man hanging his hide out over the edge in a hurtling piece of machinery. And such unmentionable payoffs it brought you! One, which he had started receiving even *before* this morning, was a look. It was a look of fraternal awe, of awe in the presence of *manly honor,* that came over the faces of other men at a base when a test pilot or combat pilot headed for the aircraft for a mission when the odds were known to be evil. Shepard had rated that look before, particularly when testing overpowered, overweight jet fighters in their first carrier landings. It was the look that came over another man when one's own righteous stuff triggered *his* adrenalin. And this morning, every step of the way, from the crew quarters in Hangar S to the gantry deck outside the capsule, where Glenn and the technicians had been waiting to help him inside, men had beamed that look straight at him—and then they had *broken into applause.* Just as he was getting ready to go onto the gantry and take the elevator up to the capsule, the entire ground crew had started applauding. They had that warm and humid smile on their faces and tears glistened in their eyes and they were banging their hands together and yelling things to him. Shepard had his helmet on, with the visor sealed, and he was carrying his own portable oxygen unit, which was pumping away, and so it was all happening in a muffled pantomime, but there was no mistaking what was going on. They were giving him the applause and homage . . . *up front* . . . come what may! . . . payable in advance!

From a sheerly analytical standpoint one knew that the odds in this flight, while bad enough, were no worse than the odds he had faced before in testing winged aircraft. Wernher von Braun had said repeatedly that the Redstone rocket's record of reliability was 98 percent, which was better than that of some of the Century series of supersonic jet fighters during the test stage. But the truth was that by now Shepard would have accepted far worse odds. He had accepted a great many rewards up front. He and his confreres had already been lionized, such as few pilots in history. The very top pilots, with the most righteous stuff, were content to receive that unmentionable glistening look from aviators and support personnel at their own base. Shepard had already had it beamed upon him by every sort of congressman, canned-food distributor, Associated Florists board chairman, and urban-renewal speculator, not to mention the anonymous little cookies with their trembling little custards who simply materialized around you at the Cape. He had already accepted the payment . . . *up front!*—and millions of wide-open humid eyes were now upon him. The ancient instinct of a people, their so-called folk wisdom, in the matter of the care, preparation, and recompense of single-combat warriors was indeed sound. Like his predecessors in the ancient past, he had reached the blessed state where one was far more afraid of not delivering on his end of the bargain—having been paid up front—than he was of getting killed. *Please, dear Lord, don't let me fuck up.* He was now where he belonged and had striven to be: atop the very thrust point of the danger. He was now precisely at that critical elevation that separated the great pilots, bearing their gigantic albeit invisible pictures of themselves, from the mere mortals on the terrain below. No one would ever know whether or not another type of human being would have handled this day with the same aplomb—this day when one became the first human being ever to sit up on top of an eight-story-high bullet and have a 66,000-pound Redstone rocket lit under his tail. All the racing drivers, mountain climbers, scuba divers, bobsledders, and Seabees they once considered using—what would have been the state of their souls at this moment? Well, by now it was pointless to ask. One could only say this: for the typical competitive military pilot, possessing the regulation-issue heroic self-esteem of the breed, throbbing with rude animal health, convinced of his utterly righteous stuff, and ravenous for glory—which is to say, for a man like Alan Shepard—being where he was right now was his vocation, his calling, his holy *Beruf*. He was home, upon the right stuff's highest elevations.

On top of everything else, the organism's deconditioning was very nearly complete. After all the full-dress rehearsals and simulations of this flight, complete with the sounds, the g-forces, and even the wires protruding from his body, after more than a hundred *pre-creations* of this moment, after riding up the gantry elevator over and over and fitting himself into the human holster and having them close the hatch and start the countdown, after lying in this very capsule, day after day, with the capsule communicator's voice coming over his headset and the signals of flight flashing on the instrument panel, until every inch and every second of the experience was familiar and the capsule had become more like an office than a vehicle . . . it was hard for a man to sense any difference this time in his own nervous system, even though intellectually he knew that this was *that day*. Now and then he could feel the adrenalin building up and his pulse rate increasing and his breathing speeding up and his heart palpitating slightly, and he would force himself to concentrate on the checklist, the console, the equipment connections, the radio hookup, and the rush would pass, and he would be back once more in his workshop, in his procedures trainer.

No, the one thing he had experienced all morning that was not second nature was his aching bladder. That had become the first terra incognita. *Please, dear Lord, don't let me fuck up.*

Shepard waited for another stop in the countdown—this time it was to wait for some clouds to pass over the launch area—and he announced his problem over the closed radio circuit. He said he wanted to relieve his bladder. Finally they told him to go ahead and "do it in the suit." And he did. Because his seat, or couch, was angled back slightly, the flood headed north, toward his head, carrying consternation with it. The flood set off a suit thermometer, and the Freon flow jumped from 30 to 45. On swept the flood until it hit his left lower chest sensor, which was being used to record his electrocardiogram, and it knocked that sensor out partially, and the doctors were nonplused. The news of the flood rushed through the worlds of the Life Science specialists and the suit technicians, like the destruction of Krakatoa, west of Java. There was no stopping it now. The wave rolled on, over rubber, wire, rib, flesh, and ten thousand baffled nerve endings, finally pooling in the valley up the middle of Shepard's back. Gradually it cooled, and he could feel a cool lake of urine in the valley. In any case, the discomfort in his bladder was gone and everything was still. They had not scratched the flight because of the dam break. He had not fucked up.

The next thing the medical team knew, a voice was coming over the closed loop, their private radio linkup with the Mercury capsule:

"Weh-ayl . . . I'm a wetback now."

The man was beautiful!

Imperturbable at every juncture!

Fifteen minutes, in the countdown, before they fire a seven-story bullet full of liquid oxygen underneath him, and he remains:

Smilin' Al!

The hold had now dragged on for four hours, and each engineer monitoring the panels showing the status of the various flight systems was agonizing over whether to declare, finally, that his system was "go"—after which it would be his responsibility if the system malfunctioned. By now there was agony on all sides. It was transmitted into the capsule in a thousand unspoken ways and sometimes in so many words. It was as if Shepard, lying here on his back, inserted, wired, strapped, and screwed into this tiny holster, were the ganglion, the agony junction, for a thousand tense and tortured souls on the gantry outside and on the ground below. Through it all he had remained Smilin' Al of the Cape. At T minus 6—six minutes before the completion of the sequence that would lead to the launching—there was yet another hold, and one of the doctors came on the closed telephone circuit and said to Shepard:

"Are you *really* ready?"

It was hard to figure out whether the question was addressed to the body or the soul. It wandered into the unmentionable terrain of the most righteous stuff itself, and it was Smilin' Al who handled it.

He laughed and said: "Go!"

"Good luck, old friend," said the doctor.

Farewell . . . from the valley of the woeful abyss . . .

At T minus 2 minutes and 40 seconds there was another hold. Now Shepard could hear engineers in the blockhouse agonizing over the fuel pressure in the Redstone, which was running high. He could sense what would be coming next. They were going to talk themselves into resetting the pressure valve inside the booster engine manually. That would mean postponing the launch for another two days at least. He could see it coming! They were going to scrub the whole thing, lest *they* hold themselves accountable for *his hide* if something went wrong! This was not a job for Smilin' Al. It was time for the Icy Commander to arrive and take charge. So he got on the circuit and put the glacial edge on his voice, as only he could do it, and he said:

"All right, I'm cooler than you are. Why don't you fix your little problem . . . and *light this candle.*"

Light the candle! he says. The words of Chuck Yeager himself! The voice of the rocket ace! Oddly enough, it seemed to do the trick. Realizing the astronaut's irritation, they began wrapping up the process and declaring their systems "go." It was nearly 9:30 a.m. by the time the countdown entered its last minute. Shepard's periscope began automatically retracting inside the capsule, and he remembered that he had put in the gray filter to cut out the sunlight. If he didn't remove it, he wouldn't be able to see any colors in flight. So he started moving his left hand toward the periscope, but his left forearm hit the abort handle. Shit! That was all he needed now! Fortunately, he had barely brushed it. The abort handle was the equivalent of the ejection seat cinch ring in an airplane. If the astronaut sensed some catastrophe that the automatic system had not picked up, he could turn the handle and the escape tower rocket would fire and pull the capsule free of the Redstone rocket and bring it down by parachute. That was all he needed—the world waits for the first American spaceman, and Shepard gives them an exhibition of a little man popping up a few thousand feet in a thimble and floating down by parachute . . . He could see it all in a flash . . . Another Popped Cork fiasco . . . The hell with changing the filter. He would look at the world in black and white. Who cared? *Don't fuck up.* That was the main thing.

Time seemed to speed up tremendously in the final thirty seconds of the countdown. In thirty seconds the rocket would ignite right underneath his back. In those last moments his entire life did not pass before his eyes. He did not have a poignant vision of his mother or his wife or his children. No, he thought about abort procedures and the checklist and about not fucking up. He only half paid attention to Deke Slayton's voice over his headset as he read out the final "ten . . . nine . . . eight . . . seven . . . six . . ." and the rest of it. The only word that counted, here in this little blind stuffed pod, was the last word. Then he heard Deke Slayton say it: "Fire! . . . You're on your way, José!"

Louise Shepard was not in the valley of the woeful abyss. She was here in her house in Virginia Beach—but beyond that it was hard to chart the locus of her soul at that moment. Never in the history of flight test had the wife of a pilot been put in any such bizarre position as this. Naturally all the wives had been aware that there might be some "press interest" in the reactions of the wife and family of the first astronaut—but Louise hadn't bargained for anything like what was now going on in her front yard. Every now and then Louise's daughters would peek out the window, and the yard looked like the clay flats three hours after the Marx Midway Carnival pulls in. Mobs of reporters and cameramen and other Big Timers were out there wearing bush jackets with leather straps running this way and that and knocking back their Pepsi-Colas and Nehis and yelling to each other and mainly just milling about, crazy with the excitement of being *on the scene,* bawling for news of the anguished soul of Louise Shepard. They wanted a moan, a tear, some twisted features, a few inside words from friends, any goddamned thing. They were getting desperate. Give us a sign! Give us anything! Give us the diaper-service man! The diaper-service man comes down the street with his big plastic bags, smoking a cigar to provide an aromatic screen for his daily task—and they're all over him and his steamy bag. Maybe he knows the Shepards! Maybe he knows Louise! Maybe he's been in there! Maybe he knows the layout of *chez* Shepard! He locks himself in the front seat, choking on cigar smoke, and they're banging on his panel truck. "Let us in! We want to see!" They're on their knees. They're slithering in the ooze. They're interviewing the dog, the cat, the rhododendrons . . .

These incredible maniacs were all out there tearing up the lawn and yearning for their pieces of Louise's emotional wreckage. The truth was, however, that Louise Shepard could hardly be said to be experiencing the feelings that all these people were so eager to wolf down. Louise had had her chances to become a nervous wreck over Al's flying many times, most recently at Pax River. In 1955 and 1956 Al had tested one hot new fighter plane after another. Their names were a delirium of sharp teeth, cold steel, cosmic warlords, and evil spirits: the Banshee, the Demon, the Tigercat, the Skylancer, and so on, and Al took them through not only maximum performance runs but also high-altitude tests, in-flight refueling tests, and "carrier suitability tests"—a stolid phrase that covered a multitude of ways for a test pilot to expire. Louise knew the entire world of the test pilot's wife . . . the calls from other wives saying that "something" has happened out there . . . the wait, in a little house with small children, to see if the Friend of Widows and Orphans is coming to make his duty call . . . Day after day she tries to be stoic, she tries not to think about the subject, not to pay attention to the clock when he doesn't return from the flight line on time—

Well, my God—what an improvement Project Mercury was over the daily lot of the test pilot's wife! No question about it! The worst part of the Pax River days had been the constant wondering and worrying, alone or with uncomprehending little faces around you. This morning Louise knew exactly where Al was every minute. He was hard to miss. He was on nationwide television. There he was. She had merely to look at the screen. On nationwide television they were talking about no one else. You could hear the laconic baritone voice of Shorty Powers, the NASA public affairs officer, in the Mercury Control room at the Cape periodically reporting the astronaut's status as the countdown proceeded. Then the telephone rings—and she hears that same voice, calling to speak to her, Louise. Al had spoken to Deke, and Deke had spoken to Shorty, and now Shorty—possessor of the voice that the entire nation was listening to—was talking to her personally, explaining the reasons for the delays, as Al had requested. Nor was she alone in this house. Hardly! It was getting crowded in here. In addition to the children, there were her parents, who had come from Ohio and had been here for days. A few of the other wives had arrived. Al had been stationed nearby in Norfolk when the program began, and so they had many Navy friends and neighbors whom they knew well, and quite a few of them had come by. There was quite a burble of voices building up in the living room. It didn't sound very tense in there. And of course she had merely half the newspapers in America in the front yard, plus the usual rabble of gawkers who materialize for car wrecks or roof jumps or traffic arguments, and the whole lot would have liked nothing better than to charge in and gather round if she had opened the front door to them so much as a crack. *Life* had wanted to have two writers and a photographer on the premises to record her reactions from start to finish, but she had held out against that. So they were waiting in a hotel on the beach, and it was agreed they could come inside as soon as the flight was completed. Louise hadn't even had all that much opportunity to sit in front of the television set and let the tension build. She had gotten up before dawn, in the dark, to fix breakfast for everybody who was staying in the house, and then there was the whole business of fixing coffee and whatnot for the other good folks as they arrived . . . until before she knew it she was caught up in the same psychology that works at a *wake*. She was suddenly the central figure in a Wake for My Husband—in his hour of danger, however, rather than his hour of death. The secret of the wake for the dead was that it put the widow on stage, whether she liked it or not. In the very moment in which, if left alone, she might be crushed by grief, she was suddenly thrust into the role of hostess and star of the show. It's free! It's *open house!* Anybody can come on in and gawk! Of course, the widow can still turn on the waterworks—but it takes more nerve to do that in front of a great gawking mob than it does to be the brave little lady, serving the coffee and the cakes. For someone as dignified and strong as Louise Shepard, there was no question as to how it was going to come out. As hostess and main character in this scene, what else was there for the pilot's wife to do but set about pulling everybody together? The press, the ravenous but genteel Beast out there upon the lawn, did not know it but he was covering not the Anguished Wife at Lift-off . . . but the Honorable Mrs. Commander Astronaut at Home . . . in the first wake, not for the dead, but for the Gravely Endangered . . . Louise didn't even have *time* to collapse in neurasthenic paralysis over the possible fate of her husband. It was all that the star and hostess could do to get back to the TV room in time for the final minutes of the countdown, to watch the flames roaring out of the Redstone's nozzles.

Hmmmm . . . with all the world wondering about the state of her soul at this moment . . . what kind of face should she have on?

Over his headset Shepard heard Deke Slayton, from inside the Mercury Control Center, as it was called, saying, "We have lift-off!" And as he had done hundreds of times on the centrifuge and on the procedures trainer, he reached up and turned on the onboard clock—which would tell him when to do this and that on his flight schedule—and he said over his microphone: "Roger, lift-off, and the clock is started" . . . as he had said hundreds of times on the centrifuge and the procedures trainer. And then, with the automatic anticipation of someone who has heard the same phonograph record over and over again and is now hearing it yet once again and senses every chord and phrase before it sounds, he waited for the gradual buildup of the g-forces and for the beating noise as the rocket lifted off . . . which he had felt and which he had heard hundreds of times on the centrifuge—

Hundreds of times! Even if he had been ordered at that point to broadcast to the American people a detailed description of precisely what it felt like to be the first American riding a rocket into space, and even if he had had the leisure to do it, he could not possibly have expressed what he was feeling. For he was introducing the era of precreated experience. His launching was an utterly novel event in American history, and yet he could feel none of its novelty. He could not feel "the awesome power" of the rocket beneath him, as the broadcasters kept referring to it. He could only compare it to the hundreds of rides he had taken on the centrifuge at Johnsville. The memory of all those rides was imbedded in his nervous system. Scores of times he had sat in the gondola on the wheel, as he sat here now, in his pressure suit, with the Mercury instrument panel in front of him and the noise of a Redstone launch piped into his headset. And compared to that—what was happening at this moment was not awesome. Quite the opposite. He was braced for it, but . . . *It didn't throw you around like the centrifuge* . . . The centrifugal force of the centrifuge threw you around inside the capsule as well as increasing the speed and the g-forces . . . The rocket was smoother and easier . . . *It wasn't as noisy as the centrifuge* . . . On the wheel the recorded sound of the Redstone rocket was piped directly into the capsule. But now, since he was the figurine in the packing box, it came from outside, through layer after layer. By the time it penetrated the escape-tower cover and the capsule wall and the contoured seat behind his head and the helmet and his headset, it was no louder than the engine noises a commercial airline pilot hears on takeoff. In fact, he was far more conscious of the noises inside the capsule . . . The camera . . . There was a camera set up to record his facial expressions and eye and hand movements, and he could hear that whirring about a foot from his head . . . There was a tape recorder set up to record all the sounds inside the capsule, and he could hear its little motor running . . . And the fans and the gyros and inverters . . . it was like an extremely compact modern kitchen in here . . . with all the gadgets running at once . . . And of course the radio . . . He thought he would have to turn the volume up to maximum, as he had on the centrifuge, but he didn't have to touch it. All he had to do was talk into his voice mike and say the exact same things he had said a thousand times on the procedures trainer . . . "Altitude one thousand . . . one-point-nine g's . . ." and so on . . . and they would answer back, "Roger, we copy, you're looking good . . ." and it even *sounded* the same way over the headset.

He could still see nothing. The periscope was still retracted. He had no way to judge his speed except by the needle on the instrument panel that showed his rising altitude and the buildup of the g-forces on his body. But this was gradual. It was a very familiar feeling. He had felt it hundreds of times on the centrifuge. It was much easier than pulling four g's in a supersonic aircraft, because he didn't have to struggle to push his hands forward against the weight of the g-forces to control the trajectory of the aircraft. He didn't have to lift a finger if he didn't want to. The computers guided the rocket automatically by swiveling the nozzles. He had very little feeling of motion, just the g-forces driving him deeper and deeper into the seat he was lying on.

The rocket rose so gradually it took forty-five seconds to reach Mach 1. An F–104 fighter plane could have done it faster. When the rocket reached transonic speed, .8 Mach, a vibration started building up—just as it had for the X series of rocket planes at Edwards years before. Shepard was quite prepared for it . . . He had been through it on the centrifuge . . . so many times . . . But it was a different vibration. It didn't shake his head about as violently, but the amplitudes were more rapid. His vision began to blur. He couldn't read the instrument panel any longer. He started to report the phenomenon over his headset but thought better of it. *Some bastard would panic and abort the mission.* Christ, the vibrations, no matter how bad, were preferable to an abort. Within another thirty seconds all the vibrations were gone, and he knew he was going supersonic. With the vibration gone, he once again had no sensation of motion. He was still blind to the world outside. He was still lying on his back, looking at the instrument panel, no more than eighteen inches from his eyes, in the greenish light of the capsule. The g-forces reached six times the force of gravity, then began to tail off as the rocket and the capsule approached the weightless phase of the arc of flight. *It was different from the centrifuge—It was easier!* On the centrifuge the only way to reduce the g-load as you simulated the approach to the weightless phase was to reduce the speed of the centrifuge arm, and this always threw you forward against your straps. When the simulated moment of zero-g arrived, you were thrown against the straps quite hard. On a rocket ride, however—as all the rocket pilots at Edwards knew—your speed did not abruptly decrease as the rocket ran out of fuel. It kept sailing. Shepard was eased into weightlessness so smoothly it was as if the weight of the g-forces had simply slid off his body. Now he could feel his heart pounding. The most critical part of the flight, next to the launch itself, was only a moment away . . . the separation of the capsule from the rocket . . . He heard a muffled explosion from above . . . *just the way it sounded in the simulations* . . . and the escape rocket blew off and the capsule was now free of the rocket. The force of the rocket pulling away accelerated his speed, and he felt as if he had had a kick from below. A three-inch-long rectangular light lit up green on the instrument panel. On it were the letters JETT TOWER, for Jettison Tower. With the tower gone, the periscope would start operating, and he could look out, but he had his eyes pinned on the green light. It was beautiful. It meant everything was going perfectly. He could forget about the abort handle now. The launch phase was over. The most treacherous part of the entire ride was now behind him. All that endless training in the procedures trainer . . . "Abort! . . . Abort! . . . Abort!" . . . he could put that behind him and forget about it. Small portholes were on either side, above his head, and he could see only the sky through them. He was now slightly more than a hundred miles up. The sky was almost navy blue. It was not the much-talked-of "blackness of space." It was the same dark-blue sky that pilots begin to see at 40,000 feet. It looked no different. The capsule was now automatically turning so that its blunt end, the end beneath his back, which had a heat shield for the reentry into the earth's atmosphere, was aimed toward the target area. He was facing back toward Florida, toward the Cape. He could not see the earth out of the high portholes at all, however. He wasn't even particularly interested in looking. He kept his eyes pinned on the instrument panel. The gauges told him he was weightless. After reaching this point so many times on the procedures trainer, he knew he must be weightless. But he felt nothing. He was so tightly strapped and stuffed into this little human holster there was no way he could float as he had in the cargo bays of the big C-131s. He didn't even experience the tumbling sensation he felt when riding backseat in the F-100s at Edwards. *It was all milder!—easier!* No doubt he should say something to the ground about the sensation of weightlessness. It was the great unknown in space flight. *But he didn't feel anything at all!* He noticed a washer floating in front of his eye. The washer must have been left in there by a workman. It was just floating there in front of his left eye. That was the only evidence his five senses had to show that he was weightless. He tried to grab the washer with his left hand, with his glove, but he missed it. It floated away, and he couldn't move his hand far enough to get it. He had no sensation of speed at all, even though he knew he was going Mach 7, or about 5,180 miles an hour. There was nothing to judge speed by. There were no vibrations at all in the capsule. Since he was now completely beyond the earth's atmosphere and was no longer attached to the rocket, there was no rushing sound whatsoever. It was as if he were standing still, parked in the sky. The sounds of the interior of the capsule, the rising and falling and whirring and moaning of the inverters and the gyros . . . the cameras, the fans . . . the busy little kitchen—they were exactly the same sounds he had heard over and over in the simulations inside the capsule on the ground at the Cape . . . The same busy little kitchen in operation, whirring and buzzing and humming along . . . There was nothing new going on! . . . He knew he was in space, but there was no way to tell it! . . . He looked out the periscope, the only way he had of looking at the earth. *The goddamned gray filter!* He couldn't see any colors at all! He had never changed the filter! The first American to ever fly this high above the earth—and it was a black-and-white movie. Nevertheless, they'll want to know about it—

"What a beautiful view!" he said.

He could hear Slayton say: "I bet it is."

In fact, there was a cloud cover over most of the East Coast and much of the ocean. He was able to see the Cape. He could see the west coast of Florida . . . Lake Okeechobee . . . He was up so high he seemed to be moving away from Florida ever so slowly . . . And the inverters moaned up and the gyros moaned down and the fans whirred and the cameras hummed . . . He tried to find Cuba. Was that Cuba or wasn't that Cuba? Over there, through the clouds . . . Everything was black and white and there were clouds all over . . . There's Bimini Island and the shoals around Bimini. He could see that. *But everything looked so small!* It had all been bigger and clearer in the ALFA trainer, when they flashed the still photos on the screen . . . The real thing didn't measure up. It was *not realistic.* He couldn't see anything but a medium-gray ocean and light-gray beaches and dark-gray vegetation . . . There were the Bahamas, Grand Bahama Island, Abaco Island . . . or were they? The pale-gray cloud cover and the medium-gray water and the pale-gray beaches and the medium-gray . . . the grays ran together like goulash . . . He didn't have time to fool with it. He had a long checklist. He was supposed to try out the roll, yaw, and pitch controls. They were all operating automatically up to now. He switched over to manual one at a time and tried using the hand controller.

"Okay," he said into his microphone, "switching to manual pitch."

And the capsule nosed up and then down.

"Pitch is okay," he said. "Switching to manual yaw."

All this he seemed to have said a hundred, a thousand times before, in the procedures trainer. And the capsule yawed from side to side, and he had *felt* it a thousand times before—on the ALFA trainer. Only the sound was different. Every time he turned the manual control handles, hydrogen-peroxide jets discharged outside the capsule. He knew it was happening, because the capsule would pitch, roll, yaw, just as it was supposed to, but he couldn't hear the jets. On the ALFA trainer he had always been able to hear the jets. The real thing was not nearly so realistic as that. The capsule pitched and yawed just like the ALFA . . . *no difference* . . . but he couldn't hear the jets at all because of the humming, moaning, and whining of the inverters and gyros and fans . . . the busy little kitchen.

Whuh?—before he knew it the ride was practically over. It was time to prepare himself for re-entry into the earth's atmosphere. He was descending the arc . . . just like a mortar shell on the way down . . . The ground was starting the countdown for the firing of the retro-rockets that would slow the capsule down for re-entry through the atmosphere. There was no need for them on this flight, because the capsule was going to come back down in any event like a cannonball, but they would be essential in orbital flights, and this flight was supposed to test the system. The retro-rockets fired automatically. He didn't have to move a finger. There was another muffled rocket firing, not a loud noise at all . . . The little kitchen kept humming and whining away . . . He was riding backward, still facing Florida. The rocket firing drove him backward into his seat with a force of about five g's. It was much more sudden than the transition from six g's to zero-g on the way up. *Man, that wasn't like the centrifuge.* The centrifuge jerked you around at this point. In the next instant he was weightless again, as he knew he would be. The retrofire was like pumping the brakes once. But it had slowed down the capsule, and soon the g-forces would start building up. They would build up gradually, however. He knew the interval exactly . . . At this point an explosive device was supposed to go off and jettison the rigs that held the retro-rockets. He heard the little dull explosions outside the capsule and then he saw one of the straps of the rig go by his window. He could feel the twist from the torque of the retro-rocket nozzles, but it corrected itself automatically. Now he was supposed to practice controlling the attitude of the capsule while it was slowing down. So he swung it this way and that way . . . like a Ferris-wheel seat . . . Oh, shit! The dial showed the g-forces were building up slightly . . . One-half g . . . This meant he had about forty-five seconds to check out the *stars* . . . He was supposed to look at the stars and the horizon and see if he could get a fix on particular constellations. Eventually this would have some bearing on space navigation. But mainly there would be millions down there wanting to know that the first American in space had been up there in the neighborhood of the stars . . . Oh, they would all want to know how the stars looked from outside the earth's atmosphere. They weren't supposed to twinkle when you looked at them from up here. They would just be little shining balls in the blackness of space . . . Except that it wasn't black, it was blue . . . The *stars,* man! . . . He kept staring out the windows trying to make out the stars . . . He couldn't see a damned thing, not one goddamned star. If the capsule were pitched up this way or rolled over that way, he would get zaps of sunlight in his face. If he pitched it or rolled it this way—he still couldn't see a damned thing. The light in the interior of the capsule was too bright. All he could make out was the navy-blue sky. He couldn't even see the horizon. There were supposed to be spectacular color bands at the horizon when viewed from above the atmosphere. But he couldn't see them out the portholes. He could see them through the periscope—but the periscope was a black-and-white movie.

What inna namea—the g-forces were building up too fast! This wasn't the way it happened on the centrifuge! The g-forces were coming up so fast, driving him back so deep into his seat, he knew he couldn't complete the maneuvers he was supposed to make on the fly-by-wire system. Whether he did them or not on this flight would make no difference as far as his own safety was concerned. This was practice for orbital flights. In orbital flights the one thing the astronaut would be able to *do* would be to hold the capsule at the proper attitude, the proper angles, if the automatic system malfunctioned. Nevertheless—he was behind on the checklist! *Falling behind* put you on the threshold of *fucking up* . . . Soon he wouldn't be able to control the capsule's attitude with the manual controls or the fly-by-wire. The g-forces would be too severe for him to use his arms. So he switched back to automatic, forgetting to shut off one of the manual buttons when he did so . . . He was behind on the checklist! The simulation had crossed him up! He was supposed to have more time than this! He was in no danger whatsoever—but how could real life vary this much from the simulation! Looking for those motherless *stars* and *color bands* had thrown him off!—eaten up time! But even so, the buildup of the g-forces came smoothly. The capsule began to swing from side to side as it came through the denser and denser layers of atmosphere, but it was not nearly as rough a ride as the same interval in the centrifuge. He was lying on his back again. If he looked up, he looked straight toward the sky. He tensed his calves and the muscles of his abdomen to counteract the g-forces . . . just as he had a thousand times on the centrifuge . . . He forced his breath out in grunts as the g's pressed down on his chest . . . He grunted out the g readings as they rose on the dial . . . "Uhhsix . . . uh seven . . . uhh eight . . . uhh nine . . ." Then he kept repeating, kept grunting out the word "Ohhkay . . . Ohhkay . . . Ohhkay . . . Ohhkay . . . Ohhkay . . . Ohhkay . . . Ohhkay" . . . to let the ground know that the g-forces were on top of him but that he was all right. He could see nothing except indigo out the portholes. He didn't even bother looking. He kept staring at the instrument panel at the big lights that would show the parachutes were coming out. They would come out automatically. The first green light came on. The drogue chute, the parachute that pulled the main chute out, had deployed. Now Shepard could see it through the periscope. He could see the needle pass 20,000 on his altimeter. *Why doesn't it—* At 10,000 feet the main chute came out, as he could see through the periscope. Then it filled and the jolt slammed him back into his seat once more. A *kick in the butt! For reassurance!* He knew he had it made. The capsule swayed from side to side under the parachute, but there was nothing to it. He was lying on his back. The landing bag was right underneath his back. He started undoing his knee straps. Get out of all these goddamned straps, hoses, and wires. That was the main thing. He took his face seal hose and oxygen exhaust hoses off his helmet. Through the periscope he could see he was coming closer and closer to the water. It was sunny out there. He was near Bermuda. The ships were all nearby. He could hear them clearly on the radio. Then the blunt end of the capsule hit the water. It was right under his back. The impact drove him back into his seat yet again. *It was just like they said it would be!* It was about the same jolt you get when you land on the deck of an aircraft carrier. No more than that. The capsule keeled over to the right. The right-hand window was underwater. But he could look out the other one and see his yellow dye marker on the surface of the water. It was released automatically. It made a big yellow stain, so that the rescue helicopters could spot the capsule better. Shepard was busy trying to get himself out of the rest of the rig. *Get free!* He broke the neck ring seal off his helmet so that he could get the helmet off. He kept looking at the window that was underwater. The capsule was supposed to right itself. The goddamned window stayed underwater. Gurgling! He kept looking all around it for signs of water seeping in. He couldn't see any but he could hear it. It had finally started seeping in on the ape's flight. The goddamned capsule was not made for the water. What a way to blow the whole thing this would be. Bobbing around in the goddamned water in a bucket. The capsule slowly came up vertical. The helicopter was already overhead radioing for instructions. He had it made as long as he didn't fuck it up getting out of the capsule. He opened the door at the top of the capsule, the neck, with a cable device and pulled himself up. A tremendous burst of sunlight hit him, sunlight on the open sea. It was barely quarter of ten in the morning. He was forty miles from Bermuda on a sunny day out in the Atlantic. The noise of the helicopter overhead obliterated every other sound. The crewmen were lowering a rescue sling, which looked like an old-fashioned horse collar, the kind they used to put on dray horses. It was a big helicopter, an industrial type. It had already hooked onto the capsule and pulled it up out of the water. There were some Marine Corps crewmen inside the helicopter. The noise was overwhelming. They kept staring at him and smiling. It was that glistening look.

It took only seven minutes to reach the helicopter's mother ship, the aircraft carrier *Lake Champlain.* The capsule swayed back and forth below the helicopter. It was very sunny. It was a perfect day in May out near Bermuda. The helicopter began descending to the flight deck of the carrier. Shepard looked down and he could see hundreds of faces. They were all looking up toward the helicopter. The entire crew of the ship seemed to be on the deck. Their faces were all turned up toward him in the sunlight. Hundreds of faces turned up in the sun. They covered the whole aft section of the deck. They were massed in between the moored airplanes. They were all looking up toward the helicopter and moving, surging toward the spot where the helicopter would come down. He could make out a whole force of masters-at-arms down there trying to hold them back behind the ropes. As the helicopter came close to the deck, he could see the faces more clearly, and they had that look. Hundreds of faces already had that glistening look.

XI. The Unscrewable Pooch

Glenn and the others now watched from the sidelines as Al Shepard was hoisted out of their midst and installed as a national hero on the order of a Lindbergh. That was the way it looked. As soon as his technical debriefings had been completed, Shepard was flown straight from Grand Bahama Island to Washington. The next day the six also-rans joined him there. They stood by as President Kennedy gave Al the Distinguished Service Medal in a ceremony in the Rose Garden of the White House. Then they followed in his wake as Al sat up on the back of an open limousine waving to the crowds along Constitution Avenue. Tens of thousands of people had turned out to watch the motorcade, even though it had been arranged with barely twenty-four hours' notice. They were screaming to Al, reaching out, crying, awash with awe and gratitude. It took the motorcade half an hour to travel the one mile from the White House to the Capitol. Al sometimes seemed to have transistors in his solar plexus. But not now; now he seemed truly moved. They adored him. He was on . . . the Pope's balcony . . . Thirty minutes of it . . . The next day New York City gave Al a ticker-tape parade up Broadway. There was Al on the back ledge of the limousine, with all that paper snow and confetti coming down, just the way you used to see it in the Movietone News in the theaters. Al's hometown, Derry, New Hampshire, which was not much more than a village, gave Al a parade, and it drew the biggest crowd the state had ever seen. Army, Navy, Marine, Air Force, and National Guard troops from all over New England marched down Main Street, and aerobatic teams of jet fighters flew overhead. The politicians thought New Hampshire was entering Metro Heaven and came close to renaming Derry "Space-town U.S.A." before they got hold of themselves. In the town of Deerfield, Illinois, a new school was named for Al, overnight, just like that. Then Al started getting tons of greeting cards in the mail, cards saying "Congratulations to Alan Shepard, Our First Man in Space!" That was already printed on the cards, along with NASA's address. All the buyers had to do was sign them and mail them. The card companies were cranking these things out. Al was that much of a hero.

Next to Gagarin's orbital flight, Shepard's little mortar lob to Bermuda, with its mere five minutes of weightlessness, was no great accomplishment. But that didn't matter. The flight had unfolded like a drama, the first drama of *single combat* in American history. Shepard had been the tiny underdog, sitting on top of an American rocket—*and our rockets always blow up*—challenging the omnipotent Soviet Integral. The fact that the entire thing had been televised, starting a good two hours before the lift-off, had generated the most feverish suspense. And then he had gone through with it. He let them light the fuse. *He hadn't resigned.* He hadn't even panicked. He handled himself perfectly. He was as great a daredevil as Lindbergh, and he was purer: he did it all for his country. Here was a man . . . with the right stuff. No one spoke the phrase—but every man could feel the rays from that righteous aura and that primal force, the power of physical courage and manly honor.

Even Shorty Powers became famous. "The voice of Mercury Control," he was called; that, and "the eighth astronaut." Powers was a lieutenant colonel in the Air Force, a onetime bomber pilot, and all during Shepard's flight he had come on the air from the flight control center at the Cape saying, "This is Mercury Control . . ." and reporting the astronaut's progress with a baritone coolness of the combat pilot's righteous sort, and people loved it. After the capsule splashed down, Powers had quoted, or seemed to have quoted, Shepard as saying everything was "A-Okay." In fact, this was a Shorty Powers paraphrase borrowed from NASA engineers who used to say it during radio transmission tests because the sharper sound of A cut through the static better than O.

Nevertheless, "A-Okay" became shorthand for Shepard's triumph over the odds and for astronaut coolness under stress, and Shorty Powers was looked up to as the medium who communicated across the gulf between ordinary people and star voyagers with the right stuff.

Bob Gilruth's status rose sharply, too. After a solid year of flak and grief, Gilruth had finally earned the eminence of riding in one of the limousines in Shepard's triumphal motorcade through Washington. James E. Webb was sitting next to him, and they were looking out at the thousands of people who were smiling and crying and waving and cheering and taking pictures. "If it hadn't worked," said Webb, "they'd be asking for your head." As it was, Gilruth and Mercury and NASA were, all at once, names that stood for American technological competence. *(Our boys no longer botch it and our rockets don't blow up.)*

None of this was lost on the President. His opinion of NASA had now swung around 180 degrees. Webb was aware of that. Three weeks before, after Gagarin's flight, when Kennedy had summoned Webb and Dryden to the White House, the President had been in a funk. He was convinced that the entire world was judging the United States and his leadership in terms of the space race with the Soviets. He was muttering, "If somebody can just tell me how to catch up. Let's find somebody—anybody . . . There's nothing more important." He kept saying, "We've got to catch up." *Catching up* became an obsession. Finally, Dryden told him that it looked hopeless to try to catch up with the mighty Integral in anything that involved flights in earth orbit. The one possibility was to start a program to put a man on the moon within the next ten years. It would require a crash effort on the scale of the Manhattan Project of the Second World War and would cost anywhere from twenty to forty billion dollars. Kennedy found the figure appalling. Less than a week later, of course, the Bay of Pigs debacle had occurred, and now his "new frontier" looked more like a retreat on all fronts. Shepard's successful flight was the first hopeful note Kennedy had enjoyed since then. For the first time he had some confidence in NASA. And the tremendous public response to Shepard as the patriotic daredevil, challenging the Soviets in the heavens, gave Kennedy an inspiration.

One morning Kennedy asked Dryden, Webb, and Gilruth to come to the White House. They sat down in the Oval Office, and Kennedy said: "All over the world we're judged by how well we do in space. Therefore, we've got to be first. That's all there is to it." After this buildup Gilruth figured Kennedy was going to tell them to cut the Redstone suborbital flights short and move straight to the series of orbital flights using the Atlas rocket. They were still considering six and possibly ten more suborbital flights, like Shepard's, using the Redstone rocket. Gilruth had thought of moving straight to the orbital flights, although it was a daring proposition, given the problems they had been having on tests of the Mercury-Atlas system. So they were all absolutely startled when Kennedy said: "I want you to start on the moon program. I'm going to ask Congress for the money. I'm going to tell them you're going to put a man on the moon by 1970."

On May 25, twenty days after Shepard's flight, Kennedy appeared before Congress to deliver a message on "urgent national needs." This was, in fact, the beginning of his political comeback from the Bay of Pigs disaster. It was as if he were starting his administration over and delivering a new inaugural address.

"Now is the time to take longer strides," he said, "time for a great new American enterprise, time for this nation to take a clearly leading role in space achievement, which in many ways may hold the key to our future on earth." He said that the Russians, thanks to "their large rocket engines," would continue to dominate the competition for some time, but that this should only make the United States step up its efforts. "For while we cannot guarantee that we shall one day be first, we can guarantee that any failure to make this effort will make us last. We take an additional risk by making it in full view of the world; but as shown by the feat of Astronaut Shepard, this very risk enhances our stature when we are successful." Then he said: "I believe this nation should commit itself to achieving the goal, before this decade is out, of landing a man on the moon and returning him safely to the earth. No single space project in this period will be more impressive to mankind or more important for the long-range exploration of space; and none will be so difficult or expensive to accomplish."

Congress was not about to quibble over expenses. NASA was given a $1.7 billion budget for the next year, and that was merely the start. It was made clear that NASA could have practically as much as it wanted.

Shepard's flight had made quite a hit. An amazing period of "budgetless financing" began. It was astonishing. Suddenly money was in the air. Businessmen of all sorts were trying to bestow it directly upon Shepard. Within a few months Leo DeOrsey, who was still the boys' unpaid business manager, had counted up half a million dollars' worth of proposals from companies who wanted Shepard to endorse products. One congressman, Frank Boykin of Alabama, wanted the government to give Shepard a house. Shepard turned it all down, but it did make one stop and think.

If Smilin' Al's experience was any indication, then this astronaut business was becoming even more of a Fighter Jock Heaven than it had been in the first year of Project Mercury. Eisenhower had never paid much attention to the astronauts personally. He had looked upon them as military volunteers for an experiment, and that was that. But Kennedy now made them an integral part of his Administration and included them in its social as well as its official life.

The other fellows had gone along for Al's trip to the White House, but the wives had remained at the Cape. When they flew back down to Patrick, the wives were out at the field to meet their plane. And they all had a single question: *"What's Jackie like?"*

Jackie Kennedy's exotic face and Sixth Floor Designer Collection clothes were in every magazine . . . They all had a curious sensation. In one corner of their souls they were still junior-level military officers & wives who saw people like Jackie Kennedy only on the pages of magazines and newspapers. And at the same time they were beginning to realize that they were part of the strange world where Those People, the people who do things and run things, actually exist.

"What's Jackie like?"

Soon enough they would all meet her. They would go to private lunches at the White House, where there were so many servants there seemed to be one behind every chair. They played you like man-to-man basketball. Jack Kennedy was very warm toward the fellows. He courted them. That glistening look would come over his face from time to time. The business of *manly honor* cut through everything at last, and even the President would become merely another awed male in the presence of the right stuff. As for Jackie, she had a certain Southern smile, which she had perhaps picked up at Foxcroft School, in Virginia, and her quiet voice, which came through her teeth, as revealed by the smile. She barely moved her lower jaw when she talked. The words seemed to slip between her teeth like exceedingly small slippery pearls. Her excitement, if any, over the prospect of lunch with seven pilots and their fraus may not have been great. But she couldn't have been kinder or more attentive. At one point she invited Rene Carpenter back for a private visit, and they talked like any two friends, about all sorts of things, including the problems of raising children in the modern age. One had only to think of any other seven pilots' wives in a squadron . . . All of a sudden the Honorable Mrs. Astronaut existed on a plateau, upon the upper reaches of American protocol, where the perquisites included *Jackie Kennedy.*

And for the fellows, it was pure heaven. None of this altered the Edwards-style perfection of their lives. It merely added something new and marvelous to the ineffable contrasts of this astronaut business. Within hours after lunch at the White House or waterskiing in Hyannis Port you could be back at the Cape, back Drinking & Driving in that marvelous Low Rent rat-shack terrain, back in your Corvette spinning out on the shoulders of those hardtack Baptist roadways and pulling into the all-night diner for a little coffee to stabilize the system for the proficiency runs ahead. And if you had switched to your Ban-Lon shirts and your go-to-hell pants, they might not even recognize you in there, which would be all the better, and you could just sit there and drink coffee and have a couple of cigarettes and listen to the two policemen in the next booth with the Dawn Patrol radio sets in their pockets, and a little voice packed in static would be coming out of the radios saying, "Thirty-one, thirty-one [garble, garble] . . . man named Virgil Wiley refuses to return to his room at the Rio Banana," and the policemen would look at each other as if to say, "Well, shit, is that anything to have to rise up from over a plate of french fries and death balls for?"—and then they'd sigh and start getting up and buckling on their gunbelts, and about the time they would head out the door, in would come the Hardiest Cracker, the Aboriginal Grit, an old guy drunk as a monkey and ricocheting off the doorframe and sliding in bowlegged over a counter stool and saying to the waitress:

"How you doing?"

And she says: "So-so, how you doing?"

"I ain't doing any more," he says. "It's dragging in the mud and it won't come up"—and since this doesn't get a rise out of her, he says it again: "It's dragging in the mud and it won't come up," and she just clamps a burglar-proof look of aloofness across her face—and all this was bound to make you smile, because here you were, listening to the merry midnight small talk of the hardiest hardtack crackers of the most Low Rent stretch of the Cape, and just twelve hours ago you were leaning across a table in the White House, straining to catch the tiny shiny pearls of tinytalk from the most famous small talker in the world—and somehow you belonged and thrived in both worlds. Oh, yes, it was the perfect balance of the legendary Edwards, the fabled Muroc, in the original Chuck Yeager and Pancho Barnes days . . . now brought forward into the billion-volt limitless-budget future.

The truth was that the fellows had now become the personal symbols not only of America's Cold War struggle with the Soviets but also of Kennedy's own political comeback. They had become *the* pioneers of the New Frontier, recycled version. They were the intrepid scouts in Jack Kennedy's race to beat the mighty Integral to the moon. There was no way they could be regarded as ordinary test pilots, much less test subjects, ever again.

For Gus Grissom that was a very fortunate thing.

Gus was assigned to the second Mercury-Redstone flight, scheduled for July. He would be in a newer capsule, one in which certain changes had been made—all of them in response to the astronauts' insistence that the astronaut function more like a pilot. It had been too late to renovate the capsule Shepard used, but Grissom's had a window, not just portholes, and a new set of hand controllers designed to let the astronaut control the capsule's attitude in a manner that was more like an aircraft pilot's, and a hatch with a set of explosive bolts that the astronaut could blow in order to get out of the capsule after the splashdown. Nevertheless, the flight would be a repetition of Shepard's, a suborbital lob three hundred miles out into the Atlantic. Gus himself encouraged certain changes in the flight plan. Since he would be taking the next flight, he had sat in on Shepard's debriefing sessions on Grand Bahama Island. Nobody, not even within NASA, was about to criticize Al openly for anything he did, but there was implied criticism of what he did toward the end of the flight when the g-forces built up faster than he expected them to and he was desperately staring out of his two portholes trying to find some stars. Some character from the Flight Systems Division kept asking him if he hadn't left a manual control button on after he shifted over to automatic control. This would have wasted hydrogen peroxide, the fuel that operated the attitude-control jets. It didn't particularly matter on a fifteen-minute suborbital flight, but it could have made a difference if it had been an orbital flight. Al kept saying he didn't think he had left the button on but he really couldn't say for sure. And this character kept coming back in and asking the question all over again. This was the first indication that the fellows had about a major truth concerning space flights. You didn't "take" the capsule off the ground, you didn't bring it up to altitude, you didn't alter its course, and you didn't land it; i.e., you didn't *fly* it—and so your performance was not going to be rated on how well you *flew* the craft, as it would be in flight test or combat. You could be rated only on how well you covered the items on your checklist. Therefore, the fewer items you had on your checklist, the better shot you had at a "perfect" flight. Each flight was so expensive there would always be people on the ground—engineers, doctors, and scientists—who wanted to load up your checklist with all sorts of things to try, their little "experiments." The way you handled this problem was to allow "operational" items on your list and growl, gruff, and otherwise balk over the rest. Testing the attitude-control system was acceptable, because that had "operational" written all over it. That was *like* flying an airplane. By the time he took off, Gus's checklist had been pared down to the point where he could concentrate on the new hand controller that had been installed.

Gus stayed at the Holiday Inn until practically the day before his flight, maintaining an even strain. He cut down a little on the waterskiing, which was his main form of exercise, and on the nocturnal proficiency runs on the highways, in order to avoid getting banged up on the eve of the flight, but otherwise life went on pretty much as usual at the Cape, here in Fighter Jock Heaven.

One night just before the flight, when he was in the cocktail lounge evening out the strain a bit, who should Gus run into but Joe Walker. NASA had given Joe a few days off from Edwards to attend the launch, and so here he was. By this time, July 1961, Walker and Bob White had been flying up a storm in the X–15. In April, White had set a new speed record of Mach 4.62, which was just over 3,000 miles an hour, and in May Joe Walker had topped that by going Mach 4.95, and White had come right back in June and flown Mach 5.27. The X–15 now had the Big Engine, the XLR–99, with its 57,000 pounds of thrust. The True Brothers were ready to go all-out toward their goal of exceeding Mach 6 and an altitude of more than fifty miles . . . in *piloted* flight. *Piloted!* These developments could be found in the press . . . if one cared to look for them . . . but they were obscured by Gagarin's flight, followed by Shepard's flight . . . the single combat for the heavens. As a matter of fact, Joe Walker had taken the X–15 to Mach 4.95, the highest speed in the history of aviation, on the same day that Kennedy addressed Congress to propose the race for the moon . . . Next to the notion of a moon voyage Walker's Mach 4.95 was pretty pedestrian stuff. But surely the truth would dawn on them eventually! It was with that in mind . . . the simple truth! . . . that Joe Walker happened to run into Gus Grissom at the Holiday Inn's cocktail lounge.

Both Gus and Joe had knocked back a few, it being after dark, after all, and Joe starts in with a little Yeager-style country-boy banter about how Gus and his pals had better hurry up or him and his boys would pass them on the way up. Oh, yeah, says Gus, how's that? Well, says Joe Walker, we've got a 57,000-pound rocket engine now, and the Redstone that shoots your little peapod up there only puts out 78,000, so we're almost up with you—and we *fly* the damn thing. We actually *fly* it and we *land* it. Joe Walker meant to keep it light and just rag Grissom a little bit, but he couldn't hold back a note in his voice concerning where things actually stood in the true scheme of things, on the *real* pyramid of flying competition. Everybody is looking at Grissom, the astronaut, to see what he's going to say. Grissom, who is a tough little nut when he wants to be, stares at Walker . . . and then he breaks into a grin and starts a kind of gruff-gus chuckle. Oh, I'll be looking over my shoulder the whole time, Joe, and if you come by, I swear I'll wave.

And so much for Joe Walker and the True Brothers! It was all right there in that scene, the *new* simple truth. Grissom didn't even *feel* angry. There was nothing that Joe Walker could say or do—and nothing that even Chuck Yeager himself could say or do—that would change the new order. The astronaut was now at the apex of the pyramid. The rocket pilots were already . . . the old guys, the eternal remember-whens . . . Oh, it didn't even have to be *said!* It was in the air, and everyone knew it. Hell, when they started flying jets and rocket planes at Muroc, somewhere there must have been the old guys, the bitter old bastards, the remember-whens, who could just fly the hell out of a propeller plane and were still insisting that that was what it was all about. Flying wasn't a competition like baseball or football. No, in flying any major advance in technology could change the rules. The Mercury rocket-capsule system—the word "system" was now on everybody's lips—was the new cutting edge. No, Gus had no need to get excited any more over Joe Walker or anybody else at Edwards.

Gus seemed like a pretty relaxed man all the way around. He would get a little irritated at the engineering sessions he sat in on during the last couple of weeks before the flight and would give them a few gus-gruff growls if they seemed to want to tinker with this and that at the last minute, but that seemed to be sheer eagerness to get on with the flight. There was even a bit of the old boondock Edwards broomstick-and-baling wire spirit about the whole thing. Just two nights before the flight it dawned on one of the doctors that they had never made provisions for a urine receptacle for Gus, to avoid the sort of thing Shepard had experienced. That was a hell of a note. They figured they could make do with an ordinary rubber condom for the receptacle. But what would hold it in place and keep it from coming off? Dee O'Hara, the nurse, helped out. She drove into Cocoa Beach and bought a panty girdle, and they rigged that up with the condom. The goddamned girdle gave you a hell of a tight grip on the groin, but Gus figured he could get by with it. All in all, he seemed pretty loose, a test pilot of the old school. He even had a foretaste of the mental atmosphere of the real thing, just as Shepard had. On July 19 he was inserted into the capsule, and the hatch was sealed, when the flight was canceled because of bad weather. The flight finally took place on July 21. Judging by his pulse rate and respiration, which were transmitted via his body sensors, Gus was more nervous than Shepard during the countdown. These rates, taken by themselves, didn't mean a great deal, however, and no one would have thought twice about it except for what happened at the end of the flight. The flight itself was very nearly a duplicate of Shepard's, except that Grissom's capsule had a window, not just a periscope, giving him a much better view of the world, and he had a more sophisticated hand controller. His pulse stayed up around 150 throughout the five minutes he was weightless—Shepard's pulse had never reached 140, not even during liftoff—and went up to 171 during the firing of the retro-rockets before the re-entry through the earth's atmosphere. The informal consensus among the program's doctors was that if an astronaut's pulse rate went above 180, the mission should be aborted. The capsule splashed down almost precisely on target, just as Shepard's had, within three miles of the recovery ship, the carrier *Randolph*. The capsule hit the water, then keeled over on one side, just as Shepard's had, and took its own sweet time righting itself. Grissom thought he heard a gurgling noise inside the capsule—as had Shepard—and began looking for water seeping in, but didn't see any. The recovery helicopter, designated Hunt Club I, was over the capsule within less than two minutes. Grissom was still in the seat, resting on his back, as he had been at the outset of the flight, and the capsule was bobbing around in the water.

Over his microphone Grissom said, "Okay, give me how much longer it'll be before you get here."

The helicopter pilot, a Navy lieutenant named James Lewis, said, "This is Hunt Club I. We are in orbit now at this time, around the capsule."

Grissom said, "Roger, give me about another five minutes here, to mark these switch positions here, before I give you a call to come in and hook on. Are you ready to come in and hook on any time?"

Lewis said, "Hunt Club I, roger, we are ready any time you are."

There was a chart on which the astronaut was supposed to record the switch positions (on or off) with a grease pencil.

Five and a half minutes later Grissom radioed Lewis in the helicopter again:

"Okay, Hunt Club, this is Liberty Bell. Are you ready for the pickup?"

Lewis said, "This is Hunt Club I, this is affirmative."

Grissom said, "Okay, latch on, then give me a call and I'll power down and blow the hatch, okay?"

"This is Hunt Club I, roger, will give you a call when we're ready for you to blow."

Grissom said, "Roger, I've unplugged my suit so I'm kinda warm now . . . so . . ."

Lewis said, "One, roger."

"One, roger."

"Now if you tell me to, ah, you're ready for me to blow, I'll have to take my helmet off, power down, and then blow the hatch."

"One, roger, and when you blow the hatch, the collar will already be down there waiting for you, and we're turning base at this time."

"Ah, roger."

As the helicopter pilot, Lewis, looked down on the capsule, it shaped up as a routine retrieval, such as he and his co-pilot, Lieutenant John Reinhard, had practiced many times. Reinhard had a pole with a hook on it, like a shepherd's crook, that he was going to slip through a loop at the neck of the capsule. The crook was attached to a cable. The helicopter could hoist up to 4,000 pounds in this fashion; the capsule weighed about 2,400 pounds. Lewis had swung out and was making a low pass toward the capsule when suddenly he saw the capsule's side hatch go flying off into the water. But Grissom wasn't supposed to blow the hatch until he told him he had hooked on! And Grissom—there was Grissom scrambling out of the hatch and plopping into the water without even looking up at him. Grissom was swimming like mad. Water was pouring into the capsule through the hatch and the damned thing was sinking! Lewis wasn't worried about Grissom, because he had practiced water egress with the astronauts many times and he knew their pressure suits were more buoyant than any life preserver. They even seemed to enjoy playing around in the water in the suits. So he gunned the helicopter down to the level of the water to try to snare the capsule. By now only the neck of the thing is visible above the water. Reinhard goes to work with the shepherd's crook, leaning out of the helicopter, desperately trying to hook on. He finally hooks on, as the capsule disappears under the water and starts sinking like a brick. Lewis is now down so low all three wheels of the helicopter are in the water. The helicopter is like a fat man squatting over a tree stump, trying to pull it out of the ground. Full of water, as it is, the capsule weighs 5,000 pounds, 1,000 over the helicopter's capacity. Lewis already has a red-light warning of impending engine failure—so he signals for a second helicopter, which is already nearby, to pick up Grissom. He finally pulls the capsule up out of the water, but he can't make the helicopter move forward toward the carrier. He's just hanging there in the air like a hummingbird. Red lights are lighting up all over the panel. He's about to lose the ship as well as the capsule. So he cuts the capsule loose. It drops and disappears forever. The water is three miles deep at that point.

They turn away finally. Grissom is still in the water. He's waving. He seems to be saying, "I'm okay." The second helicopter is moving in to lower the horse collar.

In fact, Gus's waves were saying, "I'm drowning!—you bastards—I'm drowning!"

As soon as Gus scrambled out of the hatch, he had begun swimming for his life. The goddamned capsule's going under! His suit caught momentarily on some sort of strap outside the capsule, probably leading to the dye canister. It was like a parachute!—it would pull him under!—He'd drown! The drowning man . . . No question about it . . . By this point he was neither an astronaut nor a pilot. He was the drowning man. Get away from the death capsule!—that was the idea. Then he calmed down a little. He was swimming around in the ocean under the roar of the helicopter blades. He wasn't sinking, after all. The pressure suit kept him bobbing in the water, as high as his armpits. He looked up. The horse collar was hanging out of the helicopter. The horse collar that gets him out of this! But they were pulling away from him!—they were going to the capsule! He could see the man named Reinhard leaning out of the helicopter trying to snag the loop of the capsule. Only the neck of the capsule was out of the water. He started swimming back to the capsule. It was hard swimming in the pressure suit, but it kept him up. He bobbed right up to his armpits when he stopped swimming. Little swells kept breaking over his head and he swallowed some water. He felt out of control. He was floundering around in the middle of the ocean. He looked up again and there was another helicopter. He kept waving and waving, but nobody seemed to pay any attention. And now he wasn't bobbing up so high any more. The pressure suit was losing its buoyancy. It was getting heavier . . . starting to drag him down . . . The suit had a rubber diaphragm that rolled up around his neck like a turtleneck sweater to keep the water from seeping down inside the suit. It didn't fit tight enough . . . air was escaping . . . No!—it was the oxygen inlet valve! He had completely forgotten! The valve allowed oxygen into his suit while he was in flight. He had unhooked the tube but forgotten to close the valve. The oxygen was bubbling out down there somewhere . . . the suit was becoming dead weight, pulling him down . . . He reached down and closed the valve underwater . . . But now his head kept going under and he had to fight to get to the surface and then the swells broke over his head and he swallowed more water and he'd look up at the helicopters and wave and they'd just wave back—the bastards!—how could they not know! In the window of one of the helicopters was a man with a camera, merrily taking pictures of him—they were waving and taking snapshots! The stupid bastards! They were going crazy over the goddamned capsule and he was drowning before their very eyes . . . He kept going under. He'd fight his way back up and swallow some more water and wave. But that drove him back under. The suit—he seemed to be packed in two hundred pounds of wet clay . . . The *dimes!*—and all that other shit! Christ, the dimes and those goddamned trinkets! Down there in his knee pocket . . . He'd had the bright idea of carrying a hundred one-dollar bills on the flight as souvenirs, but he didn't have a spare hundred dollars to his name—so he decided on two rolls of fifty dimes each—and he had put in three one-dollar bills for good measure and a whole bunch of little models of the capsule—and now this big junkheap of travel sentiment stuffed in his knee pocket was taking him under . . . *Dimes! . . . Silver deadweight!*

Deke! . . . Where was Deke! . . . Surely Deke would be here! . . . He had done as much for Deke. Somehow Deke would materialize and save him. Deke and Wally and him had been down at Pensacola practicing water egress, and somehow Deke, in his whole pressure suit, with his helmet on, had fallen off his raft and was going under and couldn't do a goddamn thing about it, but he and Wally had been nearby with their swimming flippers on, and they had swum straight to him and held him up until one of the Navy swabbos could reach them with the raft, and it was no sweat, because they had been by his side, and surely . . . *Deke!* . . . *Or somebody! Deke!*

Cox . . . That face up there!—it's Cox . . . Deke wasn't here and wasn't going to be here. But Cox!—Cox, whom he hardly knew, was his sole redeemer now. Cox was a Navy man in the second helicopter. Gus knew that face. Cox wasn't a stupid bastard. Cox had picked up Al Shepard! Cox had picked up the goddamned chimpanzee! Cox knew how to get people out of here! . . . *Cox!* . . . He could see Cox leaning out of the helicopter lowering the horse collar. There was a hell of a roar everywhere from the two helicopters. But Cox! Cox and his helicopter were just suspended there. They weren't coming any closer, and Gus's head kept going under. The wash from the helicopter propellers was driving him back. The closer his redeemer came in the helicopter, the farther he was driven back. *The sharks—they can smell panic!* And he was sheer panic, 160 pounds of it, plus a hundred pounds of death dimes! Lost at last at 2,800 fathoms in the middle of the Atlantic Ocean! But helicopters can drive off sharks with their prop wash! Cox would rout the sharks and save him—but Cox got no closer, even though the horse collar was now touching the water. He was still about ninety feet away, across the billows. Now he could see it, now he couldn't. The swells kept washing over him. But it was the only thing left. He swam for it. He couldn't get his legs to come up. So he fought toward the horse collar with his arms. He had no strength left. Everything pulled him down. He couldn't get enough breath. There was nothing but furious noise . . . blazing water . . . The water kept getting in his mouth. He would never make it. But the horse collar! Cox was up there! There was the horse collar. It was in front of him. He grabbed it and hung on. He was supposed to sit in it as if he were sitting on a swing. The hell with that. He flopped through the hole like a dead flounder landing on the fish-market scales. He hung on with his arms. He felt as if he weighed a ton. The suit was full of water. And it had already dawned on him: *I lost the capsule.*

As soon as Cox and his co-pilot pulled Grissom up into the helicopter, they could see that he was in a bad way. He looked funny. He was gasping for breath and he was shaking. His eyes kept darting around. He found what he was looking for: a Mae West, a life preserver. He grabbed it and started trying to strap it on. He was having a hell of a time with it because he was shaking so. His arms would fly one way and the straps would fly the other. The engines made a terrific noise. They were heading back to the carrier. Grissom was still struggling with the straps. He obviously thought they were going to crash at any moment. He thought he was going to drown. He gasped. He battled the Mae West all the way back to the carrier. What the hell had happened to the man? First he had blown the hatch before the lead helicopter could hook on and then he had floundered around out in the ocean and now he was preparing to abandon ship in a goddamned helicopter on a perfectly calm sunny morning out near Bermuda.

Once they got over the carrier *Randolph,* Grissom calmed down a bit. The same sort of awed faces that had welcomed Alan Shepard were craning up at the helicopter. But Grissom hardly noticed them. His head was in a very dark cloud.

When he went below deck, he was still shaking. He kept saying, "I didn't do anything. The damned thing just blew."

Within an hour they had started the preliminary debriefing, and Grissom kept saying, "I didn't do anything, I was just lying there—and it just blew."

A couple of hours later, at the formal debriefing on Grand Bahama Island, Grissom was much calmer, although he looked exhausted and drawn. He was grim. He was a very unhappy man. His pulse was still up to 90. Normally, at rest, it was 68 or 69. He kept saying, "I didn't touch it, I was just lying there—and it blew."

According to Gus, here was what happened. Once he knew the helicopters were nearby, he felt secure in the capsule and therefore asked for five minutes to finish getting unhooked and record his switch positions. While the capsule was still descending under the parachute, he had opened his face plate and disconnected the visor seal hose. Once the capsule was in the water, he disconnected the oxygen hose to his helmet, unfastened the helmet from the pressure suit, undid his chest strap, lap belt, shoulder harness, and knee straps, disconnected the wire leading to the biomedical sensors, and rolled the rubber neck dam up around his neck. His pressure suit was still attached to the capsule by the oxygen inlet hose, which he needed for cooling the suit, and his helmet still had its radio wiring leads hooked up; but all he had to do was take the helmet off and he would be free of the wires. Then—all in keeping with the checklist—he removed the emergency knife that was clamped onto the hatch and put it in the survival kit, which was a canvas bag about two feet long containing an inflatable life raft, shark repellant, a desalinization rig, food, a signal light, and so on. Before leaving the capsule via the hatch, as Gus recounted it, he had one more chore to perform. He was supposed to take out a chart and a grease pencil and mark the positions of all the switches on the instrument panel. Since he still had on his pressure-suit gloves, making it hard to grip the grease pencil, this took him three or four minutes. Then he armed the explosive hatch by removing the cover from the detonator, which was a button about three inches in diameter, and removed the safety pin, which was like the safety catch on a revolver. Once the cover and pin were removed, five pounds of pressure on the detonator button would blow the bolts and propel the hatch out into the water. Now he radioed Lewis in the helicopter to come in and hook on. He unhooked the oxygen hose to his pressure suit and settled back on the seat and waited for Lewis to tell him he had hooked on to the capsule. Once he got the word from Lewis, he would blow the hatch. While he was lying there, he said, he started wondering if there were some way he could retrieve the knife from the survival kit before he blew the hatch and left the capsule. He figured it would make a terrific souvenir. This thought was running idly through his mind, he said, when he heard a dull thud. He knew immediately that it was the hatch blowing. In the next instant he was looking straight out the hatchway at the brilliant blue sky over the ocean, and water was pouring in. There was not even time to grapple for the survival kit. He took his helmet off and grabbed the right side of the instrument panel and thrust his head through the hatchway and wriggled out.

"I had the cap off and the safety pin out," Gus said, "but I don't think that I hit the button. The capsule was rocking around a little, but there weren't any loose items in the capsule, so I don't see how I could have hit it, but possibly I did."

As the day wore on, and the formal debriefing got underway, Gus discounted even the possibility that he had hit the button. "I was lying there, flat on my back—and it just blew."

Nobody was about to *accuse* Gus of anything, but the engineers kept rolling their eyes at each other. The explosive hatch was new to the Mercury capsule, but explosive hatches had been in use on jet fighters since the early 1950's. When a pilot pulled his cinch ring and ejected, the hatch blew and a TNT charge rocketed the pilot and his seat-parachute rig through the opening. The pilot and anyone who might be riding backseat routinely armed their hatches and the TNT charges out on the runway before takeoff. This was the equivalent of Gus's removing the detonator cover and the safety pin.

Of course, any apparatus rigged up with explosive charges had the potential of exploding at the wrong time. Later on, NASA put a hatch assembly through every test the engineers could dream up to try to make the hatch blow without hitting the detonator button. They subjected it to trial by water, trial by heat; they shook it, pounded it, dropped it on concrete from a height of one hundred feet—and it never *just blew.*

There were many conjectures uttered very quietly, very privately.

And at Edwards . . . the True Brothers . . . well, my God, as you can imagine, they were . . . *laughing!* Naturally they couldn't say anything. But now—surely!—it was so obvious! Grissom had just screwed the pooch!

In flight test, if you did something that stupid, if you destroyed a major prototype through some lame-brain mistake such as hitting the wrong button—you were through! You'd be lucky to end up in Flight Engineering. Oh, it was obvious to everybody at Edwards that Grissom had just *fucked it,* screwed the pooch, that was all. It was doubtful that he had hit the detonator on purpose, because even if he were feeling a little panicky in the water (you have to be *afraid* to panic, old buddy), he wasn't likely to ask for trouble by blowing the hatch before the helicopter hooked on and was overhead with the horse collar. But if a man is beginning to panic, logic goes first. Maybe the poor bastard just wanted out, and—bango!—he punched the button. But what about the business of the knife? He said he wanted to take the knife as a souvenir. So he may have been trying to fish the knife out of the survival kit. The capsule is rocking in the swells . . . he bangs into the detonator—that's all it would have taken. Oh, there was no question that he had hit the damn button some way. The only thing they liked about his entire performance was the way he said, "I was lying there—and it just blew," and the way he stuck to it. There, Gus old boy, you showed the instincts of the true fighter jock! Oh, you learned many of the lessons well! After you've done some forbidden hassling and your ship flames out and you have to eject and your F–100 goes *kaboom!* on the desert floor . . . naturally you come back to base and say: "I don't know what happened, sir—it just flamed out on me!" I was minding my own business! The demons did it! And go easy on the details. A broad stroke of vagueness—that's the ticket.

"I was lying there—and it just blew" . . . oh, that was rich. And then the brethren sat back and waited for the Mercury astronaut to *get his*, the way any one of them would have *gotten his*, had a comparable *fuckup* occurred at Edwards.

And . . . nothing happened.

From first to last the publicity that came out of NASA, out of the White House, from wherever, told of what a severe disappointment it had been to brave little Gus to lose the capsule through a malfunction after so successful a flight. *Little* Gus he became. The sympathy that welled up was terrific. Only five feet six with a round face. It was amazing that so much courage could be packed into sixty-six inches. And we almost lost him through drowning.

The True Brothers were incredulous . . . the Mercury astronauts had an official immunity to three-fourths of the things by which test pilots were ordinarily judged. They were by now ablaze with the superstitious aura of the single-combat warrior. They were the heroes of Kennedy's political comeback, the updated new frontier whose symbol was a voyage to the moon. To announce that the second one, Gus Grissom, had prayed to the Lord: "Please, dear God, don't let me fuck up"—but that his prayer had not been answered, and the Lord let him *screw the pooch*—well, this was an interpretation of that event that was to be avoided at all cost. NASA was no more anxious to have to call Grissom on the carpet than Kennedy was. NASA had just been handed a *carte blanche* for a moon project. Just six months before, the organization had been in live danger of losing the space program altogether. So nothing about this flight was going to be called a failure. It was possible to argue that Grissom's flight had been a great success . . . There had just been a small problem immediately afterward. As for public opinion, the loss of the capsule didn't really matter very much. The fact that the engineers needed the capsule to study the effects of heat and stress and to retrieve various types of automatically recorded data—this certainly created no national gloom. Get the man up and bring him down alive; that, not engineering, was at the heart of the single combat. So the possibility that Gus might have blundered was never brought up again. Far from having a tarnished record, he was a hero. He had endured and overcome so much. He was back solidly in the rotation for whatever great flights might come up in the future . . . as if by magic.

In the days after the flight Gus looked gloomier and gruffer than ever. He could manage an official smile when he had to and an official hero's wave, but the black cloud would not pass. Betty Grissom looked the same way after she and the two boys, Mark and Scott, joined Gus in Florida for the celebration. Some celebration . . . It was as if the event had been poisoned by the gus-grim little secret. Betty also had the sneaking suspicion that everyone was saying, just out of earshot: "Gus blew it." But her displeasure was a bit more subtle than Gus's. They . . . NASA, the White House, the Air Force, the other fellows, Gus himself . . . were not keeping their side of the compact! Nobody could have looked at Betty at that time . . . this pretty, shy, ever-silent, ever-proper Honorable Mrs. Astronaut . . . and guessed at her anger.

They were violating the Military Wife's Compact!

By now Betty knew what to expect from Gus personally; which is to say, she seldom saw him. In one 365-day period he had been with her a total of sixty days. About six months before, Betty had had to go into the hospital near Langley for exploratory surgery. There was a good chance that she would require a hysterectomy.

Betty had a real siege in the hospital. She was there for twenty-one days. She was there for so long she had to get some of her relatives to fly in from Indiana to look after the boys. Gus managed to make it to see her in the hospital exactly once and he didn't quite make it through the entire visiting hour. He got a call right there in the hospital asking him to return to the base, and he left.

Betty seldom speculated, even to herself, on what Gus did during the 80 percent of the year that he was not with her. She had worked that out in her mind. The compact took care of it. If Gus was occasionally the Complete Fighter Jock Away from Home, that did not violate the compact . . . And now it was time for the other part of the compact to take effect. It was her time to be the Honorable Mrs. Captain Second American in Space. They *owed her* every bit of it.

Louise Shepard, over in Virginia Beach, hadn't known what was going to happen when Al went up, and so her place was invaded by reporters and sightseers. They practically tore the yard to pieces just by milling around and tramping through the shrubbery to press their noses up against the window. Gus was not having any of that. Gus saw to it that the local police were out patrolling in front of the house early in the morning, before dawn. Betty was inside the house in front of the TV set with Rene Carpenter, Jo Schirra, Marge Slayton, the children. Outside slavered the Animal. There were a lot of reporters on the sidewalk and back in the driveway of the house next door, but the palace guard kept them all under control. Betty actually felt pretty good. It was the Danger Wake business again. She was the hostess and star of the drama. She almost missed the final countdown. She was in the kitchen turning off the flame under some soft-boiled eggs for somebody.

After the flight all sorts of neighbors and NASA people at Langley came rushing in, congratulating her and bringing more food and making a fuss over her. But Betty knew enough about flight tests to know that the loss of the capsule could have some grim results. A call came in from Gus on Grand Bahama Island. There were a lot of people still in the house, but she had to ask the question, anyway.

"You didn't do anything wrong, did you . . ."

"I did not do anything wrong," he said very slowly. You could almost see the black gus gruff look over the telephone. "That hatch just blew."

"I'm glad."

She started telling him about all the people who were telephoning congratulations.

"That's good," said Gus. "Say, by the way, the motel lost two pairs of my slacks in the laundry, and I need shirts. Will you bring me some when you come down to the Cape?"

The laundry? He wanted her to remember to bring the laundry.

Betty and the boys arrived at the Cape on one of those blinding hot July days that made all of Cocoa Beach feel like a fried concrete parking lot. They were led out to a runway at Patrick Air Force Base along with a lot of NASA and military dignitaries to meet Gus's plane as it came in from Grand Bahama Island. There was a big canopy set up nearby. Under the canopy there would be a press conference. Betty stood out there on the slab with James Webb and some other NASA brass, and she slowly began to realize that . . . *they were reneging!*

This was going to be it!—a reception out on this brain-frying slab! There was going to be no trip to the White House. Webb—not John Kennedy—was going to give Gus the Distinguished Service Medal . . . under a dreadful Low Rent tent here on the slab. There was going to be no parade in Washington, no ticker-tape parade in New York—not even a parade in Mitchell, Indiana. *That* . . . Betty would have loved. To come back to Mitchell and parade down Main Street . . . But Gus would be getting nothing, just a medal from James E. Webb. They couldn't do this to her!—they were reneging.

But they did, and it was even worse than she feared. The plane comes in, taxis up to the ramp, a big cheer goes up, Gus steps out—and some NASA functionaries take her and the children by the elbows and thrust them forward at Gus like religious objects . . . Behold, the Wife, the Children . . . and Gus can hardly even look at Betty as someone he knows. She's merely the ceremonial Solid Backing on the Home Front trundled forward on the concrete slab. Gus mutters hello, hugs the two boys, and they trundle the Wife and the Children back, and then Gus is marched over to the canopy, where they have the press conference. The reporters keep harping on the blown hatch and the lost capsule. The dismal bastards—they haven't gotten the message yet. They haven't picked up the proper moral tone. But being part of the great colonial animal, the Victorian Gent, they would get it all straight in a few days and never mention the damnable hatch again . . . But for now they gave the event another shot of the poisonous secret . . . Was that what was responsible for this wretched, shabby, mean little ceremony? Gus struggled with the questions and sweated under the canopy. He kept saying, "I was just lying there minding my own business when the hatch blew. It just blew." Betty could see he was getting angrier and angrier, gruffer and grimmer and darker about the eyes. He hated talking to reporters, as it was. Her heart went out. They were making him squirm. And *this* was the Big Parade! This was what she got out of the compact after all this! It was a travesty. She was . . . the Honorable Mrs. Squirming Hatch Blower!

The day only got hotter. After the little ceremony, with Webb waxing sonorous, they drove Gus and Betty and the boys to the VIP guesthouse at Patrick Air Force Base. This was supposed to be a big deal. They were told that these were secret quarters where they would be completely screened off from the press and the gawkers. The VIP guesthouse . . . Betty looked around. Even the military VIP quarters here at Cocoa Beach were Low Rent. This VIP guesthouse was like some musty cabin court from the late 1930's. She looked out the window. Over there was the beach, that amazing hot-brick Cocoa Beach. But between the guesthouse and the beach was Route A1A, with cars roaring back and forth in the screaming heat of mid-July. She would never even make it across the highway to the beach with the children. Well, they could watch TV—but there was no TV; and no pool. Then she looked in the kitchen and opened the refrigerator. It was stuffed with food, everything you could imagine. For some reason this made her furious. She could see the afternoon shaping up and the rest of the day and tomorrow, too. She would stay here with the children, cooking and risking her life dragging them to the worst beach in Florida . . . and Gus would no doubt go to the space center or into town . . .

Town meant the Holiday Inn, where the other fellows and their wives would be. That's where they would be celebrating and having the good times.

Listen, while you're getting settled, I think I'll—

Suddenly Betty was furious: *She was not staying in this place!* Gus didn't know what had gotten into her.

She said she wanted to go to the Holiday Inn. That was where everybody would be. She told Gus to call the Holiday Inn and get a room.

She gruffed it out with such fury that Gus called the Holiday and pulled the strings and got them a room. If Gus had managed to park her here in this faded VIP mausoleum and vanish, so that she could sit here in the heat of the slab watching the hours go by while he tooled around the pool at the Holiday as the big shot—she would have slit a wrist. That was how grim it was. That was how shabbily they had treated her. That was how grossly they had welshed on the compact. Now . . . they *truly owed her.*

XII. The Tears

Since the pooch proved to be unscrewable, officially, and Gus Grissom's flight was therefore on the record as a success, NASA was suddenly in great shape. John Kennedy was happy. "We have started our long voyage to the moon." That was the idea. Neither Shepard's nor Grissom's suborbital flight measured up to Yuri Gagarin's orbit of the earth, but the fact that NASA had completed two successful manned flights seemed to mean that the United States was battling back successfully in the competition for the heavens.

Naturally, true to form, that was the moment the anonymous and uncanny Chief Designer, D-503, Builder of the Integral, chose to show the world who actually ruled the heavens.

Just sixteen days after Grissom's flight, which is to say, on August 6, 1961, the Soviets sent *Vostok* 2 into orbit with a cosmonaut named Gherman Titov aboard. Titov circled the earth for an entire day, completing seventeen full orbits, and landing where he had started, on Soviet soil. Three times he came over the United States, 125 miles overhead. Once again, all over the country, politicians and the press seemed profoundly alarmed, and the awful vision was presented: suppose the cosmonaut were armed with hydrogen bombs and flung them as he came over, like Thor flinging thunderbolts . . . one here, one there . . . Toledo disappears off the face of the earth . . . Kansas City . . . Lubbock . . . Titov's flight seemed so awesome it made the Shepard and Grissom flights look terribly insignificant. The Integral and its Chief Designer could apparently do anything they wanted, and at any time.

Seven days later, August 13, 1961, Nikita Khrushchev began the steps that led to the building of a wall, precisely like a penitentiary wall, through the middle of an entire city, Berlin, to prevent the population of East Berlin from crossing over to the West. But the world was still blinking at the radiance of the day-long space flight. "They're a bit brutal—but you have to admit they're geniuses. Imagine keeping a man in space for twenty-four hours!"

So far as NASA was concerned, the Titov flight put an end to the Mercury-Redstone program then and there. The next astronaut in line to ride on top of the Redstone, John Glenn, was now assigned to attempt an orbital flight, using the Atlas rocket, which had done so poorly in unmanned tests. Later there were those who speculated that NASA had been "saving Glenn for the big one" all along. But Glenn did not have that kind of status within NASA. He had learned that to his bitter regret. No, he had only the invisible Chief Designer, Builder of the Integral, to thank for the fact that he was assigned to be the first American to orbit the earth.

After Titov's flight the phrase *the space gap* began to be repeated throughout the American press. *Space gap* was a superstitious condition. It began to seem of urgent importance for NASA to put a man into space before the sands stopped flowing in the hourglass on the last day of the year 1961. The great cowboy rush of the winter of 1960–61 started all over again. The hell with fastidious precautions . . . For example, the Soviets revealed that Titov had suffered from nausea throughout his flight. Later they changed that to say that he had suffered nausea after "prolonged" flight. They would have probably not revealed even that much, except that they decided to participate in international scientific conferences in order to publicize their space feats. It also came out—although few specifics were given—that the Soviet manned space-flight program, from selection of their cosmonauts (from among military pilots) and their training (centrifuge rides, parabolic rides in jet fighters, and so forth) to capsule design and launching and retro-rocket systems, was remarkably similar to NASA's. Everyone at NASA regarded this as vastly reassuring. *We're on the right track, anyway!* Of course, the Soviet rockets were far more powerful. That was the given. And if a cosmonaut of the Integral had suffered nausea in orbit, then astronauts probably would, too. But there was no time to worry about that now. Find out about it the way Titov did: *up there. Más allá!* Over the next hill!

In September NASA successfully launched a Mercury-Atlas capsule with a dummy astronaut aboard and brought it back on target into the Atlantic, near Bermuda, after one orbit of the earth. The press speculated that Kennedy would pressure NASA to put an astronaut on the next flight, but Hugh Dryden and Bob Gilruth managed to hold out for an additional test. They wanted to send a chimpanzee into orbit with the Atlas rocket first.

This time, out at Edwards, the True Brothers didn't even derive a chilly smile from the fact that once more in the exalted Project Mercury an ape would be taking *the first flight.* An ape would make the first orbit of the earth for the U.S.A. The prestige of Project Mercury had by now rendered such considerations meaningless. On October 11, at Edwards, Bob White had made an extraordinary flight in the X–15—and the country hardly noticed. White took the X–15 up to 217,000 feet with the Big Engine—and the press merely nodded perfunctorily. So a man had just flown very high in an airplane; how interesting; and that was that. The fact that White was on top of a rocket, the same sort of rocket as the Redstone or the Atlas, the fact that his flight to 217,000 feet was in effect *piloted space flight*—none of this was likely to impress Kennedy or the public amid the panic over Titov and the *space gap.* White had gone forty miles up, ten miles short of the arbitarily set boundary of "space." The XLR–99, the Big Engine, had delivered 57,000 pounds of thrust, just 21,000 short of the thrust of the Redstones that took Shepard and Grissom aloft. White's speed reached Mach 5.21, or 3,647 miles an hour; Shepard's and Grissom's rocket velocities were only slightly greater, about 5,180 miles per hour. White was weightless for three minutes during his tremendous arc over the top, as compared to Shepard's and Grissom's five minutes. White saw all the things that Shepard and Grissom saw (and Shepard, only barely) . . . including the entire blue band of atmosphere at the horizon of the earth. Above all, White was a *pilot.* He controlled his plane's ascent. He used hydrogen-peroxide thrusters to control his attitude once the air became too thin for aileron control—the same system of hydrogen-peroxide thrusters that Shepard and Grissom had used—and he did it all without benefit of any automatic backup. And he brought the ship back down through the earth's atmosphere himself . . . and *landed* it himself on the holy plateau of Edwards . . . on the dome of the world. A rocket *pilot* (quoth the brethren), but the national press barely noticed.

So it was with a mainly academic fascination that the boys at Edwards followed the second Project Mercury chimpanzee flight. For nine months the veterinarians at Holloman Air Force Base had been putting their colony of chimpanzees through the operant conditioning regimen in preparation for an orbital flight. The training included all the things that had gone into the training for the first suborbital flight, the centrifuge runs, the weightless parabolas, the procedures-trainer sessions, the heat-chamber and altitude-chamber sessions, plus some intelligence tests. In one test the ape had to be able to judge time intervals. The signal light would go on, and he had to wait twenty seconds before pulling the lever or he would receive the ever-cocked electrical shock. In another the animal was required to *read* the instrument panel and throw a switch. Three symbols would flash on the panel, two of which would be identical, such as two triangles and one square, and the animal had to pull the lever under the odd one or receive the shock in the soles of his feet.

By the beginning of November, twenty veterinarians had moved into Hangar S at the Cape with five chimpanzees. One of them was Ham, thinner and more strung out than ever but still an ace in the procedures trainer, his life dedicated to the avoidance of the invisible volts. Ham was not regarded as the pick of the lot, however. The brightest and quickest member of the colony was a male who had been brought from Africa to Holloman Air Force Base in April of 1960, when he was about two and a half years old. He was known as Number 85. Number 85 had fought the veterinarians and the process of operant conditioning like a Turkish prisoner of war. He fought them with his hands, his feet, his teeth, his saliva, and his cunning. He would shake off each jolt of electricity and give them a hideous grin. When he couldn't take the shocks anymore, he would cooperate temporarily, and his hands would fly across the procedures trainer console like E. Power Biggs's at the organ—and *then* he would turn on the vets, making another desperate thrash toward freedom. He was like the slave who wouldn't break. Finally, they shut him up inside a metal box and let him thrash about in there for a week with his feces and urine for company. When they let him out, he was, at last, a different ape. He had had enough. He didn't want any more of the box. His operant conditioning could now begin in earnest. The box was certainly not the course that the good vets of Holloman would have chosen, had the times been normal. No, they had chosen this course in the name of the battle for the heavens and under the pressure of national urgency; Number 85 was the ape that the MA–5 mission (the fifth test of the Mercury-Atlas vehicle) required. He was the quickest study in the universe of the Simia satyrus. They took him up in jet fighters to get him used to the accelerations, the noise, and the disorienting sensations of high-speed flight. They put him in the gondola of the human centrifuge at the University of Southern California and ran him through entire profiles of the proposed first American orbital mission, until he was used to the seven or eight g's he would experience on ascent and on re-entry. Under high g's or low g's Number 85 could operate a Mercury console like no ape that had ever lived. He was so good they used him as the test subject for a laboratory experiment that simulated a fourteen-day orbital mission. For fourteen days Number 85 was on the procedures trainer performing the same tasks he would perform in the 4 1/2-hour MA–5 mission. For MA–5 they had added rewards as well as the punishment of the shocks in the feet. Number 85 had two tubes positioned near his mouth. Out of one came banana-flavored pellets, if he did his tasks correctly, and from the other he could take sips of water. Number 85 could do the tasks so handily, including reading the odd-symbol panel, he could have kept the tubes popping banana pellets and water until he was sated or bilious. He was outstanding.

By now, November of 1962, he had been through 1,263 hours of training—the equivalent of 158 eight-hour days. For the equivalent of 43 days he had been strapped in one simulator or conditioning apparatus or another, whether the centrifuge, the jets, or the procedures trainer. He was a marvel. The only problem was his blood pressure. Back in June of 1960, two months after his training began, they had put a blood-pressure cuff on him and obtained a reading of from 140 to 160 systolic. This was certainly high, but it was hard to tell with Number 85. He had fought every medical examination as if it were an assault. It took two or three people just to restrain him. Three months later they were getting readings from 140 to 210; by now they were running from 190 to 210. The blood pressures of all five chimpanzees out back of Hangar S had mounted steadily over the past two years, although none was so elevated as Number 85's. Well, maybe it was the pressure cuff, which he didn't see very often and which probably struck him as a big black restraining mechanism. After all, Number 85 was excitable. Perhaps they would find out more during the flight. There had been no way to read the blood pressure of the other ape, Number 61, during his Mercury-Redstone flight. But for this one they put catheters in a main artery and a main vein of Number 85's legs to provide pressure readings before the launch and throughout the flight. They also put a catheter in his urethra to collect urine.

Number 85 went through procedures trainer drills out in the vans behind Hangar S right up to the eve of the flight. He was still the pick of the litter. He must have been wound up tighter than a window-shade spring, judging by the systolic readings.

Just before the flight his name was announced to the press as Enos. Enos meant *man* in Hebrew.

The flight did not attract much interest. The public, like the President, was impatient with the tests, especially since it was already November 29 when the ape was launched and it was becoming clear that there would be no manned launch before the year was over. The year would end without a manned flight. Number 85 was supposed to make three orbits of the earth. The launch went perfectly, with Number 85 pulling his levers a mile a minute. The Atlas rocket delivered 367,000 pounds of thrust, nearly five times what Shepard and Grissom had experienced, but neither the noise nor the vibrations fazed Number 85 in the slightest. He had heard and felt worse in the centrifuge runs with their piped-in sound. And since he had no window, he did not know he was leaving the earth, and for that matter the noise, the vibrations, or departure from this globe, was preferable to the box. He kept working his levers as fast they could light up the panel. The capsule went into a perfect orbit. Throughout the first orbit Number 85 performed like a dream, not only hitting the levers on cue and in complicated sequences, but also taking six-minute rest periods when signaled to . . . or at least lying motionless, the better to avoid the juice.

During the second orbit the wiring went haywire. When Number 85 did the odd-symbol exercise, he started getting electrical shocks in his left foot even when he pulled the correct lever. He kept pulling the correct levers, nonetheless. He was unstoppable. His suit started overheating. He didn't even slow down. Now the automatic attitude controls began malfunctioning, so that the capsule kept rolling over forty-five degrees before the thrusters on either side would correct it. It kept rolling back and forth. Didn't throw Number 85 out of his routine for a second. He kept reading the lights and flipping the levers. It would have to get a whole lot worse than this before it was as bad as the box.

Because the rolling was using up too much hydrogen peroxide—they had to be sure there was enough left to position the capsule correctly, blunt end down, for re-entry—they brought the capsule down after two orbits, into the Pacific, off Point Arguello, California. Number 85 bobbed around and the capsule bobbed around in the ocean for an hour and fifteen minutes before a ship arrived to retrieve them. The capsule had an explosive hatch, but it did not "just blow." Nor had Number 85 thrown up (like Titov) because of weightlessness or erratic motion. He had been weightless for a full three hours during the flight. He was calm when they removed him from the capsule. There was evidence, however, that he had had a merry old time for himself out there in the water. He hadn't just cooled his heels. The little bastard had ripped through the belly panel of his restraint suit and removed most of the biomedical sensors from his body and damaged the rest, including those that had been inserted under the skin. He had also yanked the urinary catheter out of his penis. To just pull it out like that must have hurt like hell. What came over him?

The flight had been a great success, all things considered. But one thing bothered the NASA Life Sciences people. The animal's blood pressure had been badly elevated. It had run from 160 to 200 throughout—even when his pulse rate was normal and he was watching his lights and pulling his levers with great efficiency. Was this some sort of morbid and unforeseen effect of prolonged weightlessness? Were astronauts in earth orbit going to be candidates for apoplexy? The Holloman veterinarians hastened to reassure them that Number 85—er, Enos—had registered high-blood-pressure readings for two years now. It seemed to be the nature of the beast. The NASA folk nodded . . . although 200, systolic, was awfully goddamned high . . .

Privately, the situation had the Holloman scientists thinking, and not about space flight, either. The readings they had gotten from Number 85 in the past with the cuff may or may not have been reliable. But there was no mistaking the readings of the catheters during the flight and just before it. Once they were inserted, Number 85 wasn't even aware of them. They were giving true readings. His blood pressure had not gone up because of the stresses of the flight. He took the flight with the utmost aplomb; his heart and respiratory rates and body temperatures were actually below the readings obtained during the centrifuge runs. In fact, his blood pressure had not *gone up* at all. It had been up there all along. A theory with implications for man-on-earth, not man-in-space, was beginning to form . . . Number 85, smartest of the Simia satyrus, prince of the lower primates, had swallowed so much rage over the past two years, thanks to the operant-conditioning process, it had begun pumping out through his arteries . . . until every heartbeat was about to blow his eardrums out for him . . .

There was even a press conference at which the chimpanzee appeared. "Enos" he was, of course. At the press conference Bob Gilruth announced that John Glenn would be the pilot for the first manned orbital flight, with Scott Carpenter as the backup pilot. Deke Slayton would take the second flight, with Wally Schirra as the backup pilot. All the while the astronaut who took *the first flight* was sitting right there at the table (quoth the brethren, sotto voce). Number 85 stole the show, which was only just. He took the flashbulbs and all the talk and hubbub without blinking or even fidgeting, as if he had been waiting all along for his moment in the spotlight. Of course, the ape had been, as it were, overtrained for such a moment and was long past being able to let such things alter his behavior. Number 85 had been in rooms full of bright lights and large numbers of human beings before. Noise, vibrations, oscillations, weightlessness, space flight, fame—what earthly difference did it make compared to the shocks and the box?

At the outset neither Glenn nor his wife, Annie, foresaw the sort of excitement that was going to build up over his flight. Glenn regarded Shepard as the winner of the competition, since he looked at it the way pilots had always looked at it on the great ziggurat of flying. Al had been picked for *the first flight*, and there was no getting around that. He had been the first American to go into space. It was as if he were the project pilot for Mercury. The best Glenn could hope for was to play Scott Crossfield to Shepard's Chuck Yeager. Yeager had broken the sound barrier and become the True Brother of all the True Brothers, but at least Crossfield had gone on to become the first man to fly Mach 2 and, later on, the first man to fly the X-15.

Not even when reporters began arriving in New Concord, Ohio, his old hometown, and pushing his parents' doorbell and roaming and foaming over the town like gangs of strays, looking for anything, scraps, morsels of information about John Glenn—not even then did Glenn fully realize just what was about to happen. The deal with *Life* kept all but *Life* reporters away from him, and so the other fellows were out trying to root up whatever they could. That seemed to be the explanation. The Cape hadn't turned into a madhouse yet. As late as December, Glenn could go out to the strip on Route A1A in Cocoa Beach with Scott Carpenter, who was training with him as back-up man, out to that little Kontiki Village joint, whatever the name of it was, and listen to the combo play "Beyond the Reef." John got a kick out of "Beyond the Reef." By early January, however, it was madness to try to get to the Kontiki joint or anywhere else in Cocoa Beach. There were now reporters all over the place, all of them rabid for a glimpse of John Glenn. They would even pile into the little Presbyterian church when John went there on Sunday and turn the service into a sort of muffled melee, with the photographers trying to keep quiet and muscle their way into position at the same time. They were really terrible. So John and Scott now stuck pretty much to the base, working out on the procedures trainer and the capsule itself. At night, in Hangar S, John would try to answer the fan mail. But it was like trying to beat back the ocean with a hammer. The amount of mail he was getting was incredible.

Nevertheless, the training regimen created a curtain around John, and he didn't really have as clear an idea of the storm of publicity . . . and the *passion* of it all . . . as his wife did. At their house in Arlington, Virginia, Annie was getting the whole storm, and she had practically no protection from it and few happy distractions. John's flight was first scheduled for December 20, 1961, but bad weather over the Cape kept forcing postponements. He was finally set to go on January 27. He was inserted into the capsule before dawn. Annie was in a state. She was petrified. This had little to do with fear for John's life, however. Annie could take that kind of pressure. She had been through the whole course of worrying over a pilot. John had flown in combat in the Pacific Theater in the Second World War and then in Korea. In Korea he was hit seven times by flak. Annie had also been through just about everything that Pax River had to inflict upon a pilot's wife, short of the visit of the Friend of Widows and Orphans at her own door. But one thing she had never done. She had never had to step outside after one of John's flights and say a few words on television. She knew that would be coming up when John flew, and she was already dreading the moment. Some of the other wives were at her house for the Danger Wake, and she asked them to bring some tranquilizers. She wouldn't need them for the flight. She would take them before she had to step outside for the ordeal with the TV people. With her ferocious stutter . . . the thought of millions of people, or even hundreds, or even five . . . seeing her struggling on television . . . She had been in front of microphones with John before, and John always knew how to step in and save the day. She had certain phrases she had no trouble with—"Of course," "Certainly," "Not at all," "Wonderful," "I hope not," "That's right," "I don't think so," "Fine, thank you," and so on—and most of the questions from the television reporters were so simpleminded she could handle them with those eight phrases, plus "yes" and "no"—and John or one of the children would chime in if any amplification was called for. They were a great team that way. But today she would have to solo.

Annie could see the impending catastrophe easily enough. All she had to do was look at the television screen. Any channel . . . it didn't matter . . . she could count on seeing some woman holding a microphone covered in black foam rubber and giving a declamation on the order of:

"Inside this trim, modest suburban home is Annie Glenn, wife of Astronaut John Glenn, sharing the anxiety and pride of the entire world at this tense moment but in a very private and very crucial way that only she can understand. One thing has prepared Annie Glenn for this test of her own courage and will sustain her through this test, and that one thing is her faith: her faith in the ability of her husband, her faith in the efficiency and dedication of the thousands of engineers and other personnel who provide his guidance system . . . and her faith in Almighty God . . ."

In the picture on the screen all you could see was the one TV woman, with the microphone in her hand, standing all by herself in front of Annie's house. The curtains were pulled, somewhat unaccountably, inasmuch as it was nine o'clock in the morning, but it all looked very cozy. In point of fact, the lawn, or what was left of it, looked like Nut City. There were three or four mobile units from the television networks with cables running through the grass. It looked as if Arlington had been invaded by giant toasters. The television people, with all their gaffers and go-fers and groupies and cameramen and couriers and technicians and electricians, were blazing with 200-watt eyeballs and ricocheting off each other and the assembled rabble of reporters, radio stringers, tourists, lollygaggers, policemen, and freelance gawkers. They were all craning and writhing and rolling their eyes and gesturing and jabbering away with the excitement of the event. A public execution wouldn't have drawn a crazier mob. It was the kind of crowd that would have made the Fool Killer lower his club and shake his head and walk away, frustrated by the magnitude of the opportunity.

Meanwhile, John is up on top of the rocket, the Atlas, a squat brute, twice the diameter of the Redstone. He's lying on his back in the human holster of the Mercury capsule. The count keeps dragging on. There's hold after hold because of the weather. The clouds are so heavy they will make it impossible to monitor the launch properly. Every day for five days Glenn has psyched himself up for the big event, only to have a cancellation because of the weather. Now he has been up there for four hours, four and a half, five hours—he has been stuffed into the capsule, lying on his back, for five hours, and the engineers decide to scrub the flight because of the heavy cloud cover.

He's drained. He makes his way back to Hangar S, and they start taking the suit off and unwiring him. John is sitting there in the ready room with just the outer covering of the suit off—he still has on the mesh lining underneath and all the sensors attached to his sternum and his rib cage and his arms—and a delegation from NASA comes trooping in to confront him with the following message from on high:

John, we hate to trouble you with this, but we're having a problem with your wife.

My wife?

Yes, she won't cooperate, John. Perhaps you can give her a call. There's a phone hookup right here.

A call?

Absolutely befuddled, John calls Annie. Annie is inside their house in Arlington with a few of the wives, a few friends, and Loudon Wainwright, the writer from *Life,* watching the countdown and, finally, the cancellation on television. Outside is the bedlam of the reporters baying for scraps of information about the ordeal of Annie Glenn—and resenting the fact that *Life* has exclusive access to the poignant drama. A few blocks away, on a quaint Arlington side street, in a limousine, waits Lyndon Johnson, Vice-President of the United States. Kennedy had appointed Johnson his special overseer for the space program. It was the sort of meaningless job that Presidents give Vice-Presidents, but it had a symbolic significance now that Kennedy was presenting manned space flight as the very vanguard of his New Frontier (version number two). Johnson, like many men who have had the job of Vice-President before him, has begun suffering from publicity deprivation. He decides he wants to go inside the Glenn household and console Annie Glenn over the ordeal, the excruciating pressure of the five-hour wait and the frustrating cancellation. To make this sympathy call all the more memorable, Johnson decides it would be nice if he brought in NBC-TV, CBS-TV, and ABC-TV along with him, in the form of a pool crew that will feed the touching scene to all three networks and out to the millions. The only rub—the only rub, to Johnson's way of thinking—is that he wants the *Life* reporter, Wainwright, to get out of the house, because his presence will antagonize the rest of the print reporters who can't get in, and they will not think kindly of the Vice-President.

What he does not realize is that the only ordeal that Annie Glenn has been going through has been over the possibility that she was going to have to step outside at some point and spend sixty seconds or so stammering a few phrases. And now . . . various functionaries and secret-service personnel are calling on the telephone and banging on the door to inform her that the Vice-President is already in Arlington, in a White House limousine, waiting to pull up and charge in and pour ten minutes of hideous Texas soul all over her on nationwide TV. Short of the rocket blowing up under John, this is the worst thing she can imagine occurring in the entire American space program. At first Annie is trying to deal with it gracefully by saying that she can't possibly ask *Life* to leave, not only because of the contract, but because of their good personal relationship. Wainwright, being no fool, doesn't particularly care to get caught in the middle like this and so he offers to bow out, to leave. But Annie is not about to give up her *Life* shield at this point. Her mind is made up. She's getting angry. She tells Wainwright: "You're *not leaving* this house!" Her anger does wonders for her stutter. It flattens it right out temporarily. She's practically ordering him to stay. Annie's stutter often makes people underestimate her, and Johnson's people didn't realize that she was a Presbyterian pioneer wife living in full vitality in the twentieth century. She could deal with any five of them with just a few amps from the wrath of God when she was angry. Finally, they're getting the picture. She's too much for them. So they start trying to bend arms at NASA to get someone to *order* her to play ball. But it has to be done very rapidly. Johnson is sitting out there a few blocks away in his limousine, fuming and swearing and making life hell for everyone within earshot, wondering, in so many words, why the fuck there isn't anybody on his staff who can deal with a *housewife,* f'r chrissake, and his staff is leaning on NASA, and NASA is bucking the problem up the chain, until in a matter of minutes it's at the top, and the delegation is trooping into Hangar S to confront the astronaut himself.

So there's John, with half his mesh underlining hanging off his body and biosensor wires spouting from out of his thoracic cage . . . there's John, covered with sweat, drawn, deflated, beginning to feel very tired after waiting for five hours for a hundred tons of liquid oxygen and RP-1 kerosene to explode under his back . . . and the hierarchy of NASA has one thing on its mind: keeping Lyndon Johnson happy. So John puts in the call to Annie, and he tells her: "Look, if you don't want the Vice-President or the TV networks or anybody else to come into the house, then that's it as far as I'm concerned, they are *not* coming in—and I will back you up all the way, one hundred percent, and you tell them that. I don't want Johnson or any of the rest of them to put so much as *one toe* inside our house!"

That was all that Annie needed, and she simply became a wall. She wouldn't even discuss the matter any further, and there was no question any longer about Johnson getting in. Johnson, of course, was furious. You could hear him bellowing and yelling over half of Arlington, Virginia. He was talking about his aides. *Pansies! Cows! Gladiolas!* Webb could scarcely believe what was going on. The astronaut and his wife had shut the door in the Vice-President's face. Webb had a few words with Glenn. Glenn wouldn't back down an inch. He indicated that Webb was *way out of line.*

Way out of line! What the hell was this? Webb couldn't figure out what was happening. How could the number-one man, himself, the administrator of NASA, be *way out of line?* Webb called in some of his top deputies and described the situation. He said he was considering changing the order of the flight assignments—i.e., putting another astronaut in Glenn's place. This flight required a man who could comprehend the broader interests of the program better. His deputies looked at him as if he were crazy. He'd never get away with it! *The astronauts* wouldn't stand for it! . . . They had their differences, but on something like this the seven would stand together like an army . . . Webb was beginning to see something he had never quite figured out before. The astronauts were not *his* men. They were in a category new to American life. They were single-combat warriors. If anything, *he* was *their* man.

One could imagine what would happen if Webb tried to exercise his authority nonetheless . . . Here comes the showdown . . . the seven Mercury astronauts on the TV . . . explaining that in the very moments when their lives are on the line, he, Webb, is meddling, trying to curry favor with Lyndon Johnson, being vindictive because John Glenn's wife, Annie, would not let the hideous handwringing Texan into her living room to emote all over her on nationwide television . . . He sits in his office suite in Washington while their hides are up on the tip of the rocket . . . One could see the lines drawn in just that way. Webb would be issuing denials, furiously . . . Kennedy would be the umpire—and it wasn't too hard to figure out which way the decision would go. The changing of the assignments was never mentioned again.

Not long thereafter an old friend visited Webb in his corner office, and Webb unburdened himself.

"Look at this office," he said, making a grand gesture across a room with all the trappings of Cabinet-level rank known to the General Services Administration syllabus. "And I . . . *cannot* . . . *get* . . . *a* . . . *simple* . . . *order* . . . *carried out!*"

But in the next moment his mood changed. "All the same," he said, "I love those guys. They're putting their lives on the line for their country."

Dryden and Gilruth decided to postpone the launch for at least two weeks, to the middle of February. Glenn made a statement to the press about the delays. He said that anybody who knew the first thing about "the flight test business" expected delays; they were all part of it; the main thing was not to involve people who became "panicky" when everything didn't go just right . . . Glenn went home to Arlington for a three-day weekend. While he was there, President Kennedy invited him to the White House for a private get-together. He did not invite Webb or Johnson to join them.

On February 20 Glenn was once again squeezed inside the Mercury capsule on top of the Atlas rocket, lying on his back, whiling away the holds in the countdown by going over his checklist and looking at the scenery through the periscope. If he closed his eyes it felt as if he were lying on his back on the deck of an old ship. The rocket kept creaking and twisting, shaking the capsule this way and that. The Atlas had 4.3 times as much fuel as the Redstone, including 80 tons of liquid oxygen. The liquid oxygen, the "lox," had a temperature of 293 degrees below zero, so that the shell and tubing of the rocket, which were thin, kept contracting and twisting and creaking. Glenn was at the equivalent of nine stories up in the air. The enormous rocket seemed curiously fragile, the way it moved and creaked and whined. The contractions created high-frequency vibrations and the lox hissed in the pipes, and it all ran up through the capsule like a metallic wail. It was the same rocket lox wail they used to hear at dawn at Edwards when they fueled the D–558–2 many years before.

Through the periscope Glenn could see for miles down the Banana River and the Indian River. He could just barely make out the thousands of people along the beaches. Some of them had been camping out along there in trailers since January 23, when the flight was first scheduled. They had elected camp mayors. They were having a terrific time. A month in a Banana River trailer camp was not too long to wait to make sure you were here when an event of this magnitude occurred.

There were thousands of them, off on the periphery as Glenn looked out. He could only see them through the periscope. They looked very small and far way and far below. And they were all wondering with a delicious shudder what it must be like to be in his place now. How frightened is he! *Tell us! That's all we want to know!* The fear and the gamble. Never mind the rest. Lying on his back like this, with his legs jackknifed up above him, stuffed blind into the holster, with the hatch closed, he couldn't help but be aware of his own heartbeat from time to time. Glenn could tell that his pulse was slow. Out loud, if the subject ever came up, everyone said that pulse rates didn't matter; it was a very subjective thing; many variables; and so on. It had only been within the past five years that biosensors had even been put on pilots. They resented them and didn't care to attach any importance to them. Nevertheless, without saying so, everyone knew that they provided a rough gauge of a man's emotional state. Without saying so—not a word!—everyone knew that Gus Grissom's pulse rate had been *somewhat panicky*. It kept jumping over 100 during the countdown and then spurted up to 150 during the lift-off and stayed that high throughout his weightless flight, then jumped again, all the way to 171, just before the retrorockets went off. No one—certainly not out loud—no one was going to draw any conclusions from it, but . . . it was not a sign of the right stuff. Add to that his performance in the water . . . In his statement about people who get panicky over the flight test business, Glenn had said you had to know how to control your emotions. Well, he was as good as his word. Did any yogi ever control his heartbeat and perspiration better! (And, as the biomedical panels in the Mission Control room showed, his pulse never went over 80 and was holding around 70, no more than that of any normal healthy bored man having breakfast in the kitchen.) Occasionally he could feel his heart skip a beat or beat with an odd electrical sensation, and he knew that he was feeling the tension. (And at the biomedical panels the young doctors looked at each other in consternation—and then shrugged.) Nevertheless, he was aware that he was feeling no fear. He truly was not. He was more like an actor who is going out to perform in the same play yet once again—the only difference being that the audience this time is enormous and highly prestigious. He knew every sensation he would feel once the event began. The main thing was not to . . . "foul up." Please, dear God, don't let me foul up. In fact, there was little chance that he would forget so much as a word or a single move. Glenn had been the backup pilot—everyone said *pilot* now—for both Shepard and Grissom. During the charade before the first flight, he had gone through all of Shepard's simulations, and he had repeated most of Grissom's. And the simulations he had gone through as prime pilot for the first orbital flight had surpassed any simulations ever done before. They had even put him in the capsule on top of the rocket and *moved the gantry* away from the rocket, because Grissom had reported the odd sensation of perceiving the gantry as *falling over,* as he witnessed the event through his periscope, just before lift-off. Therefore, this feeling would be *adapted out* of Glenn. They put him in the capsule on top of the rocket and instructed him to watch the gantry move away through his periscope. *Nothing* must be novel about the experience! On top of all that, he had Shepard's and Grissom's descriptions of variations from the simulations. "On the centrifuge you feel thus-and-such. Well, during the actual flight it feels like that but with this-and-that difference." No man had ever lived an event so completely ahead of time. He was socketed into the capsule, lying on his back, getting ready to do precisely what his enormous Presbyterian Pilot self-esteem had been dying to do for fifteen years: demonstrate to the world his righteous stuff.

Exactly that! The Presbyterian Pilot! Here he is!—within twenty seconds of lift-off, and the only strange thing is how little adrenalin is pumping when the moment comes . . . He can hear the rumble of the Atlas engines building up down there below his back. All the same, it isn't terribly loud. The huge squat rocket shakes a bit and struggles to overcome its own weight. It all happens very slowly in the first few seconds, like an extremely heavy elevator rising. They've lit the candle and there's no turning back, and yet there's no surge inside him. His pulse rises only to 110, no more than the minimum rate you should have if you have to deal with a sudden emergency. How strange that it should be this way! He has been more wound up for a takeoff in an F-102.

"The clock is operating," he said. "We're underway."

It was all very smooth, much smoother than the centrifuge . . . just as Shepard and Grissom said it would be. He had gone through the same g-forces so many times . . . he hardly noticed them as they built up. It would have bothered him much more if they had been less. Nothing novel! No excitement, please! It took thirteen seconds for the huge rocket to reach transonic speed. The vibrations started. It was just as Shepard and Grissom said: it was much gentler than the centrifuge. He was still lying flat on his back, and the g-forces drove him deeper and deeper into the seat, but it all felt so familiar. He barely noticed it. He kept his eyes on the instrument panel the whole time . . . All quite normal, every little needle and switch in the right place . . . No malevolent instructor feeding *Abort* problems into the loop . . . As the rocket entered the transonic zone, the vibration became intense. The vibrations all but obliterated the roar of the engines. He was entering the area of "max q," maximum aerodynamic pressure, in which the pressure of the shaft of the Atlas forcing its way through the atmosphere at supersonic speed would reach almost a thousand pounds per square foot. Through the cockpit window he could see the sky turning black. Almost 5 g's were driving him back into his seat. And yet . . . *easier than the centrifuge* . . . All at once he was through *max q*, as if through a turbulent strait, and the trajectory was smooth and he was supersonic and the rumble of the rocket engines was more muffled than ever and he could hear all the little fans and recorders and the busy little kitchen, the humming little shop . . . The pressure on his chest reached 6 g's. The rocket pitched down. For the first time he could see clouds and the horizon. In a moment—*there it was*—the Atlas rocket's two booster engines shut down and were jettisoned from the side of the shaft and his body was slammed forward, as if he were screeching to a halt, and the g-forces suddenly dropped to 1.25, almost as if he were on earth and not accelerating at all, but the central sustainer engine and two smaller engines were still driving him up through the atmosphere . . . A flash of white smoke went up past the window . . . *No! The escape tower was firing early—but the*

JETTISON TOWER *light wasn't on!* . . . He didn't see the tower go . . . Wait a minute . . . There went the tower, on schedule . . . The JETTISON TOWER light came on green . . . The smoke must have been from the booster rockets as they left the shaft . . . The rocket pitched back up . . . going straight up . . . The sky was very black now . . . The g-forces began pushing him back into his seat again . . . 3 g's . . . 4 g's . . . 5 g's . . . Soon he would be forty miles up . . . the last critical moment of powered flight, as the capsule separated from the rocket and went into its orbital trajectory . . . or didn't . . . *Hey!* . . . All at once the whole capsule was whipping up and down, as if it were tied to the end of a diving board, a springboard. The g-forces built up and the capsule whipped up and down. Yet no sooner had it begun than Glenn knew what it was. The weight of the rocket on the launch pad had been 260,000 pounds, most of it oxidizer and fuel, the liquid oxygen and RP-1 fuel. This was being consumed at such a furious rate, about one ton per second, that the rocket was becoming merely a skeleton with a thin skin of metal stretched over it, a tube so long and light that it was flexing. The g-forces reached six and then he was weightless, just like that. The sudden release made him feel as if he were tumbling head over heels, as if he had been catapulted off the end of that same springboard and was falling through the air doing forward rolls. But he had felt this same thing on the centrifuge when they ran the g-forces up to seven and then suddenly cut the speed. At the same moment, right on schedule . . . a loud report . . . the posigrade rockets fired, throwing the capsule free of the rocket shaft . . . the capsule began its automatic turnabout, and all the proper green lights went on in front of him, and he knew he was "through the gate," as they said.

"Zero-g and I feel fine," he said. "Capsule is turning around . . ."

Glenn knew he was weightless. From the instrument readings and through sheer logic he knew it, but he couldn't feel it, just as Shepard and Grissom had never felt it. The turnaround brought him up to a sitting position, vertical to the earth, and that was the way he felt. He was sitting in a chair, upright, in a very tiny cramped quiet little cubicle 125 miles above the earth, a little metal closet, silent except for the humming of its electrical system, the inverters, the gyros, the cameras, the radio . . . *the radio* . . . He had been specifically instructed to violate the Fighter Jock code of No Chatter. He was supposed to radio back every sight, every sensation, and otherwise give the taxpayers the juicy stuff they wanted to hear. Glenn, more than any of the others, was fully capable of doing the job. Yet it was an awkward thing. It seemed unnatural.

"Oh!" he said. "That view is tremendous!"

Well, it was a start. In fact, the view was not particularly extraordinary. It was extraordinary that he was up here in orbit about the earth. He could see the exhausted Atlas rocket following him. It was tumbling end over end from the force of the small rockets throwing the capsule free of it.

He could hear Alan Shepard, who was serving as capcom in the Mercury Control Center at the Cape. His voice came in very clearly. He was saying, "You have a go, at least seven orbits."

"Roger," said Glenn. "Understand Go for at least seven orbits . . . This is *Friendship* 7. Can see clear back, a big cloud pattern way back across toward the Cape. Beautiful sight."

He was riding backward, looking back toward the Cape. It must be tremendous, it must be beautiful—what else could it be? And yet it didn't look terribly different from what he had seen at 50,000 feet in fighter planes. He had no greater sense of having left the bonds of earth. The earth was not just a little ball beneath him. It still filled his field of consciousness. It slid by slowly underneath him, just the way it did when you were in an airplane at forty or fifty thousand feet. He had no sense of being a *star voyager.* He couldn't see any stars at all. He could see the Atlas booster tumbling behind him and beginning to grow smaller, because it was in a slightly lower orbit. It just kept tumbling. There was nothing to stop it. Somehow the sight of this colossal great tumbling cylinder, which had weighed more than the average freighter while it was on the ground and which now weighed nothing and had been discarded like a candy wrapper—somehow it was more extraordinary than the view of earth. It shouldn't have been, but it was. The earth looked the way it had looked to Gus Grissom. Shepard had seen a low-grade black-and-white movie. Through his window Glenn could see what Grissom saw, the brilliant blue band at the horizon, a somewhat wider band of deeper blues leading into the absolutely black dome of the sky. Most of the earth was covered in clouds. The clouds looked very bright, set against the blackness of the sky. The capsule was heading east, over

Africa. But, because he was riding backward, he was looking west. He saw everything after he had passed over it. He could make out the Canary Islands, but they were partly obscured by clouds. He could see a long stretch of the African coast . . . huge dust storms over the African desert . . . but there was no sense of taking in the whole earth at a glance. The earth was eight thousand miles in diameter and he was only a hundred miles above it. He knew what it was going to look like in any case. He had seen it all in photographs taken from the satellites. It had all been flashed on the screens for him. Even the view had been simulated. *Yes . . . that's the way they said it would look* . . . Awe seemed to be demanded, but how could he express awe honestly? He had lived it all before the event. How could he explain that to anybody? The view wasn't the main thing, in any case. The main thing . . . was *the checklist!* And just try explaining that! He had to report all his switch and dial readings. He had to put a special blood-pressure rig on the arm of his pressure suit and pump it up. (His blood pressure was absolutely normal, 120 over 80—*perfect stuff!)* He had to check the manual attitude-control system, swing the capsule up and down, side to side, roll to the right, roll to the left . . . and there was nothing novel about it, not even in orbit, a hundred miles above the earth. *How could you explain that!* When he swung the capsule, it felt the same as it did in a one-g state on earth. He still didn't feel weightless. He merely felt less cramped, because there were no longer any pressure points on his body. He was sitting straight up in a chair drifting slowly and quietly around the earth. Just the hum of his little shop, the background noises in his headset, and the occasional spurt of the hydrogen-peroxide jets.

"This is *Friendship 7*," he said. "Working just like clockwork on the control check, and it went through just about like the procedures-trainer runs."

Well, that was it. The procedures trainer and the ALFA trainer and the centrifuge . . . He noticed that, in fact, he seemed to be moving a little faster than he had been on the ALFA trainer. When you sat in the trainer, cranking your simulated hydrogen-peroxide thrusters, they ran films on the screen of the earth rolling by below you, just the way it would be in orbital flight. "They didn't roll it by fast enough," he said to himself. Not that it mattered particularly . . . The sensation of speed was no more than that of being in an airliner and watching a cloud bank slide by far below . . . The world demanded awe, because this was a voyage through the stars. But he couldn't feel it. The backdrop of the event, the stage, the environment, the true orbit . . . was not the vast reaches of the universe. It was the simulators. *Who could possibly understand this?* Weightless he was, in the vacuum of space, humming around the earth . . . but his center of gravity was still back in that Baptist hardtack Low Rent stretch of sand and palmetto grass in Florida.

Ahhhh—but now this was truly something. Forty minutes into the flight, as he neared the Indian Ocean, off the east coast of Africa, he began sailing into the night. Since he was traveling east, he was going away from the sun at a speed of 17,500 miles an hour. But because he was riding backward, he could see the sun out the window. It was sinking the way the moon sinks out of sight as seen on earth. The edge of the sun began to touch the edge of the horizon. He couldn't tell what part of the earth it was. There were clouds everywhere. They created a haze at the horizon. The brilliant light over the earth began to dim. It was like turning down a rheostat. It took five or six minutes. Very slowly the lights were dimming. Then he couldn't see the sun at all, but there was a tremendous band of orange light that stretched from one side of the horizon to the other, as if the sun were a molten liquid that had emptied into a tube along the horizon. Where there had been a bright-blue band before, there was now the orange band; and above it a wider dimmer band of oranges and reds shading off into the blackness of the sky. Then all the reds and oranges disappeared, and he was on the night side of the earth. The bright-blue band reappeared at the horizon. Above it, stretching up about eight degrees, was what looked like a band of haze, created by the earth's atmosphere. And above that . . . for the first time he could make out the stars. Down below, the clouds picked up a faint light from the moon, which was coming up behind him. Now he was over Australia. He could hear Gordon Cooper's voice. Cooper was serving as the capcom at the tracking station in the town of Muchea, out in the kangaroo boondocks of western Australia. He could hear Cooper's Oklahoma drawl.

"That sure was a short day," said Glenn.

"Say again, *Friendship 7*," said Cooper.

"That was about the shortest day I've ever run into," said Glenn.

Somehow that was the sort of thing to say to old Oklahoma Gordo sitting down there in the middle of nowhere.

"Kinda passes rapidly, huh?" said Gordo.

"Yessir," said Glenn.

The clouds began to break up over Australia. He could make out nothing in the darkness except for electric lights. Off to one side he could make out the lights of an entire city, just as you could at 40,000 feet in an airplane, but the concentration of lights was terrific. It was an absolute mass of electric lights, and south of it there was another one, a smaller one. The big mass was the city of Perth and the smaller one was a town called Rockingham. It was midnight in Perth and Rockingham, but practically every living soul in both places had stayed up to turn on every light they had for the American sailing over in the satellite.

"The lights show up very well," said Glenn, "and thank everybody for turning them on, will you?"

"We sure will, John," said Gordo.

And he went sailing on past Australia with the lights of Perth and Rockingham sliding into the distance.

He was over the middle of the Pacific, about halfway between Australia and Mexico, when the sun began to come up behind him. This was just thirty-five minutes after the sun went down. Since he was traveling backward, he couldn't see the sunrise through the window. He had to use the periscope. First he could see the blue band at the horizon becoming brighter and brighter. Then the sun itself began to slide up over the edge. It was a brilliant red—not terribly different from what he had seen at sunrise on earth, except that it was rising faster and its outlines were sharper.

"It's blinding through the scope on clear," said Glenn. "I'm going to the dark filter to watch it come on up."

And then—*needles!* A tremendous layer of them—Air Force communications experiment that went amok . . . Thousands of tiny needles gleaming in the sun outside the capsule . . . But they couldn't be needles, because they were luminescent—they were like snowflakes—

"This is *Friendship 7*," he said. "I'll try to describe what I'm in here. I am in a big mass of some very small particles that are brilliantly lit up like they're luminescent. I never saw anything like it. They're round, a little. They're coming by the capsule, and they look like little stars. A whole shower of them coming by. They swirl around the capsule and go in front of the window and they're all brilliantly lighted. They probably average maybe seven or eight feet apart, but I can see them all down below me also."

"Roger, *Friendship* 7." This was the capcom on Canton Island out in the Pacific. "Can you hear any impact with the capsule? Over."

"Negative, negative. They're very slow. They're not going away from me more than maybe three or four miles per hour."

They swirled about his capsule like tiny weightless diamonds, little bijoux—no, they were more like fireflies. They had that lazy but erratic motion, and when he focused on one it would seem to be lit up, but the light would go out and he would lose track of it, and then it would light up again. That was like fireflies, too. There used to be thousands of fireflies in the summers, when he was growing up. These things were like fireflies, but they obviously couldn't be any sort of organism . . . unless all the astronomers and all the satellite recording mechanisms had been fundamentally wrong . . . They were undoubtedly particles of some sort, particles that caught the sunlight at a certain angle. They were beautiful, but were they coming from the capsule? That could mean trouble. They must have been coming from the capsule, because they traveled along with him, in the same trajectory, at the same speed. But wait a minute. Some of them were far off, far below . . . there might be an entire field of them . . . a minute cosmos . . . something never seen before! And yet the capcom on Canton Island didn't seem particularly interested. And then he sailed out of range of Canton and would have to wait to be picked up by the capcom at Guaymas, on the west coast of Mexico. And when the Guaymas capcom picked him up, he didn't seem to know what he was talking about.

"This is *Friendship* 7," said Glenn. "Just as the sun came up, there were some brilliantly lighted particles that looked luminous that were swirling around the capsule. I don't have any in sight right now. I did have a couple just a moment ago, when I made the transmission over to you. Over."

"Roger, *Friendship* 7."

And that was it. "Roger, *Friendship* 7." Silence. They didn't particularly care.

Glen kept talking about his fireflies. He was fascinated. It was the first true unknown anyone had encountered out here in the cosmos. At the same time he was faintly apprehensive. *Roger, Friendship 7.* The capcom finally asked a polite question or two, about the size of the particles and so on. They obviously were not carried away by this celestial discovery.

All of a sudden the capsule swung out to the right in a yaw, out about twenty degrees. Then it was as if it hit a little wall. It bounced back. Then it swung out again in the yaw and hit the little wall and bounced back. Something had gone out in the automatic attitude control. Never mind the celestial fireflies. He was sailing over California, heading for Florida. Now all the capcoms were coming alive, all right.

President Kennedy was supposed to come on the radio as Glenn came over the United States. He was going to bless his single-combat warrior as he came over the continental U.S.A. He was going to tell him the hearts of all his fellow citizens were with him. But that all went by the boards in view of the problem with the automatic controls.

Glenn went sailing over Florida, over the Cape, starting his second orbit. He couldn't see much of anything down below, because of the clouds. He no longer cared particularly. The attitude control was the main thing. One of the small thrusters seemed to have gone out, so that the capsule would drift to the right, like a car slowly skidding on ice. Then a bigger thruster would correct the motion and bounce it back. That was only the start. Pretty soon other thrusters began acting up when he was on automatic. Then the gyros started going. The dials that showed the angle of the capsule with respect to the earth and the horizon were giving obviously wrong readings. He had to line it up visually with the horizon. Fly by wire! Manual control! It was no emergency, however, at least not yet. As long as he was in orbit, the attitude control of the capsule didn't particularly matter, so far as his safety was concerned. He could be going forward or backward or could have his head pointed straight at the earth or could be drifting around in circles or pitching head over heels, for that matter, and it wouldn't change his altitude or trajectory in the slightest. The only critical point was the re-entry. If the capsule was not lined up at the correct angle, with the blunt end and the heat shield down, it might burn up. To line it up correctly, fuel was required, the hydrogen peroxide, no matter whether it was lined up automatically or by the astronaut. If too much fuel was used keeping the capsule stable while it sailed around in orbit, there might not be enough left to line it up before the re-entry. That had been the problem in the ape's flight. The automatic attitude control had started malfunctioning and was using up so much fuel they brought him down after two orbits.

Every five minutes he had to shift his radio communications to a new capcom. You couldn't receive and send at the same time, either. It wasn't like a telephone hookup. So you spent half the time just making sure you could hear each other.

"*Friendship* 7, *Friendship* 7, this is CYI." That was the Canary Islands capcom. "The time is now 16:32:26. We are reading you loud and clear; we are reading you loud and clear. CYI."

Glenn said: "This is *Friendship* 7 on UHF. As I went over recovery area that time, I could see a wake, what appeared to be a long wake in the water. I imagine that's the ships in our recovery area."

"*Friendship* 7 . . . We do not read you, do not read you. Over."

"*Friendship* 7, this is Kano. At G.M.T. 16:33:00. We do not . . . This is Kano. Out."

"*Friendship* 7, *Friendship* 7, this is CYI Com Tech. Over."

Glen said: "Hello, Canary. *Friendship* 7. Receive you loud and a little garbled. Do you receive me? Over."

"*Friendship* 7, *Friendship* 7, this is CYI Com Tech. Over."

"Hello, Canary, *Friendship* 7. I read you loud and clear. How me? Over."

"*Friendship* 7, *Friendship* 7, this is CYI Com Tech. Over."

"Hello, CYI Com Tech. *Friendship* 7. How do you read me? Over."

"*Friendship* 7, *Friendship* 7, this is CYI, CYI Com Tech. Do you read? Over."

"Roger. This is *Friendship* 7, CYI. I read you loud and clear. Over."

"*Friendship* 7, *Friendship* 7, this is CYI Com Tech, CYI Com Tech. Do you read? Over."

"Hello, CYI Com Tech. Roger, read you loud and clear. Over."

"*Friendship* 7, this is CYI Com Tech. Read you loud and clear also, on UHF, on UHF. Standby."

"Roger. *Friendship* 7."

"*Friendship* 7, *Friendship* 7, *Friendship* 7, this is Canary capcom. How do you read? Over."

"Hello, Canary capcom. *Friendship* 7. I read you loud and clear. How me?"

Finally, the Canary Islands capcom said: "I read you loud and clear. I am instructed to ask you to correlate the actions of the particles surrounding your spacecraft with the actions of your control jets. Do you read? Over."

"This is *Friendship* 7. I did not read you clear. I read you loud but very garbled. Over."

"Roger. Cap asks you to correlate the actions of the particles surrounding the vehicle with the reaction of one of your control jets. Do you understand? Over."

"This is *Friendship* 7. I do not think they were from my control jets, negative. Over."

There—exactly five minutes to get one question out and one answer. Well, at least they finally showed an interest in the fireflies. They wondered if they might have something to do with the malfunctioning thrusters. Oh, but it was a struggle.

In any case, he was not particularly worried. He could control the attitude manually if he had to. The fuel seemed to be holding out. Everything hummed and whined and buzzed as usual inside the capsule. The same high background tones came over the radio. He could hear the oxygen coursing through his pressure suit and his helmet. There was no "sensation" of motion speed at all, unless he looked down at the earth. Even then it slid by very slowly. When the thrusters spurted hydrogen peroxide, he could feel the capsule swing this way and that. But it was like the ALFA trainer on earth. He still didn't feel weightless. He was still sitting straight up in his chair. On the other hand, the camera—when he wanted to reload it, he just parked it in the empty space in front of his eyes. It just floated there in front of him. Way down there were little flashes all over the place. It was lightning in the clouds over the Atlantic. Somehow it was more fascinating than the sunset. Sometimes the lightning was inside the clouds and looked like flashlights going on and off underneath a blanket. Sometimes it was on top of the clouds, and it looked like firecrackers going off. It was extraordinary, and yet there was nothing new about the sight. An Air Force colonel, David Simons, had gone up in a balloon, alone, to 102,000 feet, for thirty-two hours and had seen the same thing.

Glenn was now over Africa, riding over the dark side of the earth, sailing backward toward Australia. The Indian Ocean capcom said: "We have message from MCC for you to keep your landing-bag switch in off position. Landing-bag switch in off position. Over."

"Roger," said Glenn. "This is *Friendship* 7."

He wanted to ask why. But that was against the code, except in an emergency situation. That fell under the heading of nervous chatter.

Over Australia old Gordo, Gordo Cooper, got on the same subject: "Will you confirm the landing-bag switch is in the off position? Over."

"That is affirmative," said Glenn. "Landing-bag switch is in the center off position."

"You haven't had any banging noises or anything of this type at higher rates?"

"Negative."

"They wanted this answer."

They still didn't say why, and Glenn entered into no nervous chatter. He now had two red lights on the panel. One was the warning light for the automatic fuel supply. All the little amok action of the yaw thrusters had used it up. Well, it was up to the Pilot now . . . to aim the capsule correctly for re-entry . . . The other was a warning about excess cabin water. It built up as a by-product of the oxygen system. Nevertheless, he pressed on with the checklist. He was supposed to exercise by pulling on the bungee cord and then take his blood pressure. The Presbyterian Pilot! He did it without a peep. He was pulling on the bungee and watching the red lights when he began sailing backward into the sunrise again. Two hours and forty-three minutes into the flight, his second sunrise over the Pacific . . . seen from behind through a periscope. But he hardly watched it. He was looking for the fireflies to light up again. The great rheostat came up, the earth lit up, and now there were thousands of them swirling about the capsule. Some of them seemed to be miles away. A huge field of them, a galaxy, a microuniverse. No question about it, they weren't coming from the capsule, they were part of the cosmos. He took out the camera again. He had to photograph them while the light was just right.

"*Friendship* 7." The Canton Island capcom was coming in. "This is Canton. We also have no indication that your landing bag might be deployed. Over."

Glenn's first reaction was that this must have something to do with the fireflies. He's telling them about the fireflies and they come in with something about the landing bag. But who said anything about the landing bag being deployed?

"Roger," he said. "Did someone report landing bag could be down? Over."

"Negative," said the capcom. "We had a request to monitor this and to ask you if you heard any flapping, when you had high capsule rates."

"Well," said Glenn, "I think they probably thought these particles I saw might have come from that, but these are . . . there are thousands of these things, and they go out for it looks like miles in each direction from me, and they move by here very slowly. I saw them at the same spot on the first orbit. Over."

And so he thought that explained all the business about the landing bag.

They gave him the go-ahead for his third and final orbit as he sailed over the United States. He couldn't see a thing for the clouds. He pitched the capsule down sixty degrees, so he could look straight down. All he could see was the cloud deck. It was just like flying at high altitudes in an airplane. He was really no longer in the mood for sightseeing. He was starting to think about the sequence of events that would lead to the retrofiring over the Atlantic after he had been around the world one more time. He had to fight both the thrusters and the gyros now. He kept releasing and resetting the gyros to see if the automatic attitude control would start functioning again. It was all out of whack. He would have to position the capsule by using the horizon as a reference. He was sailing backward over America. The clouds began to break. He began to see the Mississippi delta. It was like looking at the world from the tail-gun perch of the bombers they used in the Second World War. Then Florida started to slide by. Suddenly he realized he could see the whole state. It was laid out just like it is on a map. He had been around the world twice in three hours and eleven minutes and this was the first sense he had had of how high up he was. He was about 550,000 feet up. He could make out the Cape. By the time he could see the Cape he was already over Bermuda.

"This is *Friendship* 7," he said. "I have the Cape in sight down there. It looks real fine from up here."

"Rog. Rog." That was Gus Grissom on Bermuda.

"As you know," said Glenn.

"Yea, verily, sonny," said Grissom.

Oh, it all sounded very fraternal. Glenn was modestly acknowledging that his loyal comrade Grissom was one of the only three Americans ever to see such a sight . . . and Grissom was calling him "sonny."

Twenty minutes later he was sailing backward over Africa again and the sun was going down again, for the third time, and the rheostat was dimming and he . . . saw *blood.* It was all over one of the windows. He knew it couldn't be blood, and yet it was blood. He had never noticed it before. At this particular angle of the setting rheostat sun he could see it. Blood and dirt, a real mess. The dirt must have come from the firing of the escape tower. And the blood . . . *bugs,* perhaps . . . The capsule must have smashed into bugs as it rose from the launch pad . . . or *birds* . . . but he would have heard the thump. It must have been bugs, but bugs didn't have blood. Or the blood red of the sun going down in front of him diffusing . . . And then he refused to think about it anymore. He just turned the subject off. Another sunset, another orange band streaking across the rim of the horizon, more yellow bands, blue bands, blackness, thunderstorms, lightning making little sparkles under the blanket. It hardly mattered anymore. The whole thing of lining the capsule up for retrofire kept building up in his mind. In slightly less than an hour the retro-rockets would go off. The capsule kept slipping its angles, swinging this way and that way, drifting. The gyros didn't seem to mean a thing anymore.

And he went sailing backward through the night over the Pacific. When he reached the Canton Island tracking point, he swung the capsule around again so that he could see his last sunrise while riding forward, out the window, with his own eyes. The first two he had watched through the periscope because he was going backward. The fireflies were all over the place as the sun came up. It was like watching the sunrise from inside a storm of the things. He began expounding upon them again, about how they couldn't possibly come from the capsule, because some of them seemed to be miles away. Once again nobody on the ground was interested. They weren't interested on Canton Island, and pretty soon he was in range of the station on Hawaii, and they weren't interested, either. They were all wrapped up in something else. They had a little surprise for him. They backed into it, however. It took him a while to catch on.

He was now four hours and twenty-one minutes into the flight. In twelve minutes the retro-rockets were supposed to fire, to slow him down for re-entry. It took him another minute and forty-five seconds to go through all the "do you reads" and "how me's" and "overs" and establish contact with the capcom on Hawaii. Then they sprang their surprise.

"*Friendship 7*," said the capcom. "We have been reading an indication on the ground of segment 5–1, which is Landing Bag Deploy. We suspect this is an erroneous signal. However, Cape would like you to check this by putting the landing-bag switch in auto position, and seeing if you get a light. Do you concur with this? Over."

It slowly dawned on him . . . *Have been reading* . . . For how long? . . . Quite a little surprise. And they hadn't told him! They'd held it back! *I am a pilot and they refuse to tell me things they know about the condition of the craft!* The insult was worse than the danger! If the landing bag had deployed—and there was no way he could look out and see it, not even with the periscope, because it would be directly behind him—if it had deployed, then the heat shield must be loose and might come off during the re-entry. If the heat shield came off, he would burn up inside the capsule like a steak. If he put the landing-bag switch in the automatic control position, then a green light should come on if the bag was deployed. Then he would know. Slowly it dawned! . . . That was why they kept asking him if the switch were in the off position!—they didn't want him to learn the awful truth too quickly! Might as well let him complete his three orbits—then we'll let him find out about the bad news!

On top of that, they now wanted him to fool around with the switch.

That's stupid! It might very well be that the bag had not deployed but there was an electrical malfunction somewhere in the circuit and fooling with the automatic switch might then cause it to deploy. But he stopped short of saying anything. Presumably they had taken all that into account. There was no way he could say it without falling into the dread nervous chatter.

"Okay," said Glenn. "If that's what they recommend, we'll go ahead and try it. Are you ready for it now?"

"Yes, when you're ready."

"Roger."

He reached forward and flipped the switch. Well . . . this was it—

No light. He immediately switched it back off.

"Negative," he said. "In automatic position did not get a light and I'm back in off position now. Over."

"Roger, that's fine. In this case, we'll go ahead, and the re-entry sequence will be normal."

The retro-rockets would be fired over California, and by the time the retro-rockets brought him down out of his orbit and through the atmosphere, he would be over the Atlantic near Bermuda. That was the plan. Wally Schirra was the capcom in California. Less than a minute before he was supposed to fire the retro-rockets, by pushing a switch, he heard Wally saying: "John, leave your retropack on through your pass over Texas. Do you read?"

"Roger."

But why? The retropack wrapped around the edges of the heat shield and held the retro-rockets. Once the rockets were fired, the retropack was supposed to be jettisoned. They were back to the heat shield again, with no explanation. But he had to concentrate on firing the retro-rockets.

Next to the launch this was the most dangerous part of the flight. If the capsule's angle of attack was too shallow, you might skip off the top of the earth's atmosphere and stay in orbit for days, until long after your oxygen had run out. You wouldn't have any more rockets to slow you down. If the angle were too steep, the heat from the friction of going through the atmosphere would be so intense you would burn up inside the capsule, and a couple of minutes later the whole thing would disintegrate, heat shield or no heat shield. But the main thing was not to think about it in quite those terms. The field of consciousness is very small, said Saint-Exupéry. *What do I do next?* It was the moment of the test pilot at last. Oh, yes! *I've been here before! And I am immune! I don't get into corners I can't get out of!* One thing at a time! He could be a true flight test hero and try to line the capsule up all by himself by using the manual controls with the horizon as his reference—or he could make one more attempt to use the automatic controls. Please, dear God . . . don't let me foul up! What would the Lord answer? (Try the automatic, you ninny.) He released and reset the gyros. He put the controls on automatic. The answer to your prayers, John! Now the dials gibed with what he saw out the window and through the periscope. The automatic controls worked perfectly in pitch and roll. The yaw was still off, so he corrected that with the manual controls. The capsule kept pivoting to the right and he kept nudging it back. The ALFA trainer! One thing at a time! It was just like the ALFA trainer . . . no sense of forward motion at all . . . As long as he concentrated on the instrument panel and didn't look at the earth sliding by beneath him, he had no sense at all of going 17,500 miles an hour . . . or even five miles an hour . . . The humming little kitchen . . . He sat up in his chair squirting his hand thruster, with his eyes pinned on the dials . . . Real life, a crucial moment—against the eternal good beige setting of the simulation. One thing at a time!

Schirra began giving him the countdown for firing the rockets. "Five, four—"

He nudged it back once more with the yaw thruster.

"—three, two, one, fire."

He pushed the retro-rocket switch with his hand.

The rockets started firing in sequence, the first one, the second one, the third one. The sound seemed terribly muffled—but in that very moment, the jolt! Pure gold! One instant, as Schirra counted down, he felt absolutely motionless. The next . . . *thud thud thud* . . . the jolt in his back. He felt as if the capsule had been knocked backward. He felt as if he were sailing back toward Hawaii. All as it should be! Pure gold! The retrolight was lit up green. It was all going perfectly. He was merely slowing down. In eleven minutes he would be entering the earth's atmosphere.

He could hear Schirra saying: "Keep your retropack on until you pass Texas."

Still no reason given! He couldn't see the pattern yet. There was only the dim sense that in some fashion they were jerking him around. But all he said was: "That's affirmative."

"It looked like your attitude held pretty well," said Schirra. "Did you have to back it up at all?"

"Oh, yes, quite a bit. Yeah, I had a lot of trouble with it."

"Good enough for government work from down here," said Schirra. That was one of Schirra's favorite lines.

"Do you have a time for going to Jettison Retro?" said Glenn. This was an indirect way of asking for some explanation for the mystery of keeping the retropack on.

"Texas will give you that message," said Schirra. "Over."

They weren't going to tell him! Not so much the thought . . . as the *feeling* . . . of the insult began to build up.

Three minutes later the Texas capcom tracking station came in: "This is Texas capcom, *Friendship* 7. We are recommending that you leave the retropackage on through the entire re-entry. This means that you will have to override the zero-point-oh-five-g switch, which is expected to occur at 04:43:53. This also means that you will have to manually retract the scope. Do you read?"

That did it.

"This is *Friendship* 7," said Glenn. "What is the reason for this? Do you have any reason? Over."

"Not at this time," said the Texas capcom. "This is the judgment of Cape Flight . . . Cape Flight will give you the reason for this action when you are in view."

"Roger. Roger. *Friendship* 7."

It was really unbelievable. It was beginning to fit—

Twenty-seven seconds later he was over the Cape itself and the Cape capcom, with the voice of Alan Shepard on the radio, was telling him to retract his periscope manually and to get ready for re-entry into the atmosphere.

It was beginning to fit together, he could see the pattern, the whole business of the landing bag and the retropack. This had been going on for a couple of hours now—and they were telling him nothing! Merely giving him the bits and pieces! But if he was going to re-enter with the retropack on, then they wanted the straps in place for some reason. And there was only one possible reason—something was wrong with the heat shield. And this they would not tell him! *Him!*—the pilot! It was quite unbelievable! it was—

He could hear Shepard's voice.

He was winding in the periscope, and he could hear Shepard's voice: "While you're doing that . . . we are not sure whether or not your landing bag has deployed. We feel it is possible to re-enter with the retropackage on. We see no difficulty at this time in that type of re-entry."

Glenn said, "Roger, understand."

Oh, yes, he understood now! If the landing bag was deployed, that meant the heat shield was loose. If the heat shield was loose, then it might come off during the re-entry, unless the retropack straps held it in place long enough for the capsule to establish its angle of re-entry. And the straps would soon burn off. If the heat shield came off, then he would fry. If they didn't want him—*the pilot!*—to know all this, then it meant they were afraid he might panic. And if he didn't even *need* to know the whole pattern—just the pieces, so he could follow orders—*then he wasn't really a pilot!* The whole sequence of logic clicked through Glenn's mind faster than he could have put it into words, even if he had dared utter it all at that moment. He was being treated like a passenger—a redundant component, a backup engineer, a boiler-room attendant—in an automatic system!—like someone who did not have that rare and unutterably righteous stuff!—as if the right stuff itself did not even matter! It was a transgression against all that was holy—all this in a single limbic flash of righteous indignation as John Glenn re-entered the earth's atmosphere.

"*Seven,* this is Cape," said Al Shepard. "Over."

"Go ahead, Cape," said Glenn. "You're ground . . . you are going out."

"We recommend that you . . ."

That was the last he could hear from the ground. He had entered the atmosphere. He couldn't feel the g-forces yet, but the friction and the ionization had built up, and the radios were now useless. The capsule was beginning to buffet and he was fighting it with the controls. The fuel for the automatic system, the hydrogen peroxide, was so low he could no longer be sure which system worked. He was descending backward. The heat shield was on the outside of the capsule, directly behind his back. If he glanced out the window he could see only the blackness of the sky. The periscope was retracted, so he saw nothing on the scope screen. He heard a *thump* above him, on the outside of the capsule. He looked up. Through the window he could see a strap. *From the retropack. The straps broke! And now what!* Next the heat shield! The black sky out the window began to turn a pale orange. The strap flat against the window started burning—and then it was gone. The universe turned a flaming orange. That was the heat shield beginning to burn up from the tremendous speed of the re-entry. This was something Shepard and Grissom had not seen. They had not re-entered the atmosphere at such speed. Nevertheless, Glenn knew it was coming. Five hundred, a thousand times he had been told how the heat shield would *ablate*, burn off layer by layer, vaporize, dissipate the heat into the atmosphere, send off a corona of flames. All he could see now through the window were the flames. He was inside a ball of fire. But!—a huge flaming chunk went by the window, a great chunk of something burning. Then another . . . another . . . The capsule started buffeting . . . The heat shield was breaking up! It was crumbling—flying away in huge flaming chunks . . . He fought to steady the capsule with the hand controller. *Fly-by-wire!* But the rolls and yaws were too fast for him . . . The ALFA trainer gone amok, inside a fireball . . . The heat! . . . It was as if his entire central nervous system were now centered in his back. If the capsule was disintegrating and he was about to burn up, the heat pulse would reach his back first. His backbone would become like a length of red-hot metal. He already knew what the feeling would be like . . . and when . . . *Now!* . . . But it didn't come. There was no tremendous heat and no more flaming debris . . . Not the heat shield, after all. The burning chunks had come from what remained of the retropack. First the straps had gone and then the rest of it. The capsule kept rocking, and the g-forces built up. He knew the g-forces by heart. A thousand times he had felt them on the centrifuge. They drove him back into the seat. It was harder and harder to move the hand controller. He kept trying to damp out the rocking motion by firing the yaw thrusters and the roll thrusters, but it was all too fast for him. They didn't seem to do much good, at any rate.

No more red glow . . . he must be out of the fireball . . . seven g's were driving him back into the seat . . . He could hear the Cape capcom:

". . . How do you read? Over."

That meant he had passed through the ionosphere and was entering the lower atmosphere.

"Loud and clear; how me?"

"Roger, reading you loud and clear. How are you doing?"

"Oh, pretty good."

"Roger. Your impact point is within one mile of the up-range destroyer."

Oh, pretty good. It wasn't Yeager, but it wasn't bad. He was inside of one and a half tons of non-aerodynamic metal. He was a hundred thousand feet up, dropping toward the ocean like an enormous cannonball. The capsule had no aerodynamic qualities whatsoever at this altitude. It was rocking terribly. Out the window he could see a wild white contrail snaked out against the blackness of the sky. He was dropping at a thousand feet per second. The last critical moment of the flight was coming up. Either the parachute deployed and took hold or it didn't. The rocking had intensified. The retropack! Part of the retropack must still be attached and the drag of it is trying to flip the capsule . . . He couldn't wait any longer. The parachute was supposed to deploy automatically, but he couldn't wait any longer. Rocking . . . He reached up to fire the parachute manually—but it fired on its own, automatically, first the drogue and then the main parachute. He swung under it in a huge arc. The heat was ferocious, but the chute held. It snapped him back into the seat. Through the window the sky was blue. It was the same day all over again. It was early in the afternoon on a sunny day out in the Atlantic near Bermuda. Even the landing-bag light was green. There was nothing even wrong with the landing bag. There had been nothing wrong with the heat shield. There was nothing wrong with his rate of descent, forty feet per second. He could hear the rescue ship chattering away over the radio. They were only twenty minutes away from where he would hit, only six miles. He was once again lying on his back in the human holster. Out the window the sky was no longer black. The capsule swayed under the parachute, and over this way he looked up and saw clouds and over that way blue sky. He was very, very hot. But he knew the feeling. All those endless hours in the heat chambers—it wouldn't kill you. He was coming down into the water only 300 miles from where he started. It was the same day, merely five hours later. A balmy day out in the Atlantic near Bermuda. The sun had moved just seventy-five degrees in the sky. It was 2:45 in the afternoon. Nothing to do but get all these wires and hoses disconnected. *He had done it.* He began to let the thought loose in his mind. He must be very close to the water. The capsule hit the water. It drove him down into his seat again, on his back. It was quite a jolt. It was hot in here. Even with the suit fans still running, the heat was terrific. Over the radio they kept telling him not to try to leave the capsule. The rescue ship was almost there. They weren't going to try the helicopter deal again, except in an emergency. He wasn't about to attempt a water egress. He wasn't about to hit the hatch detonator. The Presbyterian Pilot was not about to foul up. His pipeline to the dear Lord could not be clearer. He had done it.

Annie Glenn had already had a taste of what it was going to be like. But the other six and their wives were not ready for it. It was as if some enormous tidal wave were heading for the Cape and the entire U.S.A. from out in the Atlantic, from the vicinity of Grand Bahama Island, where John was being debriefed. Riding the crest, like Triton, was the Freckled Face God, John himself. Word got back that the sailors on the *Noa*, the ship that hauled the capsule, with John in it, out of the water, had painted white lines around his footprints on the deck after he walked from the capsule to a hatchway. They didn't want his footprints on their deck to ever disappear! Well, it just seemed like some sort of goony swabbo sentimentality. But that was only the beginning.

Al Shepard and Gus Grissom didn't know what the hell was happening. Poor Gus—all he had gotten after his flight was a medal, a handshake, a gust of rhetoric from James Webb, out on a brain-frying strip of asphalt at Patrick Air Force Base, plus a few attaboys from a crowd of about thirty. For John—well, the mobs that had showed up for the launch, for the fireworks, barely seemed to have thinned out at all. Cocoa Beach was still full of the crazy adrenalin of the event. Out-of-towners were still tooling around all over the place in their automobiles and asking where the astronauts hung out. They didn't want to miss a thing. They knew that John would be flying back to the Cape after the debriefing. The next thing they knew, Lyndon Johnson was in town. He was going to meet John at the landing strip at Patrick. Underlings like Webb would just be part of the scenery. Then they learned that the President was coming, John F. Kennedy himself. Glenn wasn't going to him, in Washington—he was coming to Glenn.

Something quite extraordinary was building up. It was a wave and a half, and the other six and their wives were more surprised than anybody else. It was ironic. They had all assumed that Al Shepard was the big winner. Al had won out in the competition for the first flight. Al had been invited to the White House to receive a medal, whereas Gus had gotten his about eight steps from the palmetto grass, because Al was the certified number-one man in this thing and had taken the first flight. But even before John got back to the Cape from Grand Bahama Island, there was a note of worshipful swooning in the air that indicated that Al had not made the first flight, after all. He had merely made the first suborbital flight, which now looked like nothing at all. He was now more like Slick Goodlin to John's Chuck Yeager. Slick Goodlin had, technically, made the first flight of the X–1. But it was Yeager who made the flight that counted, the flight in which they first tried to push the bird supersonic. As Slick Goodlin to John's Chuck Yeager—what was Al supposed to do, cheer about it? And Betty Grissom—who never even got a parade down the poor dim dowdy main street of Mitchell, Indiana—was she supposed to be tickled pink about the Glenns, who were going to be paraded up and down every high road in the United States? But there was precious little time to brood. Once John's plane touched down at Patrick on February 23, the wave became so big it simply carried everyone along with it. The fellows and the wives and the children were all out at Patrick, waiting for John's plane, and the Vice-President was on hand, along with about two hundred reporters. Johnson was right up there at the head of the mob with Annie and the two children. He had gotten next to her at last. Johnson was right beside her now, out at Patrick, oozing protocol all over her and craning and straining his huge swollen head around, straining to get at John and pour Texas all over him. The plane arrives and John disembarks, a tremendous cheer goes up, a cry from the throat, from the diaphragm, from the solar plexus, and they bring Annie and the two children forward . . . the holy icons . . . the Wife and the Children . . . the Solid Backing on the Home Front . . . and John is too much! He reaches into his pocket and pulls out a handkerchief and dabs his eye, wipes away a tear! And some little guy from NASA stretched out his hand and took the used handkerchief . . . so that it could be preserved in the Smithsonian! (With this handkerchief Astronaut John H. Glenn, Jr., wiped away a tear upon being reunited with his wife after his historic earth-orbital flight.) From that moment on, Al and Gus were also-rans, minor leaguers. And they didn't even have time to fume! The events, day by day, were becoming like something elemental, like a huge change in the weather, a shift in the templates, the Flood, the Last Day, the True Brother Entering Heaven . . .

John did not merely get a parade through Washington and a trip to the White House and the medal from the President. Oh, he got those things, all right. But he also addressed a special joint session of Congress—the Senate and the House met together to hear John, the way they had for presidents, prime ministers, kings. There was John standing up there at the podium, with Lyndon Johnson and John McCormack seated behind him, and the rest of them *looking up at him* from their seats. In absolute adoration, too! That was where the tears started! The tears—they couldn't hold them back. John's great round freckled face was shining with glory. He knew just what he was doing. He was the Presbyterian Pilot addressing the world. He said some things that nobody else in the world could have gotten away with, even in 1962. He said, "I still get a lump in my throat when I see the American flag passing by." But he pulled it off! And then he lifted his hand up toward the gallery—this was in the House side of the Capitol—and five hundred pairs of congressional eyes swung up with his hand toward the gallery, and he introduced his mom and dad from New Concord, Ohio, and a few aunts and uncles for good measure, and then his children and, finally, ". . . above all, I want you to meet my wife, Annie . . . Annie . . . *the Rock!*" Well, that did it. That turned on the waterworks. Senators and representatives were trying to clap and reach for their handkerchiefs at the same time. They were dabbing their eyes and cheering through the fluttering ends of their handkerchiefs. Their faces glistened. Some fought back the tears and a couple let go. They applauded, cheered, snuffled, wheezed . . . A couple of them said, "Amen!" They said it out loud; it just came popping out of their good hardtack cracker evangelical dissenting Protestant hearts as the Presbyterian Pilot lifted up his eyes and his hand to *the Rock* and the eternal Mother of us all . . .

And that was just the start. In a way that was nothing compared to the ticker-tape parade in New York. After all, a ceremonial joint session of Congress is a tailor-made event, a command performance. But the parade in New York was an amazing event, so amazing that anyone, even Al or Gus, could only blink and shake his head and ride the wave. John had the good sense to invite Al and Gus and "the Other Four," Wally, Scott, Deke, and Gordo, and their families to join him and Annie and the children in the parade. John could have called the shot any way he wanted it. There was nobody in NASA or in the U.S. government, with the exception of Kennedy himself, who could have arranged it any way John didn't want it. So everybody came along for the show, the whole gang.

Despite the tide of cheers and tears that had already started in Washington, none of them knew what to expect in New York. Like most military people, including those in the Brooklyn Navy Yard, they didn't really consider New York part of the United States. It was like a free port, a stateless city, an international protectorate, Danzig in the Polish corridor, Beirut the crossroads of the Middle East, Trieste, Zurich, Macao, Hong Kong. Whatever ideals the military stood for, New York City did not. It was a foreign city full of a strange race of curiously tiny malformed gray people. And so forth and so on. What they saw when they got there bowled them over. The crowds were not only waiting in the airport, which was not surprising—a little publicity was all it took to get a mob of gawkers to an airport—but they were also lining the godforsaken highway into the city, through the borough of Queens, or whatever it was, out in the freezing cold in the most rancid broken-down industrial terrain you have ever seen, a decaying landscape that seemed to belong to another century—they were out there *along the highway*, anywhere they could squeeze in, and . . . they were crying!—crying as the black cars roared by!

People were crying, right out in the open, as soon as they laid eyes on

John, and perhaps the rest of them, too. They were all swept up in the wave now. The wave was too enormous for fine distinctions. When they reached the city proper, Manhattan, and came in off the expressway, the FDR Drive, there were people hanging over the railings, twenty or thirty feet up above the ramp, and they were crying and waving little flags and pouring their hearts out.

And that was just the beginning. The parade started in lower Manhattan and headed up Broadway. Each astronaut was in an open limousine. John led the lineup, of course, with Lyndon Johnson, Vice-President, in the seat beside him. It was cold as hell, seventeen degrees, but the streets were mobbed. There must have been millions of people out there, packed from the curbings clear back to the storefronts, and there were people hanging out of all the windows, particularly along lower Broadway, where the buildings were older and they could open the windows, and they were filling the air with shreds of paper, every piece of paper they could get their hands on.

Sometimes the pieces of paper would flutter right in front of your face—and you could see that they were tearing up their telephone books, just ripping the pages out and tearing them to bits and throwing them out the window as homage, as garlands, rose petals—and it was so touching! This horrible rat-gray city was suddenly touching, warm! You wanted to protect these poor souls who loved you so much! Huge waves of emotion rolled over you. You couldn't hear yourself talk, but there was nothing you could have said, anyway. All you could do was let these incredible waves roll over you. Out in the middle of the intersections were the policemen, the policemen they had all heard about or read about, New York's Finest, big tough-looking men in blue greatcoats—and they were crying! They were right out in the intersections in front of everybody, bawling away—tears streaming down their faces, saluting, then cupping their hands and yelling amazing things to John and the rest of them—"We love you, Johnny!"—and then bawling some more, just letting it pour out. The New York cops!

And what was it that had moved them all so deeply? It was not a subject that you could discuss, but the seven of them knew what it was, and so did most of their wives. Or they knew about part of it. They knew it had to do with the presence, the aura, the radiation of *the right stuff*, the same vital force of manhood that had made millions vibrate and resonate thirty-five years before to Lindbergh—except that in this case it was heightened by Cold War patriotism, the greatest surge of patriotism since the Second World War. Neither the term nor the concept of the single-combat warrior did they know about, but the sheer patriotism of that moment—even in New York, the Danzig corridor!—was impossible to miss. We pay homage to you! You have fought back against the Russians in the heavens! There was something pure and rare about it. Patriotism! Oh, yes! Here you saw it in a million-footed form, before your very eyes! Most of the seven had been around the Kennedys at one time or another, with Jack or with Bobby, and knew the way a crowd reacted to them—but it was something different from this. Around the Kennedys you saw a fan's hysteria, involving a lot of shrieking and clutching, with people reaching out to grab souvenirs and swooning and squealing, as if the Kennedys were movie stars who happened to be in power. But what the multitudes showed John Glenn and the rest of them on that day was something else. They anointed them with the primordial tears that the right stuff commanded.

The seven righteous families were put up in suites at the Waldorf-Astoria, which as far as they knew was still the grandest hotel in America. Suites!—two bedrooms and a living room! For junior officers in the military it was an experience from a fable. They were still soaring on what they had just been through, but they were afraid to try to put the right name on it, afraid of what it might reveal about what was going through their minds. They were beginning to ask themselves the question "What, precisely, have we become?"

Henry Luce gave a dinner for them at the Tower Suite, the restaurant at the top of the Time-Life Building. After dinner on the spur of the moment, the whole bunch of them went to see a play, *How to Succeed in Business Without Really Trying,* which was a big hit at the time. John and Annie and the children, all of the other fellows and their wives and children, plus the bodyguards and some NASA people and some Time-Life people, quite an entourage—and all of it arranged at the last minute. The start of the play was held up for them. People in the audience gave up their seats, so the astronauts and their party could have the best seats in the house, a whole bloc. Just like that they gave up their seats. When John and the others walked into the theater, everybody else was already seated, because by now the play was a good thirty minutes late starting—and the audience rose and cheered until John sat down. Then a member of the cast came out in front of the curtain and welcomed them and congratulated John and praised the fellows as great human beings and humbly hoped that the little diversion about to be offered would please them. "And now the play will commence!"

Then the lights went down and the curtain went up, and you had to be pretty dense not to realize what this was: a command performance! Royal treatment, point for point, right down the line, and they were the royal families. And it didn't stop there. They had rewritten some of the lines, rewritten them in an hour or so—to make the jokes contain references to space and John's flight and putting a man on the moon and so on. When they left the theater, there were still other people outside, waiting, hundreds more people, waiting in the cold, and they started yelling in those horrible twisted rat-gray New York street voices, but everything they said, even the wisecracks, was full of warmth and admiration. Christ, if they owned even New York, even this free port, this Hong Kong, this Polish corridor—what was not theirs now in America?

Somehow, extraordinary as it was, it was . . . right! The way it should be! The unutterable aura of the right stuff had been brought onto the terrain *where things were happening!* Perhaps that was what New York existed for, to celebrate those who *had it,* whatever it was, and there was nothing like the right stuff, for all responded to it, and all wanted to be near it and to feel the sizzle and to blink in the light.

Oh, it was a primitive and profound thing! Only pilots truly had it, but the entire world responded, and no one knew its name!

Not long after that, Kennedy brought the seven astronauts to the White House for a smaller, more personal visit. Kennedy's father was there, Joseph Kennedy. The old man had had a stroke, and half his body was paralyzed, and he was sitting in a wheelchair. The President took the seven astronauts in to meet his father, and the first one he introduced him to was John. John Glenn!—the first American to orbit the earth and challenge the Russians in the heavens. The old man, Joe Kennedy, reaches up with his one good hand to shake hands with John, and suddenly he starts crying. But the thing is, only half his face is crying, because of the stroke. One half of his face isn't moving a muscle. It's set, absolutely impassive. But the other half—well, it's blubbering, that's the word for it. His eyebrow is curling down over his eye, the way it does when you're really bawling, and the tears are streaming out of the crevice where his eyebrow and his eye and his nose come together, and one of his nostrils is quivering and his lips are writhing and contorting on that side, and his chin is all pulled up and pitted and trembling—but just on the one side! The other side is just staring at John, as if he saw right through him, as if he were just another Marine colonel whose career had somehow led him briefly into the White House.

The President would lean down and put his arm around the old man's shoulders and say: "Now, now, Dad, it's all right, it's okay." But Joe Kennedy was still crying when they left the room.

Obviously if the man hadn't had a stroke, he wouldn't have burst out crying. Until his stroke he had been a bear. Nevertheless, the emotion was there, and it would have been there whether he had had a stroke or not. That was what the sight of John Glenn did to Americans at that time. It primed them for the tears. And those tears ran like a river all over America. It was an extraordinary thing, being the sort of mortal who brought tears to other men's eyes.

XIII. The Operational Stuff

July 4 was not the time of year for anyone to be introduced to Houston, Texas, although just what the right time would be was hard to say. For eight months Houston was an unbelievably torrid effluvial sump with a mass of mushy asphalt, known as Downtown, set in the middle. Then for two months, starting in November, the most amazing winds came sweeping down from Canada, as if down a pipe, and the humid torpor turned into a wet chill. The remaining two months were the moderate ones, although not exactly what you would call spring. The clouds closed in like a lid, and the oil refineries over by Galveston Bay saturated the air, the nose, the lungs, the heart, and the soul with the gassy smell of oil funk. There were bays, canals, lakes, lagoons, bayous everywhere, all of them so greasy and toxic that if you trailed your hand in the water off the back of your rowboat you would lose a knuckle. The fishermen used to like to tell the weekenders: "Don't smoke out there or you'll set the bay on fire." All the poisonous snakes known to North America were in residence there: rattlers, copperheads, cottonmouths, and corals.

No, there was no best time to be introduced to Houston, Texas, but July 4 was the worst. And it was on July 4, 1962, that the seven Mercury astronauts moved to Houston. For the prodigious effort that Kennedy's moon program would require, NASA was building a Manned Spacecraft Center on a thousand acres of cattle pasture south of Houston near Clear Lake, which was not a lake but an inlet and was about as clear as the eyeballs of a poisoned bass. The astronauts, Gilruth, most of the

Langley and the Cape personnel would move to Houston, although the Cape would continue to be the launching center. The small scale and modest appearance of Langley and the Cape had somehow been perfect for the hell-for-leather *más allâ* phase that Project Mercury had just been through. They all knew Houston would be bigger. The rest they could never have guessed.

They stepped out of the airplane at the Houston airport and started gulping in the molten air. It was 96 degrees. Not that it mattered particularly; they had been assured that their entry into Houston would be easygoing and casual, Texas-style. There would be a zippy little motorcade through Downtown, just to give the good folks a look-see . . . and then there would be a cocktail party with a few prominent local figures, during which they could let their hair down and knock back a couple of long cool ones, or whatever, and relax.

Waiting there at the airport is a lineup of convertibles, one for each astronaut and his family with his name on a big paper banner taped to the side. So off they go in a motorcade, all seven of them and their wives and children, except for Jo Schirra, who was still at Langley recovering from some minor surgery. Pretty soon they're moving through the streets of Houston at a good clip, and it seems painless enough, but then all seven cars head down a ramp, into the bowels of an arena called the Houston Coliseum.

A bone-splitting chill hits them. They shudder and shake their heads. They are down inside some vast underground parking lot. The place is air-conditioned Houston-style, which is to say, within an inch of your life. There is a whole army of frozen people waiting down there in the gloaming, endless rows of marching bands in uniform, standing there like ice sculpture, politicians waiting in yet more convertibles, too cold to open their mouths, policemen, firemen, National Guard troops, stiff and still as lead, and more bands. Then they turn right around and head out of the underground parking lot, up the ramp, and back out into the blast of sunlight, a hundred degrees of heat, and the asphalt, which was lying there heaving and rippling in the caloric waves. All at once they are at the head of a big parade through the streets of Houston. Well, not quite in the lead. In the lead convertible now is a Texas congressman, a rubicund fellow named Albert P. Thomas, an influential member of the

House Appropriations Committee, waving a ten-gallon hat, as if to say, "Look what I brought you!"

It began to dawn on the boys and their wives that these people, the businessmen and the politicians, looked upon the opening of the Manned Spacecraft Center and the arrival of the astronauts as about the biggest thing in Houston history. Neiman-Marcus and all the other high-tone stores, the great banks and museums and other grand institutions, all the class, all the Culture, were in Dallas. By Houston lights Dallas was Paris, once you set your watch to Central Standard, and Houston was nothing but oil and hard grabbers. The space program and the seven Mercury astronauts were going to make Boom Town respectable, legitimate, a part of America's soul. So the big parade began with Representative Albert Thomas's waving ten-gallon hat signaling the start of the redemption of Houston.

The seven pilots and their wives thought they had seen every sort of parade there was, but this one was *sui generis.* There were thousands of people lining the streets. They did not make a sound, however. They stood there four and five deep at the curbs, sweating and staring. They sweated a river and they stared ropes. They just stared and sweated. The seven lads, each in his emblazoned convertible, were standing up grinning and waving, and the wives were smiling and waving, and the children were smiling and looking around—everyone was doing the usual—and the crowds just stared back. They didn't even smile. They looked at them with a morose curiosity, as if they were prisoners of war or had come from Alpha Centauri and no one was sure whether or not they comprehended the local lingo. Every now and then some very old person would wave and yell something hearty and encouraging, but the rest were just planted there in the sun like tarbabies. Of course, anyone fool enough to stand around in the asphalt mush of Downtown at noon watching a parade was obviously defective to begin with. The parade plowed on, however, through wave after wave of catatonia and rippling lassitude.

After about an hour of this, the fellows and their families noticed with considerable misgiving that the parade was heading back into that hole in the ground underneath the Coliseum. The air conditioning hit them like a wall. Everybody's bone marrow congealed. It made you feel like your teeth were loose. It turned out that this was where the little cocktail party was going to take place: in the Houston Coliseum. They led them up to the floor of the Coliseum, which was like a great indoor bowl. There were thousands of people milling around and some sort of incredible smell and a storm of voices and the occasional insane cackle. There were five thousand extremely loud people on the floor eager to tear into roast cow with both hands and wash it down with bourbon whiskey. The air was filled with the stench of burning cattle. They had set up about ten barbecue pits in there, and they were roasting thirty animals. Five thousand businessmen and politicians and their better halves, fresh from the 100-degree horrors of Downtown in July, couldn't wait to sink their faces in it. It was a Texas barbecue, Houston-style.

First they took the seven brave lads and their wives and children up onto a stage that had been set up at one end of the arena, and there was a little welcoming ceremony in which they introduced them one by one, and a great many politicians and businessmen made speeches. All the while the great cow carcasses sizzled and popped and the smoke of the burning meat was wafted here and there in the chilly currents of the air conditioning. Only the extreme cold kept you from throwing up. The ganglia of the solar plexus were frozen. The wives tried to be polite, but it was a losing game. The children were squirming up on the stage and the wives were getting up and whispering to any locals they could get close to. The children were *in extremis.* They hadn't been near a bathroom for hours. The wives were frantically trying to find out where the johns were in this place.

Unfortunately, now came the part where they were just supposed to relax, eat a side of beef and a peck of kidney beans sinking in gravy, drink a little whiskey, and shake hands with the good folks and make themselves at home. So they led them back out onto the floor of the arena, cleared out a space, got some folding chairs for them and some paper plates loaded with huge joints of Texas steer, and then put a lineup of folding chairs around the whole group of them, in a circle, on the order of a stockade, and around the stockade they put a ring of Texas Rangers, facing outward, toward the crowd. The crowd was now lining up, by the hundreds, at the barbecue pits, getting great lubricated hunks of beef on paper plates . . . and more whiskey. Then they took seats in the stands, thousands of them, and looked down at the stockade floor.

This was the main event, the reception, the Big Howdy: five thousand people, VIPs one and all, sitting up in the stands of the Houston Coliseum amid the burning cattle . . . watching the astronauts *eat*.

Certain VIPs with clout, however, were allowed to enter the stockade through the ring of Rangers and greet the lads and their wives personally as they juggled the great maroon hunks of meat. It was always someone such as Herb Snout from Kar Kastle, and he would come up and say: "Hi, there! Herb Snout! Kar Kastle! Listen! We're damned glad to have you folks here, just damned glad, goddamn it!" And then he would turn to one of the wives, whose hands were so full of cow meat she couldn't budge, and he would bend down and turn on a huge Karo-syrup grin, to show his deference to the ladies, and say in a suddenly huge voice that would make the poor startled woman drop the reeking maroon all over her lap: "Hi, there, little lady! Just *damned* glad to see *you*, too!" And then he'd give a huge horrible wink that would practically implode his eye, and he'd say, "We've heard a lot of good things about you gals, a *lot* of good things"—all with this eye-wrenching wink.

After a while, there were Herb Snouts and Gurney Frinks all over the place and the huge Hereford joints were sliding down every leg and splashing in the puddles of whiskey on the floor, and five thousand spectators watched their struggling jaws, and the smoke and the babble filled the air and the children screamed for mercy and relief. Just then, when the madness seemed to have outdone itself once and for all, a band struck up and the houselights dimmed and a spotlight searched out the stage and a show began and a mighty hearty voice boomed out over the p.a. system: "Ladies and gentlemen . . . in honor of our mighty special guests and mighty fine new neighbors, we are proud to present . . . Miss Sally Rand!"

The band struck up "Sugar Blues" . . . much raunchy high-hatting of the trumpets . . . *Oh, owwwwwwww wahwahwah* . . . and out into the spotlight pranced an ancient woman with yellow hair and a white mask of a face . . . Her flesh looked like the meat of a casaba melon in the winter . . . She carried some enormous plumed fans . . . She began her famous striptease act . . . Sally Rand! . . . who had been an aging but still famous stripper when the seven brave lads were in their teens, during the Depression . . . *Oh-owwwwwwww wahwahwah* . . . and she winked and minced about and took off a little here and covered up a little there and shook her ancient haunches at the seven single-combat warriors. It was electrifying. It was quite beyond sex, show business, and either the sins or rigors of the flesh. It was two o'clock in the afternoon on the Fourth of July, and the cows burned on, and the whiskey roared *goddamned glad to see you* and the Venus de Houston shook her fanny in an utterly baffling blessing over it all.

Just three years ago Rene had still been in that dogged military wife's frame of mind in which you gladly spent three days sanding a slab of monkeypod, until your hands were raw, to save the fabulous sum of ninety-five dollars. When Scott had run up a fifty-dollar telephone bill calling her from Washington, Albuquerque, and Dayton during the testing back in 1959, it had seemed like the end of the world. Fifty dollars! That was the food budget for *a month!* That was three years ago. Now she was in the living room of her own house—custom-made, not a tract home—on a lake, underneath the live oak and the pines. She and Annie Glenn had flown down to Houston one weekend from Washington and picked out lots, just like that, but they turned out to be in the best spot in the vicinity of the Space Center, a development called Timber Cove. The Schirras and Grissoms had moved into the same neighborhood. With admirable foresight, as it turned out, they had built their houses so that they opened up in the back to look out on the water and the trees, while on the side facing the street they were practically blank walls of brick. They had barely moved the first stick of furniture in when the tour buses started arriving, plus the freelance tourists in cars. They were extraordinary, these people. Sometimes you could hear the loudspeaker inside the bus. You could hear the tour guide saying, "This is the home of Scott Carpenter, the second Mercury astronaut to fly in earth orbit in outer space." Sometimes people would get out and grab handfuls of grass from your lawn. They'd get back on the bus with the miserable little green sprouts sticking out of their fingers. They believed in magic. Sometimes people would drive up, get out, stare at the house as if waiting for something to happen, and then walk up to the door and ring the bell and say: "We hate to bother you, but do you suppose you could send one of your children out so we could have our pictures taken with him?" And yet they weren't like the fans of movie stars. There was no frenzy. They thought they were really being considerate by not asking you to come out for the snapshot session yourself. They really meant it. They had more of the attitude of being at a living shrine.

This was the first house that Rene and Scott had ever built, the first house that had really seemed theirs. They had turned the page, all right. Things happened so fast now. At one point it looked as if they were going to be *given* houses completely furnished! The best that $60,000 could buy in 1962, at wholesale! A month after John Glenn's flight, a Houston man named Frank Sharp presented to Leo DeOrsey, as the fellows' business advisor, the following proposition: To show their pride in the astronauts and the new Manned Space Center, the builders, developers, furniture dealers, and others involved in the suburban home business would give each of the seven brave lads one of the homes being built for the 1962 Parade of Homes in Sharpstown. Sharpstown was a housing tract that Frank Sharp himself had been the impresario for. The Parade of Homes was a row of model houses that contractors who expected to do business in Sharpstown were putting up by way of advertising their wares. Sharp would contribute the land, a $10,000 lot to each astronaut; the contractors would contribute the houses; and the furniture and department stores would furnish them from top to bottom. The seven astronauts and their families would live right there on Rowan Drive in the Country Club Terrace section of Sharpstown, between Richmond Road and Bellaire Boulevard, in $60,000 worth of home and hearth each. Since Sharpstown at this point was nothing but maps, signs, bunting, Englishy thatchy tweedy-sounding street names and thousands of acres of wind-swept boondocker gumbo scrubland, Astronauts Row wouldn't be a bad way to start filling in the spaces. Sharp was the Big Howdy through and through, a self-made man and already quite a prominent citizen, close to the mayor, Representative Albert Thomas, Governor John Connally, and Vice-President Lyndon Johnson. He underwrote annual golf trophies and things of that sort. He had the right credentials, or the Houston version thereof, and so DeOrsey had talked it over with the boys, and they all decided the deal was okay. It had nothing to do with the space program and didn't obligate them to anything. It was just an unencumbered *goodie*, pure and simple. None of them actually wanted to live in Sharpstown, from what they had heard about the area. It was too far from the NASA facility, for a start. So they figured they would accept the houses, shake hands with one and all, and then . . . *sell them.* John Glenn was as agreeable to this type of goodie as anybody else. It was that age-old concern of the military officer with the extras. John had been in the Marines for nearly twenty years now. He was too far down that road, had been through too many measly paychecks, to decondition himself anytime soon when it came to the perks, the extras, the irresistible and perfectly honorable and authorized goodies. Therefore, not even John, for all his quite sincere sense of morality, could comprehend the furor that was erupting. Gilruth and Webb and everybody else in the NASA hierarchy were blowing fuses over Frank Sharp's Parade of the Homes of the Astronauts. And that was only the start. It had touched off a true emergency: the *Life* deal was under review! From what Scott and the others had picked up, the President himself was considering putting an end to *all* commercial exploitation of astronaut status. The rest of the press had resented the *Life* deal from the beginning and had argued that it cast a venal shadow on the astronauts' patriotic service. Sharpstown showed you where the exploitation route could lead . . .

Sharpstown was one thing . . . but a threat to the *Life* deal—now, there you had something serious! *Unthinkable* was the word. The seven pilots, steeped in the honorable goodies tradition of the military, had begun to look upon the *Life* deal in the same way they looked upon the military pension that you rated after twenty years. It was an immutable condition of the service! It was part of the drill! Regulation issue! Covered in the manual! All holes in the argument were immediately vulcanized by the heat of the emotion. This was no time to sit around waiting for the orders to be posted on the bulletin board. It was only three weeks before Scott's flight, and he was in the thick of training, but on May 3 most of the others went to see Lyndon Johnson at his ranch in Texas to try to straighten the matter out. Webb was there, too. They had quite a conclave. Lyndon Johnson gave them some fatherly discourses on private lives and public responsibility, twisting his great hands around in the air in front of him, as if making imaginary snowballs. This pained him as much as it did them, and so on. The hell of it was that neither

Johnson nor Webb would miss a minute's sleep if the *Life* deal was canceled forthwith. They had both been burned by the astronaut-*Life* connection during the incident at the Glenns' house in January. In fact, if it hadn't been for Glenn—

Fortunately, there was no getting around John. By now, three months after his flight, John had ascended to a status that only a biblical scholar could fully appreciate. John was the triumphant single-combat warrior. He had risked his life to challenge the mighty Soviet Integral on the high ground. Through his skill and courage he had neutralized the enemy's advantage, and the tears of joy and gratitude and awe still flowed. In the Bible, first Book of Samuel, eighteenth chapter, it is written that after David slew Goliath and the Philistines fled in terror and the Israelites achieved a mighty victory, King Saul took David into the royal household and gave him the status of an adopted son. It is also written that wherever Saul and David went, people thronged the streets, and the women sang of the thousands Saul had slain and the tens of thousands David had slain. "And Saul was very wroth, and the saying displeased him; and he said, They have ascribed unto David ten thousands, and to me they have ascribed *but* thousands; and *what* can he have more but the kingdom? And Saul eyed David from that day forward." And President Kennedy eyed John Glenn. The President had begun to regale John and bring him into the Kennedy family orbit. John was the sort of man a president needed to keep squarely within his camp. A vice-president, too, for that matter. Johnson had gone out of his way to be friendly to John and Annie, and they had genuinely begun to like the man. They wound up inviting Johnson and his wife, Lady Bird, for dinner at their house in Arlington, on John's fortieth birthday. And the Johnsons accepted, just like that. Rene and Scott were also invited. "What on earth are you going to serve?" Rene said to Annie.

"My ham loaf," said Annie.

"Ham loaf!"

"Why not? Everybody likes it. I bet you Lady Bird asks for the recipe, too."

The Johnsons stayed until almost midnight. Lyndon had his coat off and his sleeves rolled up and was having a rare old time for himself. As they left, Rene heard Lady Bird ask Annie for the recipe for her ham loaf.

One day John was out on the Atlantic Ocean, beyond the Hyannis Port, Massachusetts, harbor, aboard the President's yacht, the *Honey Fitz,* when the subject of the *Life* contract came up. The President wanted to know what John thought of one particular argument against the *Life* arrangement that was frequently presented; namely, that a soldier in battle—a Marine at Iwo Jima, for example—ran just as great a risk of death as any astronaut and yet did not expect recompense from Time, Inc. John said yes, that was so, but suppose that soldier's or Marine's private life, his background, his house, his way of living, his wife, his children, his thoughts, his hopes, his dreams, became of such intense interest that the press camped on his doorstep and he had to live under glass, as it were. Then he should have the right to receive compensation. The President nodded sagaciously, and the *Life* contract was saved, right there on the *Honey Fitz.*

Well, thanks to the *Life* deal, Scott and Rene could now get mortgage money and afford to build a new house in a nice area like Timber Cove. Or thanks to that and the panting eagerness of the developers to have astronauts in their new developments. It was the best advertisement they could have. They gave the boys close to at-cost deals on the land and the houses, and they let them have the mortgage money at 4 percent, with very little down payment. And for astronauts like John and Scott, who had now flown, they couldn't do enough.

The contractors and developers and the public generally thought Scott and his flight were terrific . . . but within NASA something . . . was going on. Scott and Rene had both begun to detect it, although no one ever said anything openly. Scott had gotten all the medals and all the parades and the trip to the White House, but something was up, and not even the other wives would tell Rene what it was.

Scott had flown on May 24, three months after John. Deke Slayton had been scheduled to take the flight, but then NASA made it known that Deke had a medical problem: idiopathic atrial fibrillation. This was a condition in which the electrical firing sequence of the heart went out of sync occasionally, causing an irregular pulse and a slight lowering of the pumping capacity of the organ. *Idiopathic* meant that the causes were unknown. The condition had been discovered, said NASA, during centrifuge runs in August 1959. Slayton had been examined at the

Philadelphia Navy Hospital and the Air Force's School of Aviation Medicine at San Antonio, where the verdict—or so Slayton was told—was that the condition was a minor anomaly and not serious enough to cost him his job as astronaut. But, in fact, one of the Air Force doctors at San Antonio, a highly regarded cardiologist, had written a letter to Webb recommending that Slayton not be assigned to a flight, since atrial fibrillation, idiopathic or not, did reduce the efficiency of the heart to some degree.

Webb just kept the letter on file. Slayton was assigned, in November 1961, to the second orbital flight. Early in January Webb ordered a complete reassessment of the man's heart condition. His argument was that Slayton was an Air Force pilot on loan to NASA, and an Air Force cardiologist had recommended that he not be used for flights. Therefore, the case should be reviewed. Slayton's case now went before two boards, one made up of high-ranking NASA doctors and the other made up of eight doctors convened by the Air Force Surgeon General. Both approved Slayton for the approaching Mercury flight. Nevertheless, Webb bucked the case up to three Washington cardiologists, including Eugene Braunwald of the National Institutes of Health, as a sort of blue-ribbon panel. He also requested an opinion from Paul Dudley White, who had become famous as Eisenhower's cardiologist. Why this was happening so late in the game, three months after Deke had been assigned to the flight, no one could figure. All four doctors came to the same conclusion, apparently out of sheer common sense as much as anything else. Here was a case concerning a pilot with a minor heart defect. He could probably make a space flight or any other flight with no problems. Nevertheless, from the administrator on down, the entire space agency seemed to be agonizing and oscillating and spinning its wheels over the matter, which by now had accumulated a file as thick as your arm. So if Project Mercury had plenty of ready and willing astronauts with no cardiovascular anomalies at all, why not use one of them and have done with it? That was it, so far as Webb was concerned. It was now the middle of March. Two months ago, in his set-to with Glenn, James E. Webb had run up against *astropower* and had lost. This time he had his way. Slayton was off the flight.

The great Victorian Animal was utterly baffled. The Animal had been dutifully cranking out human-interest stories about Slayton. How could NASA decide now that he was a washout with a bad heart? There was no . . . proper emotion . . . for the event.

According to the official NASA wording, Slayton was "keenly disappointed" about the decision. That was putting it somewhat delicately. The man was furious. Slayton tried to keep a rein on himself in public statements, however, because he didn't want to jeopardize his chances of reinstatement. He was convinced that the whole thing had somehow been blown up into a specious issue and that they would all come to their senses by and by. Privately he was tying knots in the flagpole. He kept saying that Paul Dudley White had made an *operational* decision. His argument was that White and the other doctors had first delivered their medical opinion—he was fit to fly—and then they had delivered their operational opinion, which was: "Even so, why not choose somebody else?" They were entitled to their medical opinion; period. But they had made an *operational* decision! This word, *operational*, was a holy word to Slayton. He was the King of Operational. *Operational* referred to action, the real thing, piloting, the right stuff. *Medical* referred to one of the many accessories to the business at hand. You didn't call in doctors to make an operational decision. The *Life* reporters knew very well how angry Slayton was, and other reporters had strong hints of it. But the Genteel Beast could find no appropriate . . . tone . . . for it. So after a short time they just dropped it. They stuck with the NASA version: "Keenly disappointed." Few of them realized that it went beyond anger. Deke Slayton was crushed. He had not merely lost his ride on the next flight; he had lost everything. NASA had just announced that he no longer had . . . the right stuff. It could blow at any seam!—and his had blown. Idiopathic atrial fibrillation—it didn't matter! *Any* seam! His whole career, his ascension from the dour grim tundra of Wisconsin was based upon his indisputable possession of that righteous stuff. That was the most important thing he had ever possessed in this Trough of Mortal Error, and it was plenty. It was the ultimate. And it had *blown*, just like that! He felt humiliated. This thing would now be rubbed in his face everywhere he turned. He couldn't go back to Edwards now, even if he wanted to. The Air Force was not going to use a NASA reject for major flight test work. Flight test? Hell, he couldn't even fly a fighter plane by himself anymore! It was true. He could go up only in two-seaters with another pilot—someone who was still intact, with no ruptured seams from which his vital stuff had leaked. There was even the possibility that the Air Force might ground him altogether, notwithstanding the fact that the Surgeon General's board had pronounced him "fully qualified as an Air Force pilot and as an astronaut." Air Force pride was at stake. The Air Force's Chief of Staff himself, General Curtis LeMay, was taking the position that if he wasn't qualified to fly for NASA, how could he be qualified to fly for the Air Force? All this was being said about *him*, Deke Slayton, who had fought hardest of all to have *the astronaut* be treated as *a pilot*, even to the extent of insisting on airplane-style controls for the capsule, or, rather, damn it, *spacecraft*.

It wasn't likely to make him any happier, either, to know that Scott Carpenter had taken his place. Carpenter had the least flight test experience of any of them, and yet he was replacing Deke Slayton—Deke Slayton, who had stood up before the Society of Experimental Test Pilots and insisted that only an experienced test pilot could do this job correctly. Wally Schirra, a man with real flight test credentials, had been training as Deke's backup. Why had he been passed over in favor of Carpenter? The two buddies, Glenn and Carpenter, were getting the first two orbital flights . . . and Deke Slayton was being *left behind* . . . to hitch airplane rides with other pilots.

Gilruth's opinion, backed up by Walt Williams, was that Carpenter had logged far more flight training, as Glenn's backup, than Schirra could possibly hope to cram in during the ten weeks remaining before the flight. Scott was not exactly ecstatic over having been handed Deke's flight on such short notice. He had trained for six months with John, but the second orbital flight was to be quite a different proposition. NASA's experimental scientists would finally have an inning. The astronaut was supposed to deploy a multicolored balloon outside the capsule in order to study the perception of light in space and the amount of drag, if any, in the presumed vacuum of space. He was supposed to observe how water in a glass bottle behaved in a weightless state and whether or not capillary action was altered. There would be a small glass sphere for that experiment. He would have a densitometer, as it was known, to measure the visibility of a ground flare. He would be trained in the use of a hand-held camera to take weather photographs and pictures of the daylight horizon and the atmospheric band above the horizon and of various land masses, particularly North America and Africa. They had the right man. Scott was intrigued by the experiments. But the addition of all these things to the checklist, which was already undergoing last-minute changes of the operational sort, put him under increasing pressure. In making all of these sightings, for the use of the camera, the densitometer, or whatever, he would be using an entirely new manual control apparatus. It was a system in which you created one pound of thrust if you pushed the hand controller slightly and twenty-five more if you pushed it beyond a small angle. It would be either/or; there would be no turning the capsule gradually like an airplane or an automobile.

The flight went off on schedule on May 24. For the first two orbits Scott had a picnic. He was more relaxed and in higher spirits than any of the three men who had preceded him. He was enjoying himself. His pulse rate, before lift-off, during the launch, and in orbit, was even lower than Glenn's. He talked more, ate more, drank more water, and did more with the capsule than any of them ever had. He obviously loved all the experiments. He was swinging the capsule this way and that way, taking photographs a mile a minute, making detailed observations of the sunrises and horizon, releasing balloons, tending his bottles, taking readings with the densitometer, having a grand time. The only problem was that the new control system used up fuel at a terrific rate. You wanted to pitch or yaw the capsule just ever so slightly, and—*bango!*—you were over the invisible line and another outsized geyser of hydrogen peroxode squirted out of the tanks.

During the second orbit he was warned by several capcoms to start conserving fuel, so that he would have enough for re-entry, but it was not until the third and last orbit that he himself seemed to realize just how low the fuel was. For most of the final orbit he just let the capsule drift and turn in any attitude it wanted to, so as not to have to use any of the thrusters, high or low, automatic or manual. It presented no problems at all. Even when you were upside down in relation to the earth, with your head pointed straight down, there was no feeling of disorientation, no feeling of up or down. Floating in a weightless state was even more enjoyable than swimming underwater, which Scott loved.

The diminished fuel supply was very much on his mind. Still, he couldn't resist the opportunity to experiment. He reached for the densitometer, and his hand hit the hatch of the capsule, and a cloud of John Glenn's "fireflies" appeared outside the window. So he swung the capsule over in a yaw to have a look at the fireflies. To him they looked more like frost or snowflakes, so he banged on the hatch and another cloud flew up and he swung around some more to take a look and used up some more fuel. Whatever they were, they were attached to the hull of the capsule and no doubt emanated from or were created by the capsule and were not some sort of micro-galaxy, all of which aroused his curiosity, and so he banged away and pitched and yawed and tooled around some more, the better to unravel the mystery. All of a sudden it was time to prepare for re-entry, and Scott was already behind in the retro-sequence, as that part of the checklist was known. Also, the fuel situation was beginning to get a little dicey. On top of that, the automatic control system would no longer hold the capsule at the proper angle for re-entry. So he switched to fly-by-wire . . . but at the same time forgot to throw the switch that cut off the manual system. For ten minutes he was eating up fuel out of both systems. He would have to fire the retro-rockets manually, as Alan Shepard, the capcom in Arguello, California, sounded off the countdown. When Shepard called "Fire one!" the capsule's angle was off about nine degrees and Scott was late in hitting the switch. He had practically no fuel left for controlling the capsule's oscillations during re-entry. By the time he hit the denser atmosphere and the radio blackout began, Chris Kraft and the other flight control engineers feared the worst. Long after radio communication should have resumed—nothing. It looked as if Carpenter had consumed all his fuel up there playing around—and had burned up. They all looked at each other and were already thinking one step ahead: "This disaster is going to set the program back a year—or do worse than that."

Rene was following Scott's re-entry over television inside a rented house in Cocoa Beach. For two days she had been involved in a hide-and-seek operation that had finally become absolutely loony. Bridge dragnets . . . amok helicopters . . . Rene had decided that since *Life*'s accounts of the brave wives bravely enduring the ordeal of their husbands' flights were written in the first person, she was actually going to write hers. Loudon Wainwright could edit what she wrote and rewrite the rough spots, but she was going to write the whole thing herself. That being the case, she wasn't going to let herself be imprisoned inside her house at Langley by the television crews and all the rest of that madness. She had seen Annie being driven far crazier from having to play quavering pigeon for the press—and the likes of Lyndon Johnson—than by any fears she had for John. It was an undignified position to be in. Despite the attention lavished on you, you were not treated as an individual but as the anxious loyal mate of the male up on top of the rocket. After a while Rene didn't know whether it was her modest literary ambitions or her resentment of the pat role of Astronaut Wife that made her do it. *Life* rented a "safe house" for her in Cocoa Beach. *Life* did things right. They rented a backup safe house as well, in case Rene's presence in the first one was discovered. Rene called up Shorty Powers, who was NASA's official press officer in matters concerning the astronauts, and told him that she was going to the Cape for the launch but wanted privacy and was telling no one where she would be, including him. Powers was not happy. The astronauts' *Life* contract had already made his job difficult enough. He was cut off from all "personal" material about the men and their families, since that was supposed to go to *Life* exclusively. And yet when a flight was on, 90 percent of the reporters Powers had to deal with were really interested in only two questions: (1) What is the astronaut doing now and how does he feel? (Is he afraid?); and (2) What is his wife doing now and how does she feel? (Is she dying from anxiety?) One of Powers's main roles was serving the television networks—and telling them where *the wife* would be during the flight, so that they could congregate for the death-watch campout. And this time all he could tell them was that *the wife* would be at the Cape . . . somewhere . . . That did it. The networks took the situation as an insult and a challenge. Before Rene left for the Cape, a correspondent for one of the networks called her and told her they were *going to find out* where she was staying . . . They could do it the hard way, if they had to, but they'd rather do it the easy way. So she'd better just tell them. It was like something out of a gangster movie. But sure enough, when she reached the Cape, the networks had people watching every bridge and causeway into Cocoa Beach. Rene knew they would be looking for a car with a woman with four children. So she had the children lie on the floor, and they slipped through. The networks were not going to be foiled that easily. After all, how could they camp on her front lawn and film her drawn shades if they didn't even know where she was? So they hired helicopters and began scouring Cocoa Beach. They went up and down the hardtack beach, looking for congregations of four small children. They would swoop right down on children on the beach until they could read the terror in their eyes. People were running for cover, abandoning their Scotch coolers and telescopes and cameras and tripods, trying to save their children from the amok helicopters. It was crazy, utterly bananas, but by now not knowing where *the wife* was— it was like not knowing where the rocket was. Finally, Rene was sending her children over to the beach two by two, in order to foil the insane people in the network helicopters.

Came time for the launch, and now Rene and the children watched the countdown on the TV in the safe house, with Wainwright and a *Life* photographer in attendance. Then the children rushed out and watched part of the rocket's slow ascent through a telescope mounted on a garage roof. The children didn't seem at all apprehensive. Flying was what their father *did*. They were in high spirits. . . . And now they were following the re-entry, as best they could, on television. They had CBS turned on. There was Walter Cronkite. Rene knew him. Cronkite had become an astrobuff. He had more than the usual reasons to like the astronauts. It was his coverage of John Glenn's flight that, in the strange workings of the television news business, had led to his current eminence among the network anchormen. Cronkite had been explaining Scott's fuel problem as he entered the atmosphere. Then Cronkite's voice began to take on more and more concern. They didn't know where Scott was. They weren't sure he had begun his re-entry at the proper angle. All at once Cronkite's voice broke. Tears came into his eyes. "I'm afraid that . . ." There was a catch in his voice. His eyes glistened. He had the waterworks turned on. "I'm afraid . . . we may have . . . *lost an astronaut* . . ." What instincts the man had! There was the Press, the Genteel Gent, coming up with the appropriate emotion . . . *live* . . . with no prompting whatsoever! Rene's children were very quiet, staring at the screen. Yet Rene herself did not believe for a moment that Scott had perished. She was like every military pilot's wife in that respect. If he were merely missing—if no dead body had been found—then he was alive and would come through it all right. There were no two ways about it. Rene had known of a case in which a cargo plane had crash-landed in the Pacific and broken in two on impact, the rear half sinking like a brick. Some men were rescued from the front half, which stayed afloat a few minutes. And yet the wives of the men who had been in the rear of the craft refused to believe that they were lost. They were out there somewhere; it was only a matter of time. Rene had marveled at how long it had taken them to accept the obvious. But her reaction was precisely the same. Scott was all right, because there was no real proof that he wasn't. Cronkite gulped on the television screen. No tears came to her eyes at all. Scott was all right. He would turn up . . . No two ways about it.

As a matter of fact, she was correct. Scott had come through the atmosphere in good shape. The capsule began rocking violently in the dense atmosphere below 50,000 feet, and he had to release his parachute early and by hand, the automatic system being out of fuel. The capsule had overshot the target area by some 250 miles. A reconnaissance plane found him in about forty minutes, but throughout that period the impression created on television was that he might be dead. When a rescue aircraft reached Scott, they found him bobbing contentedly on a life raft beside the capsule. He was very pleased with the whole adventure. When he reached the aircraft carrier *Intrepid,* he was in terrific spirits. He talked and talked into the night. He wanted to stay up and keep talking about the grand adventure he had been through. He was really pleased about all the experiments he had been able to do, despite the overcrowded checklist he had, and about solving or at least greatly narrowing down the mystery of the "fireflies." He hadn't determined precisely what they were, but he had proved that they were produced by the spacecraft itself; they were not some extraterrestrial material, and so on . . . He could have gone on all night . . . He was content . . . a job well done . . . He felt that he had helped create one of the most important roles in astronautics: man as scientist in space . . .

Over the next two weeks Scott received a hero's homage. It was not on the scale of John's, which was understandable, but it was sweet enough. There were parades in the East and parades in the West. He rode in a motorcade through Boulder, his old hometown, and through Denver, which was just down the highway. It was a great day. The sun was out and it was a light fluffy Rocky Mountain day in May, and Rene was beside him, sitting up on the ridge of the seat back in the convertible, wearing white gloves, like a proper Navy wife, and smiling and looking absolutely beautiful and radiant. Well, Scott thought he had just shot the moon.

Back at the Cape, Chris Kraft was telling his colleagues: "That sonofabitch will never fly for me again."

Kraft was furious. The truth was, he had been quietly put out beore . . . about the seven brave lads. As he saw it, Carpenter had ignored repeated warnings from capcoms all around the world about wasting fuel, and this had almost resulted in a disaster, one that might have done irreparable damage to the program. As it was, Carpenter's performance had cast doubt on the capability of the Mercury system to carry out a long flight such as Titov's seventeen orbits. And why had this catastrophe nearly occurred? Because Carpenter had insisted on comporting himself like an Omnipotent and Omniscient Mercury Astronaut. He didn't have to pay attention to suggestions and warnings from mere groundlings. He apparently believed that the *astronaut,* the passenger in the capsule, was the heart and soul of the space program. All of the resentment that the engineers had about the top-lofty status of the astronauts now crept out of its cage . . . at least within NASA. Outside of NASA, publicly, nothing was to change. Carpenter, like Grissom before him, was an exemplary brave lad; just a little dicey scrape at the end of the flight, that was all. Very successful flight; go ahead, let him have his medals and his receiving lines.

And now that the wound had been opened, there were those who were only too pleased to see the following line develop concerning the Carpenter flight: Carpenter had not merely wasted fuel while up there playing with the capsule's attitude controls, doing his beloved "experiments." No, he had also become . . . *rattled* . . . when he finally realized he was getting low on fuel. The evidence for this was that he forgot to turn off the manual system when he switched to fly-by-wire and thereby really blew his fuel supply. And then he . . . *panicked!* . . . That was why he couldn't line the capsule up at the right angle and that was why he couldn't fire the retro-rockets right on the button . . . and that was why he hit the atmosphere at such a shallow angle. He nearly skipped off it instead of going through it . . . he nearly skipped off into eternity . . . because . . . he *panicked!* There! We've said it! That was the worst charge that could be brought against a pilot on the great ziggurat of flying. It said that a man had lost whatever stuff he had in the most awful manner. He had funked it. It was a sin for which there was no redemption. Damned eternally! Once such a verdict had been pronounced, no judgment was too vile. Did you hear his voice on the tape just before the blackout? You could *hear* the panic! In fact, they could hear no such thing. Carpenter sounded very much the way Glenn had sounded and a good deal less excited than Grissom. But if one wanted to hear panic, especially in the words that a man had to force out after the g-forces built up, if that was what one was after . . . then you could hear panic. But, then, Carpenter never had the right stuff to begin with! That much was obvious. He had given up long ago. He had opted for multi-engine planes! (Now we know why!) He had only two hundred hours in jets. He was here only through a fluke of the selection process. And so forth and so on. Certain objective data had to be ignored, of course. Carpenter's pulse rate remained lower, *during the re-entry* as well as during the launch and orbital flight, than any other astronaut's, including Glenn's. It never rose above 105, even during the most critical point of the re-entry. One could argue that pulse rate was not a dependable indication of a pilot's coolness. Scott Crossfield had a chronically rapid pulse rate, and he was in the league with Yeager. Nevertheless, it was inconceivable that a man in *a state of panic*—in a *life-or-death* emergency—in a crisis that did not last for a matter of seconds but for *twenty minutes*—it was inconceivable that such a man would maintain a heart rate of less than 105 throughout. Even a pilot's heart rate could jump to more than 105 for nothing more than the fact that some cocky bastard had cut into line ahead of him at the PX. One might argue that Carpenter had mishandled the re-entry, but to accuse him of *panic* made no sense in light of the telemetered data concerning his heart rate and his respiratory rate. Therefore, the objective data would be ignored. Once it had begun, the denigration of Carpenter had to proceed at any cost.

It served many purposes at once. It made the rest of them seem like *real pilots* after all and not mere riders in a pod. A man either *had it* or he didn't . . . in space as in the air. As every pilot knew in his secret heart—deny it, if you wish!—it required washouts to make your own righteous stuff stand out. So was Carpenter, by implication, to be designated *the washout?* Logic no longer mattered—especially since none of this could be talked of openly in any case: publicly there were to be no flaws in the manned space program whatsoever. Sheer logic would have raised the question: why pick Carpenter and not Grissom? Grissom had *lost the capsule* and had then come back with the classic pilot's response to gross error: "I don't know what happened—the machine malfunctioned." The telemetry showed that Grissom's heart was on the edge of tachycardia at times. Just before re-entry his heart rate had reached 171 beats per minute. Even after Grissom was safe and sound on the carrier *Lake Champlain*, his heart rate was 160 beats a minute, his breathing was rapid, his skin was warm and moist; he didn't want to talk about it, he wanted to go to sleep. Here was the clinical picture of a man who had abandoned himself to panic. Then why was not Grissom designated *the washout*—if anybody cared to find one? But logic had nothing to do with it. One was in the area of magical beliefs now. In his everyday life doughty little Gus *lived* the life of the right stuff. He was a staunch bearer of the Operational banner. Here Gus's fate and Deke's fate came together. Deke had said all along: You need a proven operational test pilot up there. Gus and Deke were great pals. For three years they had flown together, hunted together, drunk together; their children had played together. They were both committed to the holy word: *operational.* Schirra was with them on this particular commitment, and Shepard threw his weight toward them, too, as did Cooper.

Deke had plenty to be thankful to Shepard for. One day Al had gotten the other boys together and said, "Listen, we've got to do something for Deke. We've got to do something to give him back his pride." Shepard's suggestion was that Deke be made a sort of chief of the astronauts, with an office and a title and official duties. They all went for the idea and took it to Gilruth, and in no time Deke had the title "Coordinator of Astronaut Activities." There may have been people at NASA who figured this would be a supernumerary make-work job for the fallen astronaut; if so, they underestimated Deke. He was a far shrewder and more determined individual than his Wisconsin tundra manner let on. The job gave him something to channel his tremendous thwarted energy into. The NASA hierarchy was still a political vacuum, and Deke set about filling it . . . with a vengeance, as it were. Soon Deke was a power within NASA, a man to be reckoned with, and his motivation never varied: the more powerful he became, the better his chances of reversing the decision that prevented him from flying. Justice, simple operational justice . . . in the name of the right stuff.

Operational; the word had new clout now, and a corollary to the theory of the Carpenter flight began to develop. Carpenter's flight agenda had been loaded with Larry Lightbulb experiments. The scientists, lowest men in the NASA pecking order up to now, had been given their heads on this flight . . . and the results were there for your inspection. Carpenter had taken all this Mad Professor stuff seriously, and that was what led to his problems. He became so wrapped up in his various "observations" that he fell behind in the checklist and became rattled and then blew it. All this science nonsense could wait. Just now, in the critical operational phase of the program, the crucial period of real *flight test,* it wasn't just ding-a-ling stuff, it was dangerous. There were too many goddamned *doctors* involved in this thing, too. (Look what they did to Deke!) On top of that, they had the two psychiatrists to contend with. They were nice enough men personally, Ruff and Korchin were, but they were . . . *in the way!* What the hell was all this pissing into bags and hitting little circles with a pencil . . . after you've just hung your hide out over the edge in a space flight? They hadn't even picked up the fact that Carpenter had *panicked.* They found him exhilarated, alert, full of energy, ready to take off and do it all over again . . . The two men were not invited to continue with the program after the move from Langley to Houston. Thanks a lot, gentlemen, and don't let the doorknob hit you in the butt.

Here the outlook of Grissom, Slayton, and Schirra coincided with that of Kraft and Walt Williams. Kraft and Williams also felt that nonoperational experiments should be kept to a minimum at this stage of the space program. From now on, whenever anyone said otherwise, one had only to roll his eyeballs and turn his palms up and say: You want another Carpenter flight?

On August 11 and 12 the mighty Integral struck again, and now there was absolutely no stopping the *operational* theory. On August 11 the Soviets launched *Vostok* 3 on what at first looked as if it would be a repetition of Titov's day-long flight. But no! Exactly twenty-four hours later the Chief Designer sent up *Vostok 4*, and the two craft flew together, in tandem, within three miles of each other. Within three miles of one another in the infinity of space! The Soviets spoke of a "group flight," as if the two cosmonauts, Nikolayev and Popovich, were flying in formation. In fact, neither could change his flight path in the slightest, and their proximity was due solely to the precision with which the second *Vostok* was launched as the first came orbiting overhead—but even this seemed to be a feat of incalculable sophistication. The Genteel Beast and many congressmen seemed to be on the edge of hysteria. Entire *formations* of Soviet space warriors, hurling thunderbolts at Schenectady . . . Grand Forks . . . Oklahoma City . . . Once again the Chief Designer was toying with them! God knew what his next surprise would be . . . (It would be a big one.) Well, that settled it. No more densitometers and varicolored balloons and other White Smock accessories. (No more pilots with non-operational stuff!) Which explained the singular nature of Wally Schirra's flight on October 3.

Schirra named his capsule *Sigma* 7, and there you had it. Scott Carpenter had named his *Aurora* 7 . . . *Aurora* . . . the rosy dawn . . . the dawn of the intergalactic age . . . the unknowns, the mystery of the universe . . . the music of the spheres . . . Petrarch on the mountaintop . . . and all that. Whereas *Sigma* . . . *Sigma* was a purely engineering symbol. It stood for the summation, the solution of the problem. Unless he had come right out and named the capsule *Operational,* he couldn't have chosen a better name. For the purpose of Schirra's flight was to prove that Carpenter's need not have happened. Schirra would make six orbits—twice as many as Carpenter—and yet use half as much fuel and land right on target. Whatever did not have to do with that goal tended to be eliminated from the flight. The flight of *Sigma 7* was designed to be Armageddon . . . the final and decisive rout of the forces of experimental science in the manned space program. And that it was.

Schirra cut the jolly fun-loving figure so well that people sometimes failed to notice how formidable he could be. But his emphasis, after all, was on maintaining an *even* strain. His pranksterish, rib-shaking, wild-driving gotcha intervals gave him plenty of slack when the time came to wind things up tight and get tough. Every bit as much as Shepard, Wally had the instincts of an Academy man, a leader of men, the commander, the captain of the ship. He merely operated in a different fashion. He was cool; he had "the uncritical willingness to face danger," but he wasn't afraid to show his feelings when strategy seemed to dictate it. If it was going to be his show, he insisted on running it; and he was shrewd enough to recognize the political outlines of a situation. Having seen four flights from up close, Wally couldn't help but have noticed that the secret of a successful mission lay in a simplified checklist with white space between tasks. The fewer tasks you had, the better chance you had for a 100 percent performance. Not only that, if you could control the checklist, then you could give your flight a theme, a clear-cut goal that everyone could immediately appreciate and respond to. Wally's theme for this flight was Operational Precision, which, being translated, meant conserving fuel and landing on target. Now that the operational forces were lined up shoulder to shoulder, it was possible to keep offboard most novel items that engineers or scientists had dreamed up for the flight.

It was decided that one of Wally's major operational tests would consist of powering down all of the attitude-control systems, automatic as well as manual, and just drifting in any attitude the capsule's inertia took it, upside down (in relation to the earth), head over heels, canted this way and that, whatever. Scott heard about it and told Wally that he didn't really think it was necessary. He had drifted for most of his last orbit, in an attempt to conserve fuel for the re-entry, and had proved to his own satisfaction that you could stand the capsule right up on its aerial or let it revolve or put yourself in any other attitude and it was not the least bit disorienting or uncomfortable. Why didn't Wally put the proposed drifting time to other uses? Wally said no, he was going to devote his flight to experimenting with drifting flight and to conserving fuel, in order to prepare the way for missions of long duration.

Scott would learn that planning sessions had been held—and he had not been informed of them. It was not that Scott was supposed to take any official part in the planning for Wally's flight, and it was not unusual in flight test for a pilot to have his own particular circle of colleagues and ground crew he preferred to consult with. But be that as it may, anyone could recall how much Scott had valued John Glenn's counsel before his flight. In fact, one of Scott's concerns had been that John wasn't available more. The demands on John's time in his role as NASA's number-one hero had become enormous. But anytime John was around, Scott—and the engineers, too—wanted John at the meetings. Wally also complained that John wasn't around. He caused quite a flap when he told Walter Cronkite, in a taped interview, that John was off on the banquet circuit so much, he was lost to the program. He didn't complain about the absence of Scott, however. Scott began to conclude that Kraft and Williams were overreacting to the fact that he had overshot the target by 250 miles. The possibility that there were people—*pilots*—going around saying that he had *panicked* never even crossed his mind.

Given the goal of the flight—which was to prove that a *cool* pilot could travel twice as far as Carpenter with half as much fuel consumption and ten times the accuracy—Schirra was terrific. From the moment he got up that morning, he was about as cool and relaxed a human being who ever went out to sit on top of a rocket shaft. A few days ago Wally had played one of his patented *gotchas* on Dee O'Hara, the nurse. One of her tasks was to collect urine samples. She gives him the usual little bottle and asks him to bring in a sample and leave it on her desk. She comes into her office, and on top of her desk is not a little bottle but a huge beaker holding about five gallons of an amber liquid with a head of foam on top. It couldn't possibly be—but *could* it possibly be?—and so she puts her hands on the sides of the beaker to see if it's warm, and—

"Gotcha!"

—she wheels about, and there's Wally peeking in the doorway, him and his beaming face and a couple of the boys. He had concocted it of water, tincture of iodine, and detergent. The next day Dee O'Hara presents Wally with a clear plastic bag, a big thing, about four feet long, telling him that it's the urine receptacle for his flight, replacing the little condom device that Grissom, Glenn, and Carpenter had worn. *Gotcha!* So today, the morning of his flight, here comes Wally down the hallway in Hangar S in his bathrobe, heading for the medical room. Flopping out below the robe and dragging along the floor between his legs is the huge plastic bag. He parades right past Dee O'Hara, as if he were going to suit up that way. *Gotcha!* And he kept it up. That whole day he was the jolly Wally from beginning to end. He was amazing. He never sounded for a moment like someone under the stress of a novel form of flight test. It was like listening to a good buddy at beer call, recollecting in tranquillity. He practically out-yeagered Yeager. As soon as the escape tower blew off, marking the successful completion of the fully powered part of his ascent, Schirra saw it streaking through the sky and said: "This tower is a real sayonara."

Chris Kraft, the flight director, gave his approval for the first orbit, and Deke Slayton, the capcom at the Cape, told Schirra: "You have a go from Control Center."

Schirra said: "You have a go from me. It's real fat."

Then Slayton said: "Are you a turtle today?"

"Going to VOX recorder only," said Wally. Then he spoke into the tape recorder, whose microphone was not hooked into the open radio circuit.

"You bet your sweet ass I am," he said.

The Turtle Club was one of Wally's gotcha games. If one good buddy who played the turtle game met another good buddy in public—preferably in the company of very proper folk—and challenged him with the question "Are you a turtle?" that good buddy had to answer, "You bet your sweet ass I am," in a loud voice or else treat everybody else to a round of drinks. This was three minutes and forty-one seconds into the flight. Wally was already maintaining an even strain.

He buckled down to the task of conserving hydrogen peroxide. Ordinarily, when the booster rocket separated from the capsule, the capsule was then turned around by the automatic control system, but this consumed considerable fuel. This time Schirra nudged it around manually, using only the low thrusters, the five-pound thrusters, of the fly-by-wire system. Soon he was telling Deke at the Cape: "I'm in chimp mode right now and she's flying beautifully." He started using this phrase *chimp mode*. On the chimpanzee flights the attitude of the capsule had been controlled automatically throughout. *The chimp mode* was a little zinger for the benefit of all those on the mighty ziggurat, whether astronauts or X–15 "dream pilots," who were aware of the taunt: "A monkey's gonna make the first flight." Schirra's continual reference to *the chimp mode* as much as said: "Who cares! Here—I'll wave the bloody monkey in your face." As soon as he could, however, he went into what he called *the drifting mode*. He just let the capsule twist into any attitude it wanted, as Scott had on his final orbit.

"I'm having a ball up here drifting," said Wally. "Enjoying it so much I haven't eaten yet."

When he came over California during the fourth orbit, John Glenn, acting as capcom at Point Arguello, was instructed to have Wally say something for live broadcast over television and radio.

"Ha, ha," said Wally. "I suppose an old song, 'Drifting and Dreaming,' would be apropos at this point, but at this point I don't have a chance to dream. I'm enjoying it too much." When he was over South America, he was asked to say something in Spanish for live broadcast.

"*Buenos días*, you all," said Wally. (And the Latins loved it.)

After nearly four orbits, drifting and yakking and yukking it up, cool, relaxed, a turtle to the last, Wally had used up barely 10 percent of his hydrogen peroxide. He had already traveled one orbit farther than Carpenter or Glenn. He had floated every which way, and (as Scott had told him) there was nothing to it. There was no sensation of up or down in weightless flight. It was obvious that you could send a Mercury capsule on a seventeen-orbit flight, like Titov's, if you wanted. As Schirra came over the Cape, Deke Slayton said, "Flight would like to talk to you now."

"Flight" meant the flight director, Kraft himself, was coming onto the circuit.

"Been a real good show up there," said Kraft. "I think we are proving our point, old buddy!"

Glenn was sitting in front of a microphone in the tracking station at Point Arguello. Scott was sitting in front of a microphone in the tracking station at Guaymas. Kraft had never come on the circuit to say any such thing to either of them. Scott was beginning to see what the point, *our point*, was.

As he neared the completion of his sixth and last orbit, Wally announced that he had 78 percent of the fuel left in both the automatic and manual systems. He had flown twice as far as Glenn or Carpenter, and he could have gotten through another fifteen orbits or so if he had to. One of Kraft's lieutenants, an engineer named Gene Kranz, came on the circuit and said to Wally: "Now *that's* what I call a real engineering test flight!"

Scott picked up the message in his central nervous system even before his mind analyzed it. *Unlike the last one,* the man was saying. There was even a hint of . . . *unlike the last two.*

To complete an operational triumph Schirra now had only to land on target. Carpenter had landed 250 miles off the mark. As he began his descent toward the atmosphere, Wally told Al Shepard, the capcom on Bermuda, "I think they're gonna put me on the number-three elevator." He was talking about the number-three elevator used for moving aircraft up to the flight deck on the carrier *Kearsarge.* This was a bit of the Schirra metonymy for "squarely on target." Oh, yes! And in fact he landed just 4.5 miles from the carrier. The swabbos crowding the flight deck could see him coming down under his big parachute. Carpenter had found the capsule uncomfortably hot once he splashed down and had crawled out through the neck of the capsule and waited on his life raft for the rescue planes. Glenn had also complained of the heat. Schirra's suit had an improved cooling system, and he was willing to stay in the capsule indefinitely. He refused a helicopter's offer of a lift to the carrier. What was the rush? He stayed in the capsule while a crew of swabbos in a motor-driven whaling boat towed him back to the carrier. Once he was on the *Kearsarge* he told the doctors: "I feel fine. It was a textbook flight. The flight went just the way I wanted it to."

That became the verdict on Schirra's performance: "A textbook flight." He had done everything on the checklist. He had turned in a hundred percent performance. He had successfully proved that a man could ride around the earth six times and barely turn a hand or move a muscle and hardly use an ounce of fuel or expend an extra heartbeat and never, not for a moment, surrender to psychological stress, and ride the ship down to a designated drop in the vastness of the ocean. Sigma, summa, Q.E.D.: *Operational!*

Wally came back to celebrations in Houston and Florida and a big Wally Schirra day in his hometown of Oradell, New Jersey. The next day he went to the White House for congratulations by President Kennedy, and the President presented him with the Distinguished Service Medal. It all turned out to be rather brief and unceremonious, however, and something of a disappointment. A little chat, a few grins, a few photographs with the Chief Executive in the Oval Office, and that was it. The date was October 16. In time Wally would learn that Kennedy had just seen photographic evidence, from U-2 flights, that the Soviets had set up missile bases in Cuba. The President had kept his appointment with the astronaut only to maintain appearances, to prevent word from getting out about the critical situation that was developing.

XIV. The Club

By and by Conrad started carrying Glenn's bag, as well as his own, and facing up to the role. It was the only sensible thing to do. Otherwise, the two of them would arrive at some airport, St. Louis, Akron, Los Angeles, wherever, and it would take them five minutes to walk forty feet. The autograph seekers came in waves. Every few steps Glenn would have to put his bag down and sign some more autographs and shake some more hands. Actually he was great at it. That big sunny freckle-faced smile of his lit up the place. People came up to him as if they knew him personally and loved him. *He is my protector. He risked his life and challenged the Russians in the heavens for me.* They adored him so much it would have been hard for him to brush past them, even had he been of that sort of disposition. So he would put his bag down and sign some more autographs, and the two of them would have to stop.

If Conrad carried both bags, they could keep moving. Glenn could wave and sign autographs and shake hands and chat and beam that terrific smile at everybody in transit without seeming rude. As for Conrad, he was in absolutely no danger of having to stop and put down the bags. He was now an astronaut, officially, but not to the mobs of autograph seekers. They couldn't have cared less. He looked like some little guy who carried bags for John Glenn. What was more, that was what he felt like. That was about all the second group of astronauts was doing: chores for the first, for the one, the only, the Original Seven. Conrad, as part of his training, had been accompanying Glenn in his travels. Now that Project Mercury was drawing to a close, Glenn was supposed to make Project Apollo, the moon program, his "area of specialization." He was visiting the factories of the major contractors, just as he had in the early days of Mercury. Conrad's area of specialization, officially, was "cockpit layout and systems integration"; but mainly he was . . . with John Glenn. John Glenn visiting the factory took on the aura of the general coming to inspect the troops. He was a magnet for every sort of VIP who could get next to him, particularly congressmen and senators. There were times when senators actually pushed—elbowed! hipped! bellied!—secretaries, stenographers, and other mere gawkers out of the way to get next to Glenn's fabled hide and speak to him and grin a great deal. Standing by, all the while, would be an unknown young man, the single-combat hero's valet, apparently, his batman, as the British Army called military servants. Namely, the anonymous Lieutenant Conrad, Group II Astronaut.

Nevertheless, Conrad had made it this time, and that was the main thing. That was all that any really competitive military pilot upon the great ziggurat could focus on anymore: becoming an astronaut. By now, only three years later, any such session as he and Wally Schirra and Alan Shepard and Jim Lovell and the others had gone through at the Marriott motel in February of 1959 . . . it was hard to believe it could have ever taken place. Remember Wally that night? Wally! . . . adding up the pros and cons and agonizing over what the space program might do to his chances of commanding a squadron of F-4Hs! And now Wally—the same Wally with whom they had gone waterskiing on Chesapeake Bay, with whom they had weathered that bad string at Pax River, the old affable prankster himself—now Wally stood at the very apex of the great invisible pyramid of flying. For the seven Mercury astronauts had *become* the True Brotherhood. They were so dazzling you couldn't even *see* the erstwhile True Brethren of Edwards Air Force Base any longer.

In April, when NASA announced it was accepting applications for a second group of astronauts, both Conrad and Jim Lovell had looked like good bets this time around, since they had been among the thirty-one finalists in the original selection process. Conrad was stationed at Miramar, California, requalifying for a phase of the Navy fighter jock's training he had already been through, which was night carrier landings.

There was a good reason why night carrier landings required requalifying time; even, as in Conrad's case, after training as a test pilot at Patuxent. Night landings were a routine part of carrier operations—and perhaps the best of all examples of how a man's accumulated good works did him no good whatsoever at each new step up the great pyramid, of how each new step was an absolute test, and of how each bright new day's absolutes—chosen or damned—were built into the routine. By 1962 the Navy had already shifted over to light-beam systems using angled mirrors and Fresnel lenses at the end of the flight deck. Conrad and the rest of them going through night carrier training at Miramar did not have to depend on a landing-signal officer standing at the end of the deck in a luminescent orange suit waving a pair of luminescent orange wigwag flags. At night—to the pilot way up there in the dark—there was now a blob of light, known as the *meatball*, rising and falling on a dimly perceived little slab in the middle of the ocean. The shining blob, the motherless meatball, was rising and falling because the heaving greasy skillet did not stop wallowing in the waves simply out of respect for the night. The carrier was plowing on away from you, into the wind, and therefore into the waves, and therefore was pitching up and down—five, eight, even ten feet at a gulp. On a night when the clouds were low and the moon was obscured, when the sky was black, the ocean was black, and the deck was black, the little meatball (no more than an inch high from up there) and the lights on the ship were like a single low-wattage comet, dim and lurching through the vast blackness of the universe, and the pilot was expected to have the will, the moxie, the illustrious, the all-illuminating stuff to bring a five- or ten-ton jet fighter onto that dim drunken astral plate at 125 knots. In training he had a limited number of passes down the invisible glide path. If he couldn't bring himself to make contact with the deck for so long that his fuel ran low, then the word *bingo!* sounded over his earphones, and he had to return to land, to the training base, where the landing strip didn't move when you approached it . . . and where everybody on the flight line would know that another poor sad *bingo* was coming into a safe haven, having funked it in the business of night carrier landings. A persistent case of the bingos was enough to wash a man out of night carrier landings. That did not mean you were finished as a Navy pilot. It merely meant that you were finished so far as carrier ops were concerned, which meant that you were finished so far as combat was concerned, which meant you were no longer in *the competition,* no longer ascending the pyramid, no longer qualified for the company of those with the right stuff. To have every recommendation in the book as a test pilot, to have survived bad strings galore, meant nothing when such a thing happened. Chosen or damned! (It could blow at any seam.) There were nights when that little meatball way down yonder on the deck was jumping around like the silver BBs in those maddening games you hold in the palm of your hand, and a pilot would have to drive his F–4, a big fifteen-ton brute, down onto the deck through sheer willpower, drive it down practically like *a nail.* Anything—even the Great Kaboom!—was better than hearing *bingo* over your earphones. To bingo out of a carrier landing after eight years of military flying, after completing test-pilot school at Pax River, after becoming the top of the breed . . . now, there you had something unthinkable.

Conrad had just requalified for night carrier landings, for "all-weather carrier operations," meaning that he was now fully qualified for Navy air combat, when he received his invitation to apply for astronaut. The fact that there was nothing in the role of *astronaut* that would require one-tenth the *piloting* skill of night carrier landings did not deter Conrad, Lovell, or anyone else for a second this time around. This time Conrad went through the selection process like Lt. Straight Arrow. As before, there were thirty-odd finalists. They did not have to go through the Lovelace Clinic or the Wright-Patterson Aeromedical Center, however. Instead, they were sent to Brooks Air Force Base in San Antonio, the Air Force's medical center, for a set of physicals that were time-consuming but on the whole conventional. After five Mercury flights it was obvious that no extraordinary physical hardiness was required for the job.

For the last phase of the testing they took you straight to Olympus, which was now Houston. Part of the testing was a formal interview, across a table, with NASA engineers, plus Deke Slayton, John Glenn, and Al Shepard, concerning technical matters. But part of it seemed to be social. You were expected to go to a cocktail party and a dinner in a private dining room in the Rice Hotel in Houston, with the Mercury astronauts in attendance. Al Shepard was there for a while, and Gus

Grissom, Scott Carpenter . . . and here was Wally. You kept your brain dialed up in all sectors, trying to strike the perfect middle ground between being a good righteous beer-call buddy and showing a good sober respect for the eminence of those already in the club. *Maybe I better hold it down to just one drink.* It was like a rush party in a fraternity you desperately wanted to belong to.

Naturally Wally talked to Conrad and Jim Lovell as if he were the same old Wally, comrade-in-arms, good old Wally of Group 20. Nevertheless, the difference in their ranks existed in that room like a ray of light which beamed straight upon Walter Schirra, outstandingly successful single-combat warrior; for there were now the seven Mercury astronauts up there at the apex . . . and all the rest of the pilots in America far below.

Not that the Original Seven's national eminence altered the *true and secret* nature of things, however. The self-esteem of the fighter jock knew no limits, and the members of Group II were no exceptions to this rule. As soon as they were selected, the boys began looking around and comparing themselves—the Next Nine—with the Original Seven. Here was Neil Armstrong, who had flown the X–15. (What Mercury astronaut had done anything like that?) Here was John Young, who held two world speed-to-climb records. (What Mercury astronaut, other than Glenn, could claim any such distinction?) Here were Frank Borman, Tom Stafford, and Jim McDivitt, who had been flight test instructors at Edwards. (What Mercury astronaut had ever qualified for that, except for Slayton?) The Next Nine were really rolling now. The Original Seven were chosen to withstand stress, period. Look at Carpenter! Look at Cooper! Oh, the Next Nine felt very good about themselves. Nevertheless, the exalted *status* of the Original Seven was a fact. Once the initial euphoria of being chosen as an astronaut had subsided, Conrad and the others realized that now, for all their righteous stuff, they occupied a somewhat humiliating position in the corps of astronauts. They were like plebes, rookies, fraternity pledges. Gus Grissom had a nice grim ingratiating gruff gus way of telling them—if their paths had to cross here and there—not to get big ideas, not to go around calling themselves *astronauts.* "You're not an astronaut," he would say, "you're a trainee. You're not an astronaut until you go up." He said it without a trace of a smile. The Next Nine spent their time going to classes, like nine freshmen, like nine primary flight training candidates at Pensacola, which was bad enough, and doing scutwork for the Holy Seven, which was worse.

That was how Conrad had ended up being John Glenn's go-fer and carrying his bags. It could get cold as hell here on Olympus. There were levels upon levels, even here at the top. At the very apex there was John Glenn, and there were others among the Original Seven themselves who could not get over that fact. At the very first press conference, the one introducing the Next Nine to the public, the Original Seven were on hand, and Shorty Powers happened to introduce them in the reverse order of their flights. When he got to Shepard, he said: "And finally, this is Alan Shepard, the man who's been saying for years, 'But I was first!'" Well, that just cracked the place up. Everyone was laughing, with the single and obvious exception of Smilin' Al. He didn't even move a lip. If slow burns gave off needle rays, Shorty Powers would have had two small green holes through his frontal lobes. And you realized all at once that after Glenn's great orbital triumph Shepard—the prime pilot, the first American in space—must have felt like a forgotten man. Nobody ranked with Glenn, however, not even Webb, the administrator of NASA.

One day Glenn dropped by Webb's office in Washington and informed him that there was going to be a change in his personal agenda. He was going to make no more trips for NASA at the request of this or that congressman or senator. He was no longer going to fly halfway across the country, nor would he walk across the street, to stand on a platform to please some congressman who was looking for votes or whatever else. Glenn didn't put it as a request. He was letting Webb know how it was going to be from now on. He was just laying down the law. There was no way for Webb to take it except as a direct contradiction of his authority. Webb answered in a reasonable, if somewhat aroused, manner. Now, listen, John, we don't send you anywhere because some congressman wants you to be there. We send you because NASA wants you to be there. Congressional support is absolutely essential at this point, and this is one of the most important things you could possibly do for the program. To which Glenn says that nevertheless, he is taking no more such trips. Webb's color begins to rise, and he says that if he is instructed to carry out such duties, then he will be obliged to carry them out. To which Glenn says that Webb happens to be mistaken; he will do no such thing. All at once it's practically a shouting match.

Webb didn't push the situation to the brink. He just let the storm wear itself out; and when it was over, it was obvious that the Administrator of NASA was not a chief so long as John Glenn was in the room. Glenn did not back down or apologize. Far from it; he made it obvious who held the cards around here, and that was that.

It was John Glenn who had realized from the first that Project Mercury was like a new branch of the armed services, despite its civilian coloration. It would have simplified matters tremendously if NASA had given everybody formal rankings and had done with it. That way people such as Webb would have known where they actually stood. The seven Mercury astronauts could have been designated Single-Combat General, a category with the honors and privileges of five-star general but with none of the duties and obligations of command. After his flight John Glenn, then, would have been promoted to Galactic Single-Combat General, a category ranking slightly above the Chiefs of Staff of the Armed Services and slightly below the Commander-in-Chief. Webb, as NASA administrator, would have been a two-star general and would have known the protocol for dealing with GSC General Glenn. Newly inducted astronauts, such as Conrad, Lovell, and Young, could have been ranked as majors, with rapid promotion promised in the event of the successful completion of flights.

It would have greatly simplified matters for the wives as well. For as much as they would have denied it, had anyone confronted them with the topic, many of the wives of the Original Seven reacted to the arrival of the Next Nine and their wives . . . precisely like service wives since time was. The classic and often-told story of service wives concerned the wives of a group of Navy pilots who had just been transferred to a new base. A commander designated to give the wives an orientation lecture says: "First, would you ladies please rearrange yourselves by rank, with the highest-ranking wives sitting in the first row and so on back to the rear." It takes about fifteen minutes for the women to sort themselves out and change their seats, since very few of them know one another. Once the process has been completed, the commander fixes a stern glare upon them and says: "Ladies, I want you to know that I have just witnessed the most ridiculous performance I have ever seen in my entire military career. Allow me to inform you that no matter who your husbands are, *you* have *no* rank whatsoever. You are all equals, and you should kindly remember to conduct yourselves as such in all dealings with one another." That was not the end of the story, however. The wives stared back at their instructor with looks of utter bemusement and, as if with a single mind, said to themselves: "Who is this idiot and what planet has he been stationed on?" For the inexpressible provisions of the Military Wife's Compact were well known to all. A military officer's wife rose in rank with her husband and immediately took on all the honors and perquisites pertaining thereto, and only a fool or the sort of simple-minded jerk who was assigned to give orientation lectures to wives could fail to comprehend this.

Further, said the code, the wise wife of a junior officer was careful not to make her family's style of living so ostentatious that it outdazzled that of higher-ranking families. It was on just this point that the Next Nine began to rankle some of the wives of the Original Seven and, for that matter, some of the seven astronauts, as well. They were irritated to notice what *terrific houses* many of the Next Nine immediately bought. Just like that—in Timber Cove even!—they began gobbling up *the goodies!* For the Original Seven the ascension from the drab life of the junior officer had seemed like a glorious pioneer struggle and part of the prize for winning the contest, for being chosen as the Original Seven. They found something distasteful about the attitude of the new crowd—this notion that as soon as a man was designated an astronaut, he and his family were entitled to stride out upon the golden boulevards of the Celestial City as if they owned the place.

The Group II astronauts immediately produced an agent, their version of Leo DeOrsey, and sat down to negotiate and cut up the *Life* pie. He was Harry Batten, president of the N. W. Ayer advertising agency in Philadelphia. He was as much of a big leaguer as DeOrsey, and like DeOrsey he agreed to serve as agent without pay. It was a bit too much!

People were already treating the Next Nine like the Original Seven! The developers in Timber Cove and Nassau Bay, the second-best development, offered them big houses for small down payments and huge mortgages at low interest—a mortgage of forty to fifty thousand dollars seemed enormous in 1962—and the Next Nine took it all on without even blinking. They were *not behaving like junior officers* in the presence of the Single-Combat Generals. For, as everyone understood, tacitly, this was no mere civilian agency; this was a new branch of the service.

It was in that spirit that the "A.W.C." was started. Without anyone coming out and saying so, it was understood that Marge Slayton was the C.O.'s wife in this outfit. By now Deke had taken on his job as Coordinator of Astronaut Activities with such determination that he was about to be put in charge of crew selection—meaning that he would be the one man who had the most to say about who flew, and in what order, particularly when it came to the Next Nine. He was about to be made Assistant Director of Flight Crew Operations. He became the equivalent of a commanding officer, making Marge the C.O.'s wife. Marge organized a couple of coffee hours for all the wives, the First Seven and the Next Nine, so they could all get to know each other. By the second time they met, they all realized, without a word—nobody had to say it—what this was. This was . . . the Officers Wives Club, such as existed at every base in the land. One of the newcomers who seemed absolutely tickled to death about the coffee hours was Sue Borman, the wife of one of the Next Nine, Frank Borman. Borman had been one of the first instructors at the Air Force's new Experimental Test Pilot School at Edwards. He was a short, compact West Point man from Arizona, and his wife, Sue, was the perfect super-efficient officer's wife. She had a fireproof cheeriness about her and a determination to get things organized. They were a great team. "This is fun," she said, speaking of Marge's coffee hours for the wives. "Let's start doing this on an organized basis." So the A.W.C. began. The initials stood for Astronauts Wives Club, of course, but the full name was never used. It was a gaffe for an astronaut or an astronaut's wife to use the word *astronaut*. Marge herself was always talking about "the fellows." Besides, the full name would make the military analogy too pointed.

The A.W.C. was no great delight to most of the wives of the Original

Seven, however. Some of the newcomers, such as Sue Borman, were too gung-ho about it all. That was what they would tell themselves. In fact, the newcomers acted too *equal* about it all. There was no protocol for showing the deference due Single-Combat Generals' wives. The wives of the Original Seven began attending the monthly A.W.C. hours less and less. Betty Grissom almost never showed up; but then Betty had hated teatime functions from the very beginning. If one was going to feel ill at ease in the cheery chitchat game *and* not be treated as the Honorable Mrs. Single-Combat General . . . why bother?

For both the wives and the men there was at first an Oaken Bucket nostalgia for the early days, the Langley era, the pioneer period, the period of youth and idealism and spartan courage and yahoo cowboy disregard for the bureaucratic proprieties. You would even see engineers and support personnel who had moved from Langley and the Cape here to Houston getting misty about those old days . . . three years ago . . . The new facility, the Manned Spacecraft Center, was taking shape out in the middle of a thousand acres of absolutely flat gumbo pasture. The buildings were great squat beige cubes set at grandiose intervals from one another and connected by wide roads, veritable highways, lined with aluminum light poles. The place looked like one of those "industrial parks" that were always being touted in the real-estate sections of the Sunday newspapers.

Nevertheless, it was obvious that something of vast proportions was underway; and given a rosy enough picture of what lies down the road, a man could get over a case of homesickness soon enough. Perhaps Houston, the boom town of boom towns, was just the place for the expansion occurring in the space program. After a while you could begin to appreciate Houston's energy and its sense of the grand sweep, the risk-all plunge. And Timber Cove and Nassau Bay and the other new housing developments out by Clear Lake weren't half bad, it turned out. In fact, they were luxury itself compared to what you found around most air bases, and the locals, out here in this erstwhile farming country, were really good folks. Two-thirds of NASA was already geared up for Projects Gemini and Apollo and the great race to beat the Soviets to the moon. Just imagine what waited, here on earth, for the first man to walk on the moon . . . and the boys could imagine it. You had only to look at John Glenn. Glenn had not been the first man to fly in earth orbit or even the second, merely the first American. Yet he had ascended to a status so extraordinary it had no precedent. Some of the boys were convinced that Glenn had his eyes set on becoming President. (Nor was the notion farfetched; after all, it was David who had succeeded to the throne of Saul.) Glenn now moved in a world full of the Kennedys, the Johnsons, senators, congressmen, foreign dignitaries, heads of corporations, VIPs of every description. Next to John Kennedy himself, John Glenn was probably the best-known and most admired American in the world. Oh, the boys were aware of all that! Just ask Al Shepard!—although of course no one did.

Al was now in training as backup for Gordon Cooper's flight, which would be in May 1963. Gordo, the last in line, had drawn what was shaping up as the final flight in the Mercury series, thirty-four hours, twenty-two orbits, designed to put the United States into the game with the Soviets, who had now achieved flights of seventeen, sixty-four, and forty-eight orbits. The original planning had called for four long-duration flights, with the second one lasting three days. Shepard had been counting on that. He was desperate for an orbital flight. His suborbital flight, as well as Grissom's, now seemed terribly insignificant. As for Gus, he was already getting revved up for the Gemini program, spending a lot of time in St. Louis, where McDonnell was building the Gemini spacecraft. Gus had put the gloom of his flight behind him and was looking ahead to Gemini and Apollo. His friend Deke was in charge of crew selection for the two new programs and was throwing himself heart and soul into the job, and not just because he enjoyed the exercise of his newly found power. The main thing was that he would be on top of every flight from beginning to end, familiar with every detail of every mission. It seemed to be Deke's fervent belief that it would be only a matter of time—following the next physical or the one after that, or the one next year—before he was back on full flight status and into the rotation. And Wally—Wally was riding high. Wally's flight was still the shining example of what an *operational space flight* should be. It had taken Wally's performance to show that Cooper's twenty-two-orbit flight would be possible. Wally couldn't have been in better shape in the program. He had been as efficient and as cool as they made them.

Wally had come back from his flight and landed in the middle of the Cuban missile crisis. For a week Kennedy and Khrushchev had their showdown, which appeared to bring the world to the brink of a nuclear war, but then Khrushchev backed off and withdrew all Soviet missiles from Cuba. After that things cooled down considerably. Like everyone else, the boys noticed that negotiations for a nuclear test ban and for "cooperation in space," whatever that might prove to be, were in the news a lot. But to tell the truth, it didn't seem like much more than the usual drizzle of words. For that matter, on May 11, four days before Cooper was to go up, Lyndon Johnson drew the lines the same way they had been drawn ever since October of 1957, when the first Sputnik went up. He made a speech answering charges that some congressmen were making about the high cost of the new programs, Gemini and Apollo, and he said: "I, for one, don't want to go to bed by the light of a Communist moon." Christ, that was *worse* than Sputnik: every night, overhead, sails the silvery moon, occupied by the Russians.

As for Gordo himself, he was already on top of the world. There were those among his brethren who had their doubts about the man, but he never had a moment's doubt about himself. Once more his light shone 'round about him. Was he the last of the seven to be assigned a flight? Well, so what . . . it wasn't a contest . . . the press had dreamed up all that crap . . . Shepard, Glenn, and the others had paved the way for his endurance test. The potential hazards? They didn't bother him in the slightest. They hadn't from the beginning.

Confronted with any feasible form of manned flight, Gordo was a picture of righteous aplomb. This was a side of Cooper that Jim Rathmann understood better than any of his confreres in the corps. Gordo had been in Florida for a long stretch, preparing for his flight, and he saw Rathmann a lot. Thanks to Rathmann, he, like Gus and Wally and Al, had become crazy about automobile racing. Rathmann, in turn, had decided to learn to fly. Cooper took him up in a Beechcraft one day and told him: "Never fly under a sea gull—they'll shit on your airplane." Rathmann made the mistake of laughing, as if he thought Gordo was kidding, whereupon Gordo said, "I'll show you," and headed for a flock of sea gulls flying low over the Everglades. The first thing Rathmann knew, Cooper was down so low he could hear a sound that went *whup*

whup whup whup whup whup whup whup. It was the propellers cutting the marsh grass. For more than a mile old Gordo mowed the marsh grass to make sure he stayed under the sea gulls. Rathmann could hear it the whole time: *whup whup whup whup whup whup.* By the time they landed, Rathmann was Jell-O. But Cooper just popped open the cockpit door and stood up on the lower frame and pointed triumphantly at the roof and yelled to Rathmann: "Look here—I *told* you!"

So far as a Mercury flight was concerned, he seemed to regard it as easy enough. He had lobbied as hard as Deke Slayton himself for more *pilot* control of the spacecraft. But since you didn't have it, why get excited? Why get your bowels in an uproar? Just take the ride and relax.

Early in the morning of May 15, while it was still dark, Gordo was inserted into the little human holster atop the rocket. As usual there was a long hold before the lift-off. The doctors monitoring the biomedical telemetry began noticing something very odd. In fact, they couldn't believe it. Every objective reading of the calibrations and printouts indicated . . . the astronaut had gone to sleep! The man was up there stacking Z's on top of a rocket loaded with 200,000 pounds of liquid oxygen and RP–1!

Well, why the hell not? Gordo had had plenty of opportunity to see how the launch days always went. You hit the sack in Hangar S about ten or eleven the night before, and they woke you up about three in the morning, in the dark, and they took you out to the rocket and they laid you down on a contoured couch for two, three, four hours while they tuned in all the systems for the lift-off. You didn't have a damned thing to do, really, during most of this; so why not catch up on all the sleep you'd missed?

Throughout America, throughout the world, untold millions were by their radios and in front of television sets, waiting for the moment of lift-off, wondering, as always: *My God, what goes through a man's mind at a moment like this!* Scarcely able to believe it themselves, NASA never supplied the answer.

By the moment of lift-off there was quite a circus in progress out front of Gordo's house in Timber Cove. The Genteel Beast was outdoing itself in the death-watch department. One of the television networks had erected a stupendous aerial on the lawn of a house across the way, a gigantic thing, about eight stories high, the better to beam to the world live pictures of the house inside which Astronaut Cooper's wife, Trudy, was maintaining her anxious vigil in front of the TV set. Milling around underneath the aerial and in the street and on the sidewalks was the biggest mob of reporters and camera crewmen you could imagine. Slovenly but nevertheless seemly they were. They treated Trudy Cooper, in their coverage, as if she had stepped right out of *Life* magazine, as if she were wearing a pageboy bob and playing "Moonlight in Vermont" on the old upright in the family room to keep up the spirits of herself and the children while Gordon's life was on the line in the longest American mission yet.

That was rich. By now the very presence of the Beast itself had made any such private and personal response to the event impossible, even in households where the marriage was a lot more solid than Gordo and Trudy's. For the astronaut's wife the days of the lonely vigil by the telephone with the little ones tugging at her skirt, Edwards- or Pax River-style, were over. First had come the Danger Wake, as Louise Shepard had experienced it, with a big crowd in the house and a bigger one—the Genteel Animal—out on the front lawn. Since then the Astronaut's Wife had been converted from an individual to a performer, at least for the duration of the flight—ready or willing or good at it or not. It had become an immutable part of the drill: at the completion of the flight the astronaut's wife had to leave the house and confront the Beast and all his cameras and microphones and submit to a press conference and answer questions and be the Perfect Astronaut's Wife with merely the entire world watching. It was this grim prospect that truly lacerated one's heart while Mr. Wonderful was aloft. It was *this* that gave the test pilot's wife a royal case of nerves in the space age. For the astronaut the flight consisted of riding the rocket and, God willing, not fucking up. For the wife the flight consisted of . . . the Press Conference.

The questions they asked you were unbelievably simple-minded, and yet there was no smooth way to field them. As soon as you touched one, it popped all over your face like bubble gum.

"What is in your heart?"

"What advice do you have for other women whose husbands have to go through dangerous situations?"

"What's the first meal you plan to cook for [Al, Gus, John, Scott, Wally, Gordon]?"

"Did you feel you were with him while he was in orbit?"

Pick out one! Try answering it!

Problems of protocol had arisen. Sometimes the Genteel Animal besieged the Mother's house as well as the Wife's. John Glenn's mother had been a great hit on television. She looked and sounded like about as ideal a mother as an astronaut could possibly have. She had white hair and a marvelous smile, and when Walter Cronkite, on CBS, cut from the Cape to New Concord, Ohio, to say a few words to her, she said, "Well, Wal-ter Cron-kite!"—as if she were saying hello to a cousin she hadn't heard from in years. But whom should the networks interview first after a flight, the Wife, the Mother, or the President? Opinions varied, and this added to the tension. Regardless of the order, however, there seemed to be no way for the wife to get out of it. Even Rene, after hiding throughout Scott's flight, had dutifully turned up at the press tent on the base at the Cape for the Wife's Press Conference. By now, when the other wives came around to the house of the Wife during a flight, they were not there to hold her hand over the dangers her husband was facing. They were there to hold her hand over the television cameras she would be facing. They were there to try to buck her up for a *true ordeal.* They liked to do the Squarely Stable routine. One of the wives—Rene Carpenter was good at it—would take the role of Nancy Whoever, TV correspondent, and hold her fist up to her mouth, as if she were holding a microphone, and say:

"We're here in front of the trim, modest suburban home of Squarely Stable, the famous astronaut who has just completed his historic mission, and we have with us his attractive wife, Primly Stable. Primly Stable, you must be happy, proud, and thankful at this moment."

And then she would shift her fist over underneath the chin of another wife, and she would say:

"Yes, Nancy, that's true. I'm happy, proud, and thankful at this moment."

"Tell us, Primly Stable—may I call you Primly?"

"Certainly, Nancy, Primly."

"Tell us, Primly, tell us what you felt during the blastoff, at the very moment when your husband's rocket began to rise from the earth and take him on this historic journey."

"To tell you the truth, Nancy, I missed that part of it. I'd sort of dozed off, because I got up so early this morning and I'd been rushing around a lot taping the shades shut, so the TV people wouldn't come in the windows."

"Well, would you say you had a lump in your throat as big as a tennis ball?"

"That's about the size of it, Nancy, I had a lump in my throat as big as a tennis ball."

"And finally, Primly, I know that the most important prayer of your life had already been answered: Squarely has returned safely from outer space. But if you could have one other wish at this moment and have it come true, what would that one wish be?"

"Well, Nancy, I'd wish for an Electrolux vacuum cleaner with all the attachments—"

—and they'd all crack up at the thought of what a dim lummox the Genteel Beast really was. Still . . . that didn't make it any easier when your time came.

Gordo's flight was to last thirty-four hours, meaning that Trudy would undergo the longest siege by the Beast and have the most protracted danger wake yet. Two sets of wives came by. Louise Shepard brought most of the other Original Seven wives in her convertible. Later on, some of the Other Nine wives came by—Jim Lovell's wife Marilyn, Ed White's wife Pat, Neil Armstrong's wife Jan, and John Young's wife Barbara. Everybody tried to listen to Gordo's transmissions from the capsule over a high-frequency radio receiver Wally Schirra had loaned Trudy. It was the receiver that had been in Wally's capsule during his flight. But about all you could get out of it was static. So they went out on the patio in the back, out of sight of the Animal, and watched the television coverage of the flight, off and on, and ate devil's-food cake. In the true spirit of the wake, friends and neighbors had brought over food. During his ninth orbit, which began about 7:30 p.m., Gordo was supposed to try to go to sleep for a few hours, and Trudy decided that she and their two daughters, Jan and Cam, should try to get some rest, too. In the morning Gordo was still up there, twenty-four hours into the flight, and the Beast was still outside the door, and the danger wake was going strong. About noon, as Gordo began his last four orbits, you could tell from the television reports that his capsule was beginning to develop electrical problems. During the next-to-last orbit they became worse. It now appeared as if Gordo would have to line up the capsule for re-entry manually, without any assistance from the automatic control system at all. Trudy received a telephone call from Deke Slayton. He told her that she and the children shouldn't worry, because Gordo had practiced completely manual re-entries many times on the procedures trainer. "This is what we wanted to do anyway," he said.

Well, Gordo was going to have his hands full. Nevertheless, Trudy couldn't help but jump yet one more step forward in the retro sequence. If Gordo was beginning his re-entry, then very soon . . . she would have to step out the front door and face the Beast and his cameras and microphones and go through the press conference . . .

Meantime, aloft, Gordo was having a hell of a time for himself. Right after the lift-off he said to Wally Schirra, who was serving as the capcom, "Feels good, buddy . . . All systems go." He kept adding things such as "Working just like advertised." The Life Sciences people, who had finally been allowed a few experiments since the flight was so long, were interested in determining the limits of adaptability to weightlessness. They hoped to see what sleep would be like, although they were not sure they could learn anything about this during a thirty-four-hour flight, given the naturally high adrenal excitement of the astronaut. They needn't have worried. Ol' Gordo obliged by falling asleep during his second orbit, even though his suit was overheating and he had to adjust the temperature settings continually. One of his tasks was to provide urine samples at specified intervals. This he dutifully did. Since in a weightless condition it would be impossible to pour the sample from the urine receptacle—it would have floated about the cockpit as globules—Gordo was provided with a syringe to transfer it from the receptacle to a container. But the syringe leaked all over the place, and Gordo had the reeking amber globules floating around, anyway. So he just tried to herd them together into one big blob periodically and went on with his tasks, which included light and photographic experiments, somewhat like Carpenter's. Gordo was really something. He seemed even cooler about the whole thing than Schirra, and nobody had believed that possible. Every now and then he would look out the window and give the folks on the ground a little travelogue, Gordo-style.

"Down there's the Himalayas," he said. He seemed to like the sound of the word. "Ay-yuh . . . the Himalayas." In Oklahoma lingo Gordo it came out "Himmuh-lay-yuz."

On the nineteenth orbit, with three more to go, Gordo started getting readings of g-force buildups, as if the capsule had begun its re-entry. Sure enough, the capsule started rolling, just as it would have during a re-entry in order to increase stability. The automatic control system had begun the re-entry sequence, even though the capsule was still in orbit and hadn't slowed down in the slightest. The electrical system was shorting out. On the next orbit, the twentieth, the capsule lost all attitude readings. This meant Cooper would have to line it up manually for the re-entry. On the next-to-last orbit, the twenty-first, the automatic control system went out completely. For re-entry Cooper would not only have to establish the capsule's angle of attack by hand, using the horizon as his point of reference, he would also have to hold the capsule steady on all three axes, pitch, roll, and yaw, with the hand controller and fire the retro-rockets by hand. Meantime, the electrical malfunction had done something to the oxygen balance. Carbon dioxide started building up in the capsule and inside Cooper's suit and helmet as well.

"Well . . . things are beginning to stack up a little," said Gordo. It was the same old sod-hut drawl. He sounded like the airline pilot who, having just slipped two seemingly certain mid-air collisions and finding himself in the midst of a radar fuse-out and control-tower dysarthria, says over the intercom: "Well, ladies and gentlemen, we'll be busy up here in the cockpit making our final approach into Pittsburgh, and so we want to take this opportunity to thank you for flying American and we hope we'll see you again real soon." It was second-generation Yeager, now coming from earth orbit. Cooper was having a good time. He knew everybody was in a sweat down below. But this was what he and the boys had wanted all along, wasn't it? They had wanted to take over the complete re-entry process—become *true pilots* in this damned thing, bring her in manually—and the engineers had always shuddered at the thought. Well, now they had no other choice, and he had the controls. On top of that, during his final orbit he would have to keep the capsule at the proper angle, by eye, on the night side of the earth and then be ready to fire the retro-rockets soon after he entered daylight over the Pacific. No sweat. Just made it a little more of a sporty course, that was all—and Gordo lined up the capsule, hit the button for the retro-rockets, and splashed down even closer to the carrier *Kearsarge* than Schirra had.

No one could deny it . . . no brethren, old or new, could fail to see it . . . when the evil wind was up, Ol' Gordo had shown the world the pure and righteous stuff.

Over the next week Gordo became the most celebrated of all the astronauts aside from John Glenn himself. Ol' Gordo!—whose confreres had pictured him as forever bringing up the rear . . . There he was, sitting on the back of the open limousine, in parade after parade . . . Honolulu, Cocoa Beach, Washington, New York . . . And such parades! The ticker-tape parade in New York was one of the biggest ever, Glenn-scale, with signs along the way, saying things such as GORDO COOPER—YOU'RE SUPER-DUPER! in letters three or four feet high. Not only that, he addressed a joint session of Congress, just as Glenn had. A "textbook flight" like Schirra's was all well and good, but there was nothing like a hair-raiser to capture the imagination and stir the gourds. Gordo was also the first American to spend an entire day in space, of course, and he had put the United States back in the ball game with the Soviets. The role of single-combat warrior seemed more glorious than ever.

XV. The High Desert

By the time of Gordon Cooper's flight, Chuck Yeager had returned to Edwards Air Force Base. He was only thirty-nine, the same age, it so happened, as Wally Schirra and Alan Shepard and two years younger than John Glenn. Yeager no longer had quite the head of dark curly hair that everybody at Edwards saw in the framed photographs of him stepping out of the X-1 in October 1947. And God knows, his face had more mileage on it. This was typical of military pilots by that age and came not so much from the rigors of the job as from taking the sun rays head-on twelve months every year out on the concrete of the flight lines. Yeager had the same trim muscular build as always. He had been flying supersonic fighter aircraft as regularly, day in and day out, as any colonel in the Air Force. So in the ten years since he had made his last record-setting flight here at Edwards, that wild ride to Mach 2.4 in the X-1 A, he really hadn't changed too much. You couldn't say the same about Edwards itself.

When Yeager had departed in 1954, Pancho's had still been standing. Today the base was loaded with military and civil-service personnel, every GS-type in the manual, working for the Air Force, for NASA, even for the Navy, which had a small piece of the X-15 program. At four o'clock it was worth your life to be heading upstream during the mad rush from the air conditioners in the office buildings to the air conditioners in the tract homes in Lancaster.

This much Yeager already knew about; this was the part that was easy to take. He had been commanding a squadron of F–100s at George Air Force Base, which was only about fifty miles southeast of Edwards in the same stretch of prehistoric dry-lake terrain. Yeager and Glennis and their four children had lived at Victorville in the same sort of housing development you found in Lancaster; just a bit more barren, if anything, a little grid of Contractor Suburbans lined up alongside Interstate Highway 15. The same old arthritic Joshua trees dared you to grow a blade of grass, much less a real tree, and the cars heading from Los Angeles to Las Vegas hurtled by without so much as a flick of the eye. Not that any of this weighed upon Yeager, however. As commander of a squadron of supersonic fighters he had led training operations and readiness maneuvers over half the world and as far away as Japan. Besides, nobody stayed in the Air Force because of the glories of the domestic architecture. Where he was living was standard issue for a colonel such as himself who after twenty years was making just a little over two hundred dollars a week, including extra flight pay and living allowance . . . and without magazine contracts or any other unorthodox goodies . . .

The Air Force had brought Yeager back to Edwards two years ago to be director of flight test operations. Last year, 1962, they created the new Aerospace Research Pilot School and made him commandant. ARPS, as the school became known, was part of big plans the Air Force had for a manned space program of its own. As a matter of fact, the Air Force had envisioned a major role in space ever since the first Sputnik went up, only to be thwarted by Eisenhower's decision to put the space effort in civilian hands. They now wanted to create a military program, quite apart from NASA's, using fleets of ships such as the X–20 and various "lifting body" craft, wingless ships whose hulls would be shaped to give them aerodynamic control when they re-entered the earth's atmosphere, and the Manned Orbiting Laboratory, which would be a space station. Boeing was building the first X–20 at its plant in Seattle. The Titan 3C rocket booster it would require was almost ready. Six pilots had already been chosen to train to take it into orbit.

The X–20 and the MOL were not yet operational, of course. In the meantime, it seemed to be highly important that Air Force pilots be chosen as NASA astronauts. The prestige of the astronaut absolutely dominated flying, and the Air Force was determined to be the prime supplier of the breed. Four of the nine new astronauts selected in 1962, before ARPS was instituted, had been from the Air Force; that was not considered good enough.

To tell the truth, the brass had gone slightly bananas over this business of producing astronauts. They had even set up a "charm school" in Washington for the leading candidates. The best of the young test pilots from Edwards and Wright-Patterson flew to Washington and were given a course in how to impress the NASA selection panels in Houston. And it was dead serious! They listened to pep talks by Air Force generals, including General Curtis LeMay himself. They went through drills on how to talk on their feet—and that was the more sensible, credible part of the course. From there it got right down to the level of cotillion etiquette. They were told what to wear to the interviews with the engineers and the astronauts. They were to wear knee-length socks, so that when they sat down and crossed their legs no bare flesh would show between the top of the socks and bottom of the pant cuffs. They were told what to drink at the social get-togethers in Houston: they should drink alcohol, in keeping with the pilot code of Flying & Drinking, but in the form of a tall highball, either bourbon or Scotch, and only one. They were told how to put their hands on their hips (if they must). The thumbs should be to the rear and the fingers forward. Only women and interior decorators put the thumbs forward and the fingers back.

And the men went through it all willingly! Without a snigger! The brass's passion for the astronaut business was nothing compared to that of the young pilots themselves. Edwards had always been the precise location on the map of the apex of the pyramid of the right stuff itself. And now it was just another step on the way up. These boys were coming through Chuck Yeager's prep school so they could get a ticket to Houston.

The glamour of the space program was such that there was no longer any arguing against it. In addition to the chances for honor, glory, fame, and the celebrity treatment, all the new hot dogs could see something else. It practically glowed in the sky. They talked about it at beer call at every Officers Club at every air base in the land. Namely, the Astronaut Life. The youngsters knew about that, all right. It existed just over the rainbow, in Houston, Texas . . . the *Life* contract . . . $25,000 per year over and above your salary . . . veritable *mansions* in the suburbs, custom-designed . . . No more poor sad dried-up asbestos-shingle-roof clapboard shacks rattling in the sandstorms . . . free Corvettes . . . an enormous free lunch from one side of America to the other, for that matter . . . and the tastiest young cookies imaginable! One had only to reach for them! . . . The vision of all the little sugarplums danced above the mighty ziggurat . . . You bet! A veritable Fighter Jock's Forbidden Dream of *the goodies* had been brought to life, and all these young hot dogs looked upon it like people who believed in miracles . . .

It really made some of the older pilots shake their heads. If a man got a piece of tail every now and then, the world wasn't going to come to an end. But to dream of a goddamned aerial nookie circus . . . What was worse, however, was the *Life* contract. The way any true Blue-Suiter saw it, to let an experimental test pilot exploit his job commercially was only asking for trouble. If a man had the opportunity to fly machines with incalculable millions of dollars' worth of resources and facilities and man hours built into them, if they put him in a position to make history—that was more than enough compensation.

Yeager had flown the X–I at straight pay, $283 a month. The Blue Suit!—that was enough for him. The Blue Suit had brought him everything he had in this world, and he asked for nothing else.

And what would all of that mean to these boys, even if someone said it? Not a hell of a lot, probably. Not even the fact that the X–15 project was in its finest hour, right here, for all to see, affected the new order of things. In June the X–15, with Joe Walker at the controls, had achieved Mach 5.92, or 4,104 miles per hour, which brought the project close to the optimum speed—"in excess of Mach 6"—it had been aiming for. In July Bob White had flown to 314,750 feet, or 59.6 miles, 9.6 miles into space (50 miles was now officially regarded as the boundary line) and well above the project's goal of 280,000 feet. These and many less spectacular flights of the X–15 were bringing back data concerning heat buildup (from air friction) and stability upon which the design of all the supersonic and hypersonic aircraft of the future, commercial and military, would be based. The X–15's XLR–99 rocket had 57,000 pounds of thrust. The Mercury-Redstone had 78,000; the Mercury-Atlas had 367,000 pounds; but soon there would be the X–20, and the X–20 would have a Titan 3 rocket's 2.5 million pounds of thrust, and it would be the first ship to go into orbit with a pilot at the controls from beginning to end, a pilot who could land it anywhere he wanted, eliminating the tremendous expense and risk of the Mercury ocean-rescue operations, which involved carriers, spotter planes, helicopters, frogmen, and backup vessels strung halfway around the world.

Yeager's students had a chance to experience something close to what such space piloting would be like. They went "booming and zooming" in the F–104. The F–104 was a fighter-interceptor that had been built to counter the MiG–21, which the Russians were known to be developing. The F–104 was fifty feet long and had two razor-thin wings, each only seven feet long, set far back on the fuselage, close to the tail assembly. The pilot and his guy-in-back were in two seats way up in the nose. The F–104 was built for speed in combat, period. It could climb at speeds in excess of Mach 1 and it could achieve Mach 2.2 in level flight. The faster it went, the steadier it was; it was unstable at low speeds, however, and oversensitive to the controls, with an evil tendency to pitch up and then snap into rolls and spins. At glide speed it seemed to want to fall like a length of pipe. After practicing on an F–104 simulator, Yeager's students would take the ship up to 35,000 feet and open her up to Mach 2 (the boom), then aim her up at about forty-five degrees and try to poke a hole in the sky (the zoom). The g-forces slammed them back in their seats and they shot up like shells, and the pale-blue desert sky turned blue-black and the g-forces slid off and they came sailing over the top of the arc, about 75,000 feet up, silent and weightless—an experience like unto what the brethren themselves had known!—

—and these boys thought that was neat. Maybe it would be nice to fly the X–15 or the X–20, if you didn't make astronaut . . .

Yeager liked to take the ARPS students up for mock dogfights, hassling, just to . . . keep the proficiency up . . . Few of the lads had ever been in combat and they knew little about the critical tolerances of fighter aircraft during violent maneuvers. They knew where the outside of the envelope was, but they didn't know about the part where you reached the outside and then *stretched* her a little . . . without breaking through . . . Yeager waxed their tails with regularity, but they took that in stride. These days the way to the top—meaning the road to test-pilot astronaut—involved being very good at a lot of things without necessarily being "shit-hot," to use the beer-call expression, at anything. A balance of pilot skills and engineering; that was the ticket. Joe Walker's backup pilot in the X–15 project, Neil Armstrong, was typical of the new breed. A lot of people couldn't figure out Armstrong. He had a close blond crew cut and small pale blue eyes and scarcely a line or a feature in his face that you could remember. His expression hardly ever changed. You'd ask him a question, and he would just stare at you with those pale-blue eyes of his, and you'd start to ask the question again, figuring he hadn't understood, and—*click*—out of his mouth would come forth a sequence of long, quiet, perfectly formed, precisely thought-out sentences, full of anisotropic functions and multiple-encounter trajectories, or whatever else was called for. It was as if his hesitations were just data punch-in intervals for his computer. Armstrong had been preparing for an X–15 launch from the Smith's Ranch dry-lake bed last year when Yeager, who was director of flight test operations, told him the lake bed was still too muddy from the rains. Armstrong said the meteorological data, considering the wind and temperature factors, indicated the surface would be satisfactory. Yeager received a call from NASA asking him if he would take a small plane over to the lake bed and make a ground inspection. "Hell, no," said Yeager. "I've been flying over these lakes for fifteen years, and I know it's muddy. I'm not going to be responsible for disabling an Air Force plane." Well, would he fly a NASA plane up there? Hell, no, said Yeager; he didn't want that on his record, either. It was finally arranged that he would fly up there backseat, with Armstrong at the controls and therefore responsible for the mission. As soon as they touched down, Yeager could tell that the mud was going to suck up the landing gears like a couple of fence posts, which it did. Now they were hopelessly mired in the muck, and a range of hills blocked radio contact with the base. "Well, Neil," said Yeager, "in a few hours it'll be dark, and the temperature's going down to zero, and we're two guys standing out here in the mud wearing windbreakers. Got any good ideas?" Armstrong stared at him, and the computer interval began, and it ended, and nothing came out. A rescue team from the base, alerted by the loss of radio contact, retrieved them before nightfall—and brought back the story, which entertained the old-timers for a few days.

Nevertheless, the new breed had their share of the proper righteous stuff, same as the stick'n'rudder tigers of yore. Armstrong himself had flown more than a hundred missions off carriers during the Korean War, and had done good work in the X–15. Then you had men like Dave Scott and Mike Adams, who were two of Yeager's ARPS students. They were practicing low lift-over-drag landings one day in the F–104. In this maneuver, which simulated an X–15 landing, you gunned the afterburner for speed (and stability) and flared the flaps and tried to grease the ship onto the runway at 200 knots. As Scott and Adams neared the ground, the "eyelids" on the afterburner malfunctioned, opening too wide, cutting the thrust down to 20 or 30 percent of maximum. Visually they could tell the ship was sinking too fast. Scott, who had the controls, gunned it but got very little response. They were dropping like a brick. Adams, in back, knew that the tail would hit the runway first, due to the angle of attack they were in, if Scott couldn't regain power. He told Scott over the radio circuit that if the tail hit he was ejecting. The tail hit, and in that moment he pulled his cinch ring and ejected at zero altitude. Scott elected to stay with the ship. The belly smashed onto the runway and the ship went careening down it and off into the mesquite. When the beast finally came to a halt, Scott looked back, and the engine was jammed up into the space where Adams used to be. Both men had made the right decision. Adams had been exploded up into the air and had come down safely by parachute. Scott's ejection mechanism had been broken in the torque of the initial impact and he would have been killed had he pulled the cinch ring, either by the nitroglycerine explosion or by a partial ejection.

Yeager was tremendously impressed by those two decisions by two men in the very mouth of the Gulp. There you had it, with the ante doubled: the right stuff. And when NASA had announced several months ago that a third group of astronauts would be chosen both men immediately applied, although Adams also seemed to have a sincere interest in the X–15 project. The X–15 pilots themselves had their eyes on Houston, for that matter. Armstrong had applied as soon as civilians had been eligible and was now a Group II astronaut. He had Joe Walker's blessing, too. Walker himself had considered applying but figured that his age—he was forty-two—pretty well ruled him out.

That was the way the pyramid was now constructed. The old

argument—namely, that an astronaut would be a mere passenger monitoring an automated system—didn't have much sock to it anymore. The truth was that there you had a picture of the pilot in practically all the hypersonic vehicles of the future, whether in space or in the atmosphere. The Mercury vehicle had merely been one of the first. Way back in April of 1953, Yeager had made a speech in which he said, "Some of the proposed fighters of tomorrow will be able to find and destroy a target and even return to their home stations and land by themselves. The only reason a pilot will be needed is to take over and decide what to do if anything goes wrong with the electronic equipment." Talking about the Ships of Tomorrow had made it all seem far off. But now, ten years later, they were already bringing such systems into the hardware stage. They were even working on a system to land F-4s automatically on aircraft carriers; the pilot would take his hands off the controls and let the computers bring him down onto that heaving slab. The supersonic transports and airliners would be so automated they would give the pilot an override stick just so he could push on it every now and then and feel *like a pilot;* it would be a goddamned right-stuff security blanket. They were even developing an automatic guidance system to bring the X-15 back through the atmosphere at a precise angle of attack. Maybe the age of "the flyboys," the stick'n'rudder fighter jocks, was about finished.

All of that Yeager could accept. On the great pyramid there was no steady state. Sixteen years ago, when he came to Muroc, he was only twenty-four, and few other test pilots had ever heard of him, and most people in aviation thought "the sound barrier" was as solid as a wall. Once he flew Mach 1, however, it was a whole new ball game. And now there were cosmonauts and astronauts, and it was a whole new ball game once again. A man could do a pretty good job of being philosophical about it. What finally got to Yeager, however, was the Ed Dwight case.

It had been early this year that Yeager got word from the brass that the President, John F. Kennedy, was determined that NASA have at least one Negro astronaut in their lineup. The whole process was to take place organically, however, as if in the natural order of things. Kennedy was leaning on the Defense Department, Defense was leaning on the Air Force brass, and they tossed the potato to Yeager. The pilot who had been singled out was an Air Force captain named Ed Dwight. He was to go through ARPS and be selected by NASA. The clouds developed soon enough. Dwight was enrolled in the basic flight test course along with twenty-five other candidates. Only the top eleven students could enter ARPS's six-month space-flight course, which had limited facilities, and Dwight did not rank among the top eleven. Yeager didn't see how he could jump him over other young tigers, all of them desperate to become astronauts. Every week, it seemed like, a detachment of Civil Rights Division lawyers would turn up from Washington, from the Justice Department, which was headed by the President's brother Bobby. The lawyers squinted in the desert sunlight and asked a great many questions about the progress and treatment of Ed Dwight and took notes. Yeager kept saying he didn't see how he could simply jump Dwight over these other men. And the lawyers would come back the next week and squint some more and take some more notes. There were days when ARPS seemed like the Ed Dwight case with a few classrooms and some military hardware appended. A compromise was finally struck in which Dwight would be admitted to the space-flight course, but only if every man who ranked above him was also admitted. That was how it came to pass that the next class had fourteen students instead of eleven and included Captain Dwight. Meantime, the White House, apparently, was signaling to the Negro press that Dwight was going to be "the first Negro astronaut," and he was being invited to make public appearances. He was being set up for a fall, because the chances of NASA accepting him as an astronaut appeared remote in any event.

The whole thing was baffling. On the upper reaches of the great ziggurat the subject of race had never been introduced before. The unspoken premise was that you either had the right stuff or you didn't, and no other variables mattered. When the seven Mercury astronauts had been chosen in 1959, the fact that they were all white and all Protestant seemed to be interpreted as wholly benign evidence of their Small-Town American virtues. But by now, four years later, Kennedy, who had been supported by a coalition of minority groups in the 1960 election, had begun to raise the question of race as a matter of public policy in many areas. The phrase "white Protestant" took on a different meaning, so that it was now possible to regard the astronauts as some sort of cadre of whites of northern European racial background. In fact, this had nothing to do, *per se*, with their being astronauts. It was typical of career military officers generally. Throughout the world, for that matter, career officers came from "native" or "old settler" stock. Even in Israel, which had existed for barely a generation as an independent nation and was dominated politically by immigrants from Eastern Europe, the officer corps was made up overwhelmingly of "real Israelis"—men born or raised from an early age in the pre-war Jewish settlements of the old Palestine. The other common denominator of the astronauts was that they were all first or only sons; yet not even this had any special significance, for studies soon showed that first or only sons dominated many occupations, including scholarly ones. (In an age when the average number of children per family was barely more than two, the odds were two out of three that *any* male would be a first or only son.) None of which was going to mollify the White House, however, because the astronaut, the single-combat warrior, had become a creature with greater political significance than any other type of pilot in history.

The squinting and hassling was still going on the day the NF–104 arrived. Perhaps that was one reason the monster looked so good to Yeager. All of the world's accumulated political cunning, from Machiavelli to John McCormack, wouldn't be worth a dogscratch in the NF-104 at 65,000 feet. Two extraordinary pieces of equipment were being developed specifically for ARPS. One was a space mission simulator, a device more realistic and sophisticated than the Mercury procedures trainer or any simulator NASA had on the boards. The other was the NF–104, which was an F–104 with a rocket engine mounted over the tailpipe. The rocket engine used hydrogen peroxide and JP4 fuel and would deliver 6,000 pounds of thrust. It was like a super-afterburner. The main engine plus the regular afterburner would take you to about 60,000 feet, and then you cut in the rocket, and that would take you somewhere between 120,000 and 140,000 feet. At least that was what the engineers confidently assumed. The plan was that the ARPS students would run profiles on the space mission simulator, then put on silver pressure suits, space-flight-style, and take the NF-104 up to 120,000 feet or more in a tremendous arc, affording up to two minutes of weightlessness. During this interval they could master the use of the reaction controls, which were hydrogen-peroxide thrusters of the sort used in all vehicles above 100,000 feet, whether the X–15, the Mercury capsule, or the X–20.

The only problem was, nobody had ever wrung out the NF–104. Just how it would handle in the weak molecular structure of the atmosphere above 100,000 feet, what the limits of its performance envelope would be, nobody knew. The F–104 had been built as a high-speed interceptor, and when you tried to do other things with it, it became very "unforgiving," as the expression went. Pilots were already beginning to crunch it with the F–104 simply because the engine flamed out and they fell to the ground with about as much glide as a set of car keys. But Yeager loved the damned ship. It went like a bat. As the commandant of ARPS, he seized the opportunity to test the NF–104 as if it had his name on it.

The main reason he would be testing it would be for use in the school, but there was an extra dividend. Whoever was first to push the NF–104 to optimum performance was certain to set a new world record for altitude achieved by a ship taking off under its own power. The Soviets had set the current record, 113,890 feet, in 1961 with the E–66A, a delta-winged fighter plane. The X–2 and the X–15 had flown higher, but they had to be hauled aloft by a larger ship before their rockets were ignited. The Mercury and Vostok space vehicles were lifted to altitude by automated booster rockets, which were then disengaged and jettisoned. Of course, all aircraft records were losing their dazzle now that space flight had begun. It was getting to be like setting some sort of new record for railroad trains. Yeager hadn't tried to break a record in the skies over Edwards since December 1953, ten years ago, when he had set a new speed mark of Mach 2.4 in the X–1A and had come down the far side of the arc in the most horrendous bout with high-speed instability any man had ever survived. Now Yeager was back on the flight line again to go for broke, out by the shimmering mirage surface of Rogers Lake, under that pale-blue desert sky, and the righteous energy was flowing again. And if the good lads of the prep school could sense through him . . . and through that wild unbroken beast . . . a few volts of that righteous old-time religion . . . well, that would be all right, too.

Yeager had taken the NF–104 up for three checkout flights, edging it up gradually toward 100,000 feet, where the limits of the envelope, whatever they were, would begin to reveal themselves. And now he was out on the flight line for the second of two major preliminary flights. Tomorrow he would let it all out and go for the record. It was another of those absolutely clear brilliant afternoons on the dome of the world. In the morning flight everything had gone exactly according to plan. He had taken the ship up to 108,000 feet after cutting in the rocket engine at 60,000. The rocket had propelled the ship up at a 50-degree angle of attack. One of the disagreeable sides of the ship was her dislike of extreme angles. At any angle greater than 30 degrees, her nose would pitch up, which was the move she made just before going into spins. But at 108,000 feet it was no problem. The air was so thin at that altitude, so close to being pure "space," that the reaction controls, the hydrogen-peroxide thrusts, worked beautifully. Yeager had only to nudge the sidearm hand controller by his lap and a thruster on top of the nose of the plane pushed the nose right down again, and he was in perfect position to re-enter the dense atmosphere below. Now he was going up for one final exploration of that same region before going for broke tomorrow.

At 40,000 feet Yeager began his speed run. He cut in the afterburner and it slammed him back in his seat, and he was now riding an engine with nearly 16,000 pounds of thrust. As soon as the Machmeter hit 2.2, he pulled back on the stick and started the climb. The afterburner would carry him to 60,000 feet before exhausting its fuel. At precisely that moment he threw the switch for the rocket engine . . . terrific jolt . . . He's slammed back in his seat again. The nose pitches up to 70 degrees. The g-forces start rising. The desert sky starts falling away. He's going straight up into the indigo. At 78,000 feet a light on the console . . . as usual . . . the main engine overheating from the tremendous exertion of the climb. He throws the switch, and shuts it down, but the rocket is still accelerating. Who doesn't know this feeling if he doesn't't! The bastards are fantastic! . . . One hundred thousand feet . . . He shuts down the rocket engine. He's still climbing. The g-forces slide off . . . makes you feel like you're pitching forward . . . He's weightless, coming over the top of the arc . . . 104,000 feet . . . It's absolutely silent . . . Twenty miles up . . . The sky is almost black. He's looking straight up into it, because the nose of the ship is pitched up. His angle of attack is still about 50 degrees. He's over the top of the arc and coming down. He pushes the sidearm control to bring down the nose of the ship. Nothing happens . . . He can hear the thruster working but the nose isn't budging. It's still pitched up. He hits the thruster again . . . Shit! . . . She won't go down! . . . Now he can see it, the whole diagram . . . This morning at 108,000 feet the air was so thin it offered no resistance and you could easily push the nose down with the thrusters. At 104,000 feet the air remains just thick enough to exert aerodynamic pressure. The thrusters aren't strong enough to overcome it . . . He keeps hitting the reaction controls . . . The hydrogen peroxide squirts out of the jet on the nose of the ship and doesn't do a goddamned thing . . . He's dropping and the nose is still pitched up . . . The outside of the envelope! . . . well, here it is, the sonofabitch . . . It doesn't want to stretch . . . and here we go! . . . The ship snaps into a flat spin. It's spinning right over its center of gravity, like a pinwheel on a stick. Yeager's head is on the outer edge of the circle, spinning around. He pushes the sidearm control again. The hydrogen peroxide is finished. He has 600 pounds of fuel left in the main engine but there's no way to start it up. To relight the engine you have to put the ship nose down into a dive and force air through the intake duct and start the engine windmilling to build up the rpms. Without rpms there's no hydraulic pressure and without hydraulic pressure you can't move the stabilizer wings on the tail and without the stabilizer wings you can't control this bastard at the lower altitudes . . . He's in a steady-state flat spin and dropping . . . He's whirling around at a terrific rate . . . He makes himself keep his eyes pinned on the instruments . . . A little sightseeing at this point and it's vertigo and you're finished . . . He's down to 80,000 feet and the rpms are at dead zero . . . He's falling 150 feet a second . . . 9,000 feet a minute . . . *And what do I do next?* . . . here in the jaws of the Gulp . . . *I've tried A!—I've tried B!*—The damned beast isn't making a sound . . . just spinning around like a length of pipe in the sky . . . He has one last shot . . . the speed brakes, a parachute rig in the tail for slowing the ship down after a highspeed landing . . . The altimeter keeps winding down . . . Twenty-five thousand feet . . . but the altimeter is based on sea level . . . He's only 21,000 feet above the high desert . . . The slack's running out . . . He pops the speed brake . . . *Bango!*—the chute catches with a jolt . . . It pulls the tail up . . . He pitches down . . . The spin stops. The nose is pointed down. Now he only has to jettison the chute and let her dive and pick up the rpms. He jettisons the chute . . . and the beast heaves up again! The nose goes back up in the air! . . . It's the rear stabilizer wing . . . The leading edge is locked, frozen into the position of the climb to altitude. With no rpms and no hydraulic controls he can't move the tail . . . The nose is pitched way above 30 degrees . . . Here she goes again . . . She's back into the spin . . . He's spinning out on the rim again . . . He has no rpms, no power, no more speed chute, and only 180 knots airspeed . . . He's down to 12,000 feet . . . 8,000 feet above the farm . . . There's not a goddamned thing left in the manual or the bag of tricks or the righteousness of twenty years of military flying . . . Chosen or damned! . . . It blows at any seam! Yeager hasn't bailed out of an airplane since the day he was shot down over Germany when he was twenty . . . I've tried A!—I've tried B!—I've tried C! . . . 11,000 feet, 7,000 from the farm . . . He hunches himself into a ball, just as it says in the manual, and reaches under the seat for the cinch ring and pulls . . . He's exploded out of the cockpit with such force it's like a concussion . . . He can't see . . . *Wham* . . . a jolt in the back . . . It's the seat separating from him and the parachute rig . . . His head begins to clear . . . He's in midair, in his pressure suit, looking out through the visor of his helmet . . . Every second seems enormously elongated . . . infinite . . . such slow motion . . . He's suspended in midair . . . weightless . . . The ship had been falling at about 100 miles an hour and the ejection rocket had propelled him up at 90 miles an hour. For one thick adrenal moment he's weightless in midair, 7,000 feet above the desert . . . The seat floats nearby, as if the two of them are parked in the atmosphere . . . The butt of the seat, the underside, is facing him . . . a red hole . . . the socket where the ejection mechanism had been attached . . . It's dribbling a charcoal red . . . lava . . . the remains of the rocket propellant . . . It's glowing . . . it's oozing out of the socket . . . In the next moment they're both falling, he and the seat . . . His parachute rig has a quarter bag over it and on the bag is a drogue chute that pulls the bag off so the parachute will stream out gradually and not break the chute or the pilot's back when the canopy pops open during a high-speed ejection. It's designed for an ejection at 400 or 500 miles an hour, but he's only going about 175. In this infinitely expanded few seconds the lines stream out and Yeager and the rocket seat and the glowing red socket sail through the air together . . . and now the seat is drifting above him . . . into the chute lines! . . . The seat is nestled in the chute lines . . . dribbling lava out of the socket . . . eating through the lines . . . An infinite second . . . He's jerked up by the shoulders . . . it's the chute opening and the canopy filling . . . in that very instant *the lava*—it smashes into the visor of his helmet . . . Something slices through his left eye . . . He's knocked silly . . . He can't see a goddamned thing . . . The burning snaps him to . . . His left eye is gushing blood . . . It's pouring down inside the lid and down his face and his face is on fire . . . Jesus Christ! . . . the seat rig . . . The jerk of the parachute had suddenly slowed his speed, but the seat kept falling . . . It had fallen out of the chute lines and the butt end crashed into his visor . . . 180 pounds of metal . . . a double visor . . . the goddamned thing has smashed through both layers . . . He's burning! . . . There's rocket lava inside the helmet . . . The seat has fallen away . . . He can't see . . . blood pouring out of his left eye and there's smoke inside the helmet . . . Rubber! . . . It's the seal between the helmet and the pressure suit . . . It's burning up . . . The propellant won't quit . . . A tremendous *whoosh* . . . He can feel the rush . . . He can even hear it . . . The whole left side of the helmet is full of flames . . . A sheet of flame goes up his neck and the side of his face . . . The oxygen! . . . The propellant has burned through the rubber seal, setting off the pressure suit's automatic oxygen system . . . The integrity of the circuit has been violated and it rushes oxygen to the helmet, to the pilot's face . . . A hundred percent oxygen! Christ! . . . It turns the lava into an inferno . . . Everything that can burn is on fire . . . everything else is melting . . . Even with the hole smashed in the visor the helmet is full of smoke . . . He's choking . . . blinded . . . The left side of his head is on fire . . . He's suffocating . . . He brings up his left hand . . . He has on pressure-suit gloves locked and taped to the sleeve . . . He jams his hand in through the hole in the visor and tries to create an air scoop with it to bring air to his mouth . . . The flames . . . They're all over it . . . They go to work on the glove where it touches his face . . . They devour it . . . His index finger is burning up . . . His goddamned finger is burning! . . . But he doesn't move it . . . Get some air! . . . Nothing else matters . . . He's gulping smoke . . . He has to get the visor open . . . It's twisted . . . He's encased in a little broken globe dying in a cloud of his own fried flesh . . . The stench of it! . . . rubber and a human hide . . . He has to get the visor open . . . It's that or nothing, no two ways about it . . . It's smashed all to hell . . . He jams both hands underneath . . . It's a tremendous effort . . . It lifts . . . Salvation! . . . Like a sea the air carries it all away, the smoke, the flames . . . The fire is out. He can breathe. He can see out of his right eye. The desert, the mesquite, the motherless Joshua trees are rising slowly toward him . . . He can't open his left eye . . . Now he can feel the pain . . . Half his head is broiled . . . That isn't the worst of it . . . The damned finger! . . . Jesus! . . . He can make out the terrain, he's been over it a million times . . . Over there's the highway, 466, and there's Route 6 crossing it . . . His left glove is practically burned off . . . The glove and his left index finger . . . he can't tell them apart . . . they look as if they exploded in an oven . . . He's not far from base . . . Whatever it is with the finger, it's very bad . . . Nearly down . . . He gets ready . . . Right out of the manual . . . A terrific wallop . . . He's down on the mesquite, looking across the desert, one-eyed . . . He stands up . . . Hell! He's in one piece! . . . He can hardly use his left hand. The goddamn finger is killing him. The whole side of his head . . . He starts taking off the parachute harness . . . It's all in the manual! Regulation issue! . . . He starts rolling up the parachute, just like it says . . . Some of the cords are almost melted through, from the lava . . . His head feels like it's still on fire . . . The pain comes from way down deep . . . But he's got to get the helmet off . . . It's a hell of an operation . . . He doesn't dare touch his head . . . It feels enormous . . . Somebody's running toward him . . . It's a kid, a guy in his twenties . . . He's come from the highway . . . He comes up close and his mouth falls open and he gives Yeager a look of stone horror . . .

"Are you all right!"

The look on the kid's face! Christalmighty!

"I was in my car! I saw you coming down!"

"Listen," says Yeager. The pain in his finger is terrific. "Listen . . . you got a knife?"

The kid digs into his pocket and pulls out a penknife. Yeager starts cutting the glove off his left hand. He can't bear it any more. The kid stands there hypnotized and horrified. From the look on the kid's face, Yeager can begin to see himself. His neck, the whole left side of his head, his ear, his cheek, his eye must be burned up. His eye socket is slashed, swollen, caked shut, and covered with a crust of burned blood, and half his hair is burned away. The whole mess and the rest of his face and his nostrils and his lips are smeared with the sludge of the burning rubber. And he's standing there in the middle of the desert in a pressure suit with his head cocked, squinting out of one eye, working on his left glove with a penknife . . . The knife cuts through the glove and it cuts through the meat of his finger . . . You can't tell any longer . . . It's all run together . . . The goddamn finger looks like it's melted . . . He's got to get the glove off. That's all there is to it. It hurts too goddamned much. He pulls off the glove and a big hunk of melted meat from the finger comes off with it . . . It's like fried suet . . .

"Arrggghhh . . ." It's the kid. He's retching. It's too much for him, the poor bastard. He looks up at Yeager. His eyes open and his mouth opens. All the glue has come undone. He can't hold it together any longer.

"God," he says, "you . . . look *awful!*" The Good Samaritan, A.A.D.! Also a doctor! And he just gave his diagnosis! That's all a man needs . . . to be forty years old and to fall one hundred goddamned thousand feet in a flat spin and punch out and make a million-dollar hole in the ground and get half his head and his hand burned up and have his eye practically ripped out of his skull . . . and have the Good Samaritan, A.A.D., arrive as if sent by the spirit of Pancho Barnes herself to render a midnight verdict among the motherless Joshua trees while the screen doors bang and the pictures of a hundred dead pilots rattle in their frames:

"My God! . . . you look awful."

A few minutes later the rescue helicopter arrived. The medics found Yeager standing out in the mesquite, him and some kid who had been passing by. Yeager was standing erect with his parachute rolled up and his helmet in the crook of his arm, right out of the manual, and staring at them quite levelly out of what was left of his face, as if they had had an appointment and he was on time.

At the hospital they discovered one stroke of good luck. The blood over Yeager's left eye had been baked into a crustlike shield. Otherwise he might have lost it. He had suffered third- and second-degree burns on his head and neck. The burns required a month of treatment in the hospital, but he was able to heal without disfigurement. He even regained full use of his left index finger.

It so happened that on the day of Yeager's flight, at just about the time he headed down the runway on takeoff, the Secretary of Defense, Robert McNamara, announced that the X–20 program had been canceled. Although the Manned Orbiting Laboratory scheme remained alive officially, it was pretty obvious that there would be no American military space voyagers. The boys in Houston had the only ticket; the top of the pyramid was theirs to extend to the stars, if they were able.

Yeager was returned to flight status and resumed his duties at ARPS. In time he would go on to fly more than a hundred missions in Southeast Asia in B–57 tactical bombers.

No one ever broke the Russian mark with the NF–104 or even tried to. Up above 100,000 feet the plane's envelope was goddamned full of holes. And Yeager never again sought to set a record in the sky over the high desert.

Epilogue

Well, the Lord giveth, and the Lord taketh away. After Gordon Cooper's triumph, Alan Shepard had launched a campaign for one more Mercury flight, a three-day mission, which it would be his turn to fly. He had the backing of Walt Williams and most of the astronauts. James Webb headed them off easily, however, with President Kennedy's tacit blessing, and announced that Project Mercury had been completed. NASA and the Administration were having a hard enough time keeping the Congress in line to support the $40 billion Gemini and Apollo lunar-landing programs without prolonging the Mercury series. The spirit of two years before, when Kennedy had raised his arms toward the moon and congressmen cheered and issued forth a limitless budget, had evaporated. The space race was a . . . "race for survival"? The United States faced . . . "national extinction"? Whoever controlled outer space . . . controlled the earth? The Russians were going to . . . put a red spot on the moon? It was impossible to recall the emotion of those days. In mid-June 1963 the Chief Designer (still the anonymous genius!) put *Vostok 5* into orbit with Cosmonaut Valery Bykovsky aboard, and two days later he sent the first woman into space, Cosmonaut Valentina Tereshkova, aboard *Vostok 6*, and they remained in orbit for three days, flying within three miles of each other at one point and landing on Soviet soil on the same day—and not even then did the old sense of warlike urgency revive in the Congress.

In July, Shepard began to be bothered by a ringing in his left ear and occasional dizziness, symptoms of Meniere's disease, which affects the inner ear. Like Slayton, he had to go on inactive status as an astronaut and could only fly an aircraft with a co-pilot aboard. Slayton, meantime, had made a decision that would have been unthinkable to most military officers. He had resigned from the Air Force after nineteen years—one year short of qualifying for the twenty-year-man's pension, that golden reward that blazed out there beyond the horizon throughout the career officer's years of financial hardship. Slayton's problem was that the Air Force had decided to ground him altogether because of his heart condition. As a civilian working for NASA, he could continue to fly high-performance aircraft, so long as he was accompanied by a co-pilot. He could keep up his proficiency, he could remain on flight status, he could keep alive his hopes of proving, somewhere down the line, that he had the right stuff to go aloft as an astronaut, after all. Next to that consideration the pension didn't amount to much.

On July 19 Joe Walker flew the X–15 to 347,800 feet, which was sixty-six miles up, surpassing the record of 314,750 feet set by Bob White the year before; and on August 22 Walker reached 354,200 feet, or sixty-seven miles, which was seventeen miles into space. In addition to White and Walker, one other man had flown the X–15 above fifty miles. That was White's backup, Bob Rushworth, who had achieved 285,000 feet, fifty-four miles, in June. The Air Force had instituted the practice of awarding *Air Force Astronaut* wings to any Air Force pilot who flew above fifty miles. They used the term itself: *astronaut.* As a result, White and Rushworth, the Air Force's prime and backup pilots for the X–15, now had their astronaut wings. Joe Walker, being a civilian flying the X–15 for NASA, did not qualify. So some of Walker's good buddies at Edwards took him out to a restaurant for dinner, and they all knocked back a few, and they pinned some cardboard wings on his chest. The inscription read: "Asstronaut."

On September 28 the seven Mercury astronauts went to Los Angeles for the awards banquet of the Society of Experimental Test Pilots. Iven Kincheloe's widow, Dorothy, presented them with the Iven C. Kincheloe Award for outstanding professional performance in the conduct of flight test. The wire services devoted scarcely more than a paragraph to the occasion, and that they took from the handout. After all the Distinguished Service Medals and the parades and the appearances before

Congress, after every sort of tribute that politicians, private institutions, and the Genteel Beast could devise, the Iven C. Kincheloe Award didn't seem like very much. But for the seven astronauts it was an important night. The radiant Kinch, the great blond movie-star picture of a pilot, was the most famous of the dead rocket pilots and could have cut his own orders in the Air Force, had he lived. He would have had Bob White's job as prime pilot of the X–15 and God knows what else. There were aviation awards and aviation awards, but the Kincheloe Award—for "professional performance"—was the big one within the flight test fraternity. The seven men had finally closed the circle and brought together the scattered glories of their celebrity. They had fought for a true pilot's role in Project Mercury, they had won it, step by step, and Cooper's flight, on top of the others, had shown they could handle it in the classic way, out on the edge. Now they had the one thing that had been denied them for years while the rest of the nation worshipped them so unquestioningly: acceptance by their peers, their true brethren, as *test pilots* of the space age, deserving occupants of the top of the pyramid of the right stuff.

During the summer Kennedy had gone on television to tell the nation that a nuclear test ban agreement with the Russians had been reached. Thereupon the Soviet foreign minister, Andrei Gromyko, had proposed a corollary that would ban even the placing of nuclear weapons in earth orbit. The Soviets themselves were extinguishing the notion of the thunderbolts from space. On August 30 there had gone into service the piece of equipment by which the entire interlude would be remembered: the hotline, a telephone hookup between the White House and the Kremlin, the better to avoid misunderstandings that might result in nuclear war. When Kennedy was assassinated on November 22 by a man with Russian and Cuban ties, there was no anti-Soviet or anti-Cuban clamor in the Congress or in the press. The Cold War, as anyone could plainly see, was over.

No one, and certainly not the men themselves, could comprehend the meaning of this for the role of *the astronaut,* however. The new President, Lyndon Johnson, proved to be even more of a proponent of the space program than his predecessor. Due partly to the political genius of James E. Webb, who now came into his own, the Congress swung about and gave

NASA a blank check for the missions to the moon. Nevertheless, the fact remained: the Cold War was over.

No more manned spaceflights were scheduled until the start of the Gemini program in 1965. By then the seven Mercury astronauts would begin to feel a change in the public attitude toward them and, for that matter, toward the Next Nine and the groups of astronauts that were to follow. They would *feel* the change, but they would not be able to put it into words. What *was* that feeling? Why, it was the gentle slither of the mantle of soldierly glory sliding off one's shoulders!—and the cooling effect of oceans of tears drying up! The single-combat warriors' war had been removed. They would continue to be honored, and men would continue to be awed by their courage; but the day when an astronaut could parade up Broadway while traffic policemen wept in the intersections was no more. Never again would an astronaut be perceived as a protector of the people, risking his life to do battle in the heavens. Not even the first American to walk on the moon would ever know the outpouring of a people's most primal emotions that Shepard, Cooper, and, above all, Glenn had known. The era of America's first single-combat warriors had come, and it had gone, perhaps never to be relived.

The Lord giveth, and the Lord taketh away. The mantle of Cold Warrior of the Heavens had been placed on their shoulders one April day in 1959 without their asking for it or having anything to do with it or even knowing it. And now it would be taken away, without their knowing that, either, and because of nothing they ever did or desired. John Glenn had made up his mind to run for the Senate in Ohio in 1964. He could not have foreseen that the voters of Ohio would no longer regard him as a man with a protector's aura. But at least he would be remembered. It would have been still more impossible for his confreres to realize that the day might come when Americans would hear their names and say, "Oh, yes—now, which one was he?"

AUTHOR'S NOTE

The writing of this book would have been impossible without the personal recollections of many people, pilots and non-pilots, who were intimately involved in the beginning of the era of manned rocket flight in America. I wish there were some way to thank them properly for their generosity and for the time and effort it took them to review events that go back twenty years or more in some cases.

The NASA history office at the Johnson Space Center in Houston was unfailingly helpful, especially in giving me access to transcriptions of the post-flight debriefings of the astronauts. I should mention in particular NASA historian James M. Grimwood, the author, with his colleagues, Loyd S. Swenson, Jr., and Charles C. Alexander, of *This New Ocean: A History of Project Mercury*. Other books I would like to acknowledge are *Always Another Dawn*, by A. Scott Crossfield with Clay Blair, Jr.; *Starfall*, by Betty Grissom and Henry Still; *Across the High Frontier*, by Charles E. Yeager and William Lundgren; *The Lonely Sky*, by William Bridgeman and Jacqueline Hazard; *X–15 Diary*, by Richard Tregaskis; and *We Seven*, by the seven Mercury astronauts.

The names of four figures appearing briefly in the narrative have been changed: Bud and Loretta Jennings, Mitch Johnson, and Gladys Loring.